$125.95

MW01518324

RESIDENTIAL WIRING FOR THE TRADES

250101

H. Brooke Stauffer

Boston   Burr Ridge, IL   Dubuque, IA   Madison, WI   New York   San Francisco   St. Louis
Bangkok   Bogotá   Caracas   Kuala Lumpur   Lisbon   London   Madrid   Mexico City
Milan   Montreal   New Delhi   Santiago   Seoul   Singapore   Sydney   Taipei   Toronto

Higher Education

RESIDENTIAL WIRING FOR THE TRADES

1 2 3 4 5 6 7 8 9 0 QPD/QPD 0 9 8 7 6 5

ISBN-13 978–0–07–351081–1
ISBN-10 0–07–351081–5

Publisher, Career Education: *David T. Culverwell*
Publisher, Trades and Engineering Technology: *Thomas E. Casson*
Managing Developmental Editor: *Jonathan Plant*
Editorial Coordinator: *Lindsay M. Roth*
Marketing Coordinator: *Amy L. Reed*
Project Manager: *April R. Southwood*
Senior Production Supervisor: *Laura Fuller*
Lead Media Project Manager: *Audrey A. Reiter*
Media Technology Producer: *Daniel M. Wallace*
Designer: *John Joran*
(USE) Cover Images: Upper left: *Michael Rosenfeld/Stone;* Lower left: *Andy Sacks/Stone;*
    Upper right: *Hisham F Ibrahim/Photodisc Green;* Lower right: *Brand X Pictures*
Compositor: *Modern Graphics, Inc.*
Typeface: *11/13 Times Roman*
Printer: *Quebecor World Dubuque, IA*

National Electrical Code® and *NEC®* are registered trademarks of the National Fire Protection Association, Inc., Quincy, MA 02169

Library of Congress Cataloging-in-Publication Data

Stauffer, H. Brooke.
    Residential wiring for the trades / H. Brooke Stauffer. — 1st ed.
        p.   cm.
    Includes index.
    ISBN 978–0–07–351081–1 — ISBN 0–07–351081–5
    1. Electric wiring, Interior. 2. Dwellings—Electric equipment. I. Title.

    TK3285.S723   2007
    621.319'24—dc22   2005053118
                                                                    CIP

www.mhhe.com

Contents _____

Preface

Overview

Residential Wiring for the Trades is a practical guide for wiring homes according to *National Electrical Code*® rules. It covers *Code* requirements and actual construction practices for installing electrical systems in new one- and two-family dwellings.

This book contains explanations of many technical requirements in the *NEC*®, and the safety-based intent behind them, to give students a fuller and more comprehensive understanding of the *Code*.

Residential Wiring for the Trades combines in-depth instruction about 2005 *NEC* rules with detailed, hands-on, information about real life residential wiring practices. It is a textbook for electrician apprenticeship programs, vocational-technical schools, inspector training, and other educational, training, and certification programs.

Features

Clear text and more than 200 color illustrations provide a thorough understanding of the residential wiring requirements of the 2005 *National Electrical Code*. Students learn how to design and install residential electrical systems in one- and two-family dwellings, in accordance with all applicable *Code* rules. Important features include:

- General principles that guide the development of *NEC* requirements
- Design of residential electrical systems, including load calculations
- Specific *Code* rules that apply to different rooms and areas of dwellings
- Wiring methods used in residential construction
- *NEC* rules that apply specifically to two-family dwellings
- Important *NEC* terms are listed at the beginning of each chapter
- Official *NEC* definitions that apply to electrical systems in dwellings are extracted from Article 100 of the *NEC* and listed at the end of the book
- Detailed coverage of safety begins in Chapter 1 and is stressed throughout the book
- Chapter 17, Old Work, covers the important topic of additions and modifications to existing residential wiring systems
- Chapter Review sections test student knowledge with an array of multiple choice, true or false, and fill in the blank questions
- Challenge Questions at the end of each unit provide critical thinking situations for students to analyze

Instructor Resources

The following companion pieces are available for use with the textbook on the Instructor Productivity Center (IPC) CD ROM:

- **EZ TEST:** McGraw-Hill's EZ Test is a flexible and easy-to-use electronic testing program. The program allows instructors to create tests from book specific items. It accommodates multiple choice, true/false, matching, short answer and essay questions and instructors may add their own questions. Multiple versions of the test can be created, and any test can be exported for use with course management systems such as WebCT, BlackBoard, or PageOut. EZ Test Online is a new service and gives you a place to easily administer your EZ Test created exams and quizzes online. The program is available for Windows and Macintosh environments.

- **Test Bank:** Hundreds of multiple choice, true/false, and completion questions have been prepared for each chapter. Instructors can edit and add to this test bank using EZ Test.

- **Lecture Presentations:** Rich PowerPoint presentations have been developed for every chapter in the text, and instructors can edit and add to the PowerPoint slides to customize classroom presentations.

- **Classroom Performance System (CPS):** This wireless response system is available from McGraw-Hill for in-class testing, in-class quizzing, and classroom management. Contact your McGraw-Hill sales representative for details.

Acknowledgments

Opinions given in this book are those of the author. They represent my best understanding of the *NEC* rules for wiring one- and two-family dwellings, based on long experience of working with it and being one of many volunteers who help to shape the nation's wiring rules. In writing this book, I have also benefited from the expertise and advice of knowledgeable colleagues throughout the electrical industry.

I would like to thank the following reviewers for their comments and suggestions.

Jeff Cook
Ivy Tech State College; Richmond, IN

Orville N. Lake
Augusta Technical College; Augusta, IN

Robert W. Saxon
Bessemer State Technical College; Bessemer, AL

Steve Vietor
Riverland Community College; Albert Lea, MN

I would also like to thank Frank Petruzella as the *Training for the Trades* series advisor.

H. Brooke Stauffer
Executive Director, Standards & Safety
National Electrical Contractors Association (NECA)
Member of N.E. Code-Making Panel No. 1 (CMP-1)

About the Author _____

Brooke Stauffer is Executive Director of Standards and Safety for the National Electrical Contractors Association (NECA) in Bethesda, Maryland. He is responsible for all of NECA's regulatory activities and for developing and publishing the *National Electrical Installation Standards* (NEIS)®.

Stauffer is a member of the IEEE, NFPA, and IAEI, and has served on three different code-making panels. He is also a member of the ANSI Board of Standards Review (BSR), which approves all American National Standards, including the *National Electrical Code*®.

Before joining NECA in 1995, Stauffer was technical director for the Association of Home Appliance Manufacturers (AHAM), manager of codes and standards development with the Smart House Limited Partnership, and assistant manager of engineering at the National Electrical Manufacturers Association (NEMA). He was educated at Baltimore Polytechnic Institute and the Catholic University of America in Washington, DC.

General Principles

OBJECTIVES

After completing this chapter, the student will be able to understand the following:

- The overall structure of the *National Electrical Code® (NEC®)*
- What a dwelling unit is and how it differs from other kinds of buildings and facilities covered by the *Code*
- The relationship between the *NEC* and other building codes
- What is meant by a product or installation being "approved"
- The importance of using official *Code* terminology

INTRODUCTION

This chapter covers general principles concerning the *National Electrical Code* and how they apply to electrical installations in residential construction. The following major topics are included:

- Overview of the *NEC*
- Special rules for wiring residences
- Other codes and regulatory requirements
- Symbols for residential wiring
- Basics of residential electrical systems
- Definitions for residential wiring

SCOPE

Dwellings Included. *NFPA's Residential Wiring* is a practical guide for wiring homes according to *National Electrical Code* rules. It covers requirements and construction practices for wiring new one- and two-family dwellings. It concentrates on *Code* rules for one- and two-family dwellings built at grade—detached houses, duplexes, and townhouses.

One- and two-family dwellings form a special class of small buildings in the *National Electrical Code*. They are exempt from a number of rules that apply to larger buildings, including all types of multifamily dwellings.

Dwellings Not Included. This book doesn't cover multifamily dwellings such as "three-deckers," garden apartment buildings, high-rise apartments, and condominiums. It also doesn't cover guest rooms in hotels and motels.

Individual Dwellings in Multifamily Buildings. Each individual living unit of a multifamily apartment or condominium building (or a hotel room that qualifies as a dwelling unit because it includes a kitchenette) must be wired according to the requirements described in this book. See "Many Types of Dwelling Units" later in this chapter.

In other words, *NEC* rules for one- and two-family dwellings cover the internal wiring of apartment-type dwelling units themselves, but not the power distribution systems or common areas of those larger, multifamily buildings.

The material in this book provides useful guidance for wiring all these types of dwelling units, but it concentrates on electrical requirements for one- and two-family dwellings built at grade—detached houses, duplexes, and townhouses.

OVERVIEW OF THE *NEC*

This section provides a brief overview of the *National Electrical Code*, concentrating on those portions most used in wiring one- and two-family dwellings. The *NEC* is one of nearly 300 technical standards published by the National Fire Protection Association (NFPA), based in Quincy, Massachusetts. The *NEC* is also known by its numerical designation, NFPA 70. Throughout this book, NFPA 70 is referred to interchangeably in three ways: *National Electrical Code*, *NEC*, and *Code*.

The *National Electrical Code* is "the law of the land" where safety requirements for electrical wiring are concerned. Most states, cities, and counties adopt the *NEC* for regulatory use. State and local jurisdictions also adopt a number of other NFPA codes and standards for regulatory use, including the following:

NFPA *101*®, *Life Safety Code*®
NFPA *72*®, *National Fire Alarm Code*®
NFPA 13, *Standard for the Installation of Sprinkler Systems*
NFPA 54, *National Fuel Gas Code*

The structure, organization, and language of the *Code,* along with the process through which it is developed, are explained in much greater detail in two other books published by NFPA:

National Electrical Code® Handbook, 2005 edition
User's Guide to the National Electrical Code®, 2005 edition

Organization

The *National Electrical Code* is a comprehensive and complex document. It consists of nine chapters followed by seven annexes:

Chapter 1, General

Chapter 2, Wiring and Protection

Chapter 3, Wiring Methods and Materials

Chapter 4, Equipment for General Use

Chapter 5, Special Occupancies

Chapter 6, Special Equipment

Chapter 7, Special Conditions

Chapter 8, Communications Systems

Chapter 9, Tables

Annex A, Product Safety Standards

Annex B, Application Information for Ampacity Calculation

Annex C, Conduit and Tubing Fill Tables for Conductors and
          Fixture Wires of the Same Size

Annex D, Examples

Annex E, Types of Construction

Annex F, Cross-Reference Tables

Annex G, Administration and Enforcement

The *Code* also contains one article that isn't part of the enforceable requirements, but provides important information to users:

Article 90, Introduction

The interrelationships between the different parts of the *Code*, and how they work together, are explained in 90.3 Code Arrangement.

The rules in Chapters 1–4 of the *Code* apply generally to all electrical installations. The rules in Chapters 5–7 modify those general rules for special occupancies, equipment, or conditions.

Chapter 8 covers communications systems, including telephone and cable television wiring used in homes.

Chapter 9 contains tables relating to conductors installed in raceways, along with other tables dealing with power source limitations of Class 2, Class 3, and fire alarm circuits. Chapter 9 isn't used much in residential wiring. The great majority of houses are wired with Type NM and Type AC cable, and the power limitation tables are primarily of interest to product listing agencies.

In general, the *NEC* annexes deal with commercial and industrial construction rather than residential wiring. However, Examples D1(a), D1(b), D2(a), D2(b), and D2(c) in Annex D provide useful guidance for service load calculations for one-family dwellings.

Structure

Chapters. Chapters covering broad areas are the major divisions of the *NEC*. There are nine chapters, with Chapters 1–8 being subdivided into articles.

Articles. Each article covers a specific major subject such as Overcurrent Protection, Grounding, Nonmetallic-Sheathed Cable: Type NM, Switches, Appliances, and Solar Photovoltaic Systems. With the exception of Article 90 (the first one in the *Code*), articles are designated by three-digit numbers. The 200-series articles are in Chapter 2, the 400-series articles are in Chapter 4, and so on. Articles are divided into sections.

Sections. Each section of the *NEC* can be thought of as representing a separate rule. Sections are designated using the article number followed by a period, which is followed by another number. The following are examples:

110.12 Mechanical Execution of Work
210.8 Ground-Fault Circuit-Interrupter Protection for Personnel
680.9 Electric Pool Water Heaters

SPECIAL RULES FOR WIRING DWELLING UNITS

As we have seen, not all *Code* rules apply to residential wiring. Most house wiring situations fall within Chapters 1–4, although there are exceptions. Article 680 contains requirements for hot tubs, spas, and fountains (Figure 1.1). Elevators and wheelchair lifts found in some one- and two-family dwellings are covered by Article 620. Telephone and cable TV wiring are governed by Articles 800 and 820 in Chapter 8 on communications systems.

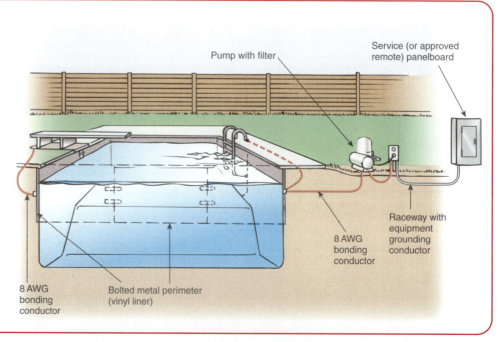

FIGURE 1.1 These are the bonding requirements for an all-metal swimming pool with vinyl liner.

Pump with filter

Service (or approved remote) panelboard

Raceway with equipment grounding conductor

8 AWG bonding conductor

8 AWG bonding conductor

Bolted metal perimeter (vinyl liner)

What's a Dwelling Unit?

NEC Article 100 defines the residences within the scope of this book as follows.

> **Dwelling Unit:** A single unit, providing complete and independent living facilities for one or more persons, including permanent provisions for living, sleeping, cooking, and sanitation.
>
> **Dwelling, One-Family:** A building that consists solely of one dwelling unit.
>
> **Dwelling, Two-Family:** A building that consists solely of two dwelling units.

Thus, the typical hotel or motel room doesn't qualify as a dwelling unit because it lacks permanent provisions for cooking. (See the Article 100 definition of *Guest Room*.) However, a hotel/motel guest room with a kitchenette would be considered a dwelling unit.

Similarly, an office building with a kitchenette or coffee room, as well as bathrooms on every floor, doesn't qualify as a dwelling unit even if it also has a nurse's office or other room with a bed where people sometimes sleep. Such different facilities scattered throughout the building don't add up to a housekeeping unit. However, an office building might include a self-contained apartment, possibly for an on-site superintendent, which would satisfy the definition of a dwelling unit.

While not all *NEC* requirements apply to residential wiring, there are a number of special rules that apply only to dwelling units. Examples of these include the following:

- Rules for locating and spacing receptacles
- Locations of lighting outlets and wall switches
- Small-appliance and laundry branch circuits
- Arc-fault circuit-interrupter (AFCI) protection for branch circuits in bedrooms
- Ground-fault circuit-interrupter (GFCI) protection for receptacles located in kitchens, bathrooms, basements, outdoors, and various other locations exposed to moisture

Many Types of Dwelling Units

The *NEC* rules for wiring dwelling units apply not only to houses but also to apartment or condominium units, extended-stay hotel/motel rooms with kitchenettes, and other similar types of housekeeping units.

Each dwelling unit in a two-family or multifamily dwelling is essentially subject to the same *Code* rules as a detached house, with some exceptions. For example, the requirements for receptacle spacing, switched lighting outlets in habitable rooms, and GFCI protection for bathroom and kitchen counter receptacles are identical for all dwelling units (Figure 1.2).

Much of the material in this book provides useful guidance for wiring all types of dwelling units, but it concentrates on *National Electrical Code* rules for one- and two-family dwellings built at grade—detached houses, duplexes, and townhouses.

A Note About Terminology. Throughout this book, the terms *dwelling, house, home,* and *residence* are used interchangeably. The term *residential wiring* is also used frequently to describe electrical installations in one- and two-family dwellings.

FIGURE 1.2 This 15-ampere duplex receptacle has integral GFCI protection. (Courtesy of Pass & Seymour/Legrand®)

OTHER CODES AND REGULATORY REQUIREMENTS

Local Electrical Codes

Although most jurisdictions adopt the wiring requirements of the current edition of the *National Electrical Code*, some localities maintain their own electrical safety codes, adopt older editions of the *NEC*, or adopt it with local variations. For example, while the *Code* permits a minimum conductor size of 14 AWG for branch circuits [310.5], some localities require minimum 12 AWG. A few places, such as Chicago and some South Florida jurisdictions, restrict the use of cable wiring methods and nonmetallic raceways, requiring that residences be wired using metallic raceways.

Even where local codes are enforced, they typically resemble the *National Electrical Code* in many respects. Thus, the guidelines and recommendations in this book can generally still be followed. However, installers should always be aware of which state and local codes are enforced in the area where they are working and follow them carefully.

Other Building Codes

The *National Electrical Code* is a building code. It is one of a series of regulatory building codes published by NFPA and its partners, the International Association of Plumbing and Mechanical Officials (IAPMO), American Society of Heating, Refrigerating, and Air-Conditioning Engineers (ASHRAE), and Western Fire Chiefs Association (WFCA). These *Comprehensive Consensus Codes®* (C3®), as they are known, cover all aspects of constructing buildings and similar structures and are designed to work together as a system.

For example, *NEC* 300.21 requires that "openings around electrical penetrations through fire-resistant–rated walls, partitions, floors, or ceilings shall be firestopped using approved methods to maintain the fire resistance rating," but does not provide detailed instructions on firestopping. Instead, the Fine Print Note (FPN) states that "assistance in complying with 300.21 can be found in building codes, fire resistance directories, and product listings." *NFPA 5000®*, *Building Construction and Safety Code®*, which is another volume in the C3 series, provides technical requirements for firestopping openings in rated partitions (Figure 1.3).

Similarly, the *Code* has no requirements for hard-wired smoke detectors in dwellings. Those requirements are found in *NFPA 72®, National Fire Alarm Code®*. But once the requirement has been established in another volume of the *Comprehensive Consensus Codes*, they must be wired following *NEC* rules. To put it another way—smoke detectors aren't within the scope of the *National Electrical Code*, but the branch circuits that supply them are.

In most cases the *NEC* is a stand-alone, self-contained document. But installers need to be aware of other building codes whose requirements may affect the installation of residential electrical systems. When in doubt, consult the authority having jurisdiction (AHJ).

Product Listing

In many places, the *Code* requires the use of listed or labeled electrical equipment. Agencies such as Underwriters Laboratories Inc. (UL) test products for safety and compliance with *NEC* requirements. AHJs typically accept the listing mark as evidence that the product complies with applicable *NEC* rules (Figure 1.4).

However, it is important to understand that when a product has been listed by

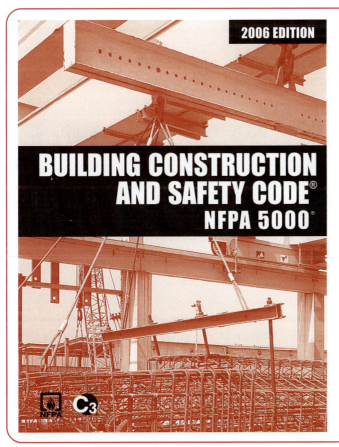

FIGURE 1.3 *NFPA 5000®, Building Construction and Safety Code®*, is the centerpiece of the *Comprehensive Consensus Codes®* (C3®) series of publications.

FIGURE 1.4 AHJs look for listing marks on electrical products to verify that they comply with *NEC* requirements. (Courtesy of Underwriters Laboratories Inc.)

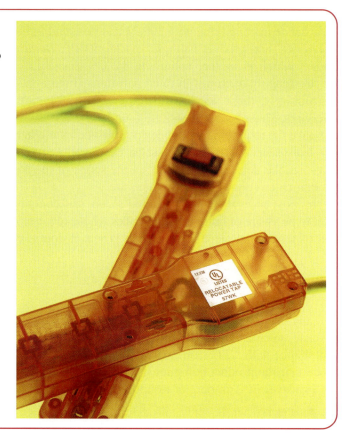

UL or another agency, the listing indicates only that it is safe and suitable when used under the conditions for which it was designed. For example, most jurisdictions permit the use of listed Type NM cable for interior wiring in dwellings. But an inspector would reject such cable if it were direct-buried to provide power to a freestanding garage. Section 334.10(A) states that NM cable is permitted in normally dry locations and in void spaces of masonry block or tile walls and isn't listed for direct burial in the earth.

There are a number of different product listing agencies, and sometimes they are referred to collectively as Nationally Recognized Testing Laboratories (NRTLs, pronounced *nertles*). However, this language does not appear in the *National Electrical Code*. Different states and localities recognize different listing agencies, and the acceptable ones in a given area can be determined by consulting the AHJ.

Listed and Labeled. Listed products typically are marked or labeled in some way, and testing agencies publish directories of their listed products (Figure 1.5). Section 110.3(B) states that listed or labeled equipment must be installed in accordance with any instructions included as part of the listing or labeling. In effect, this requirement makes the manufacturer's instructions part of the *Code* requirements that can be enforced by the AHJ. It is important for installers to keep this provision in mind.

Approvals. People sometimes talk about an electrical apparatus being "UL-approved," or a wiring method being "approved by the *Code*," but this language isn't correct. Only the AHJ can approve wiring methods and equipment installed under the rules of the *National Electrical Code*. However, listing of products by testing

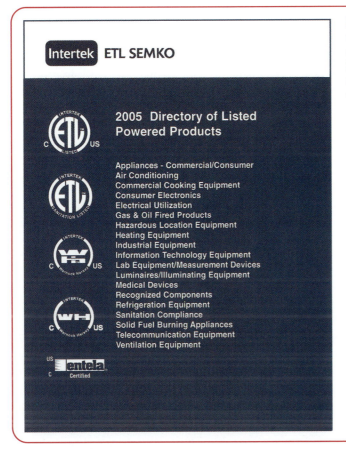

FIGURE 1.5 Product testing and listing agencies publish directories like this one. (Courtesy of Intertek Testing Services)

laboratories often provides a basis for the AHJ's approval. The following are four key definitions from Article 100.

Approved: Acceptable to the authority having jurisdiction.

Authority Having Jurisdiction: The organization, office, or individual responsible for approving equipment, materials, an installation, or a procedure.

Labeled: Equipment or materials to which has been attached a label, symbol, or other identifying mark of an organization that is acceptable to the authority having jurisdiction and concerned with product evaluation, that maintains periodic inspection of production of labeled equipment or materials, and by whose labeling the manufacturer indicates compliance with appropriate standards or performance in a specified manner.

Listed: Equipment, materials, or services included in a list published by an organization that is acceptable to the authority having jurisdiction and concerned with evaluation of products or services, that maintains periodic inspection of production of listed equipment or materials or periodic evaluation of services, and whose listing states that the equipment, material, or services either meets appropriate designated standards or has been tested and found suitable for a specified purpose.

National Electrical Installation Standards® (NEIS®)

NEIS are quality standards for electrical construction published by the National Electrical Contractors Association (NECA) (Figure 1.6). They define what is meant by installing electrical equipment "in a neat and workmanlike manner," as required by *NEC* 110.12. All of NECA's standards comply with the *National Electrical Code.* However, because they are quality and workmanship standards, the NEIS often have additional requirements over and above *NEC* minimum safety provisions. The NEIS have been adopted for regulatory use in some states and localities.

Permits and Inspections

Nearly all jurisdictions require building permits for electrical construction work. The permit fees finance inspections by the AHJ. Building permits of the type shown in Figure 1.7 contain a general description of the work authorized to be performed (for example, electrical, plumbing, structural) and normally must be posted at the job site. Typically, there are two inspections during an electrical construction job: rough inspection and final inspection.

Rough Inspection. The first inspection takes place when the service has been installed, outlets are in place, and wiring is complete—but before the walls have been "closed in" and receptacles, switches, luminaires (lighting fixtures), and other

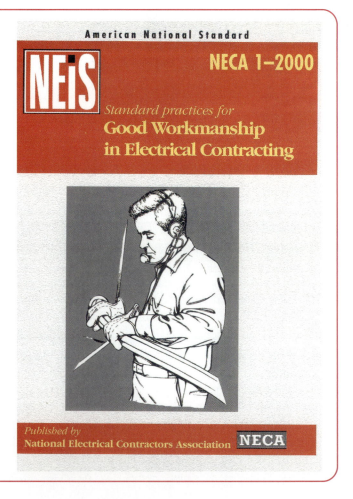

FIGURE 1.6 NECA standards for electrical construction are adopted for regulatory use in some jurisdictions. (Courtesy of NECA)

BUILDING PERMIT

Any Town, Any County
Department of Inspection Services
987-654-3210

The Department of Inspection Services hereby grants permission to

_____

_____

_____

_____

_____

Approved by _____ , Director of Inspection Services   Date: _____

PERMIT NO.: _____   By _____ , Permit Clerk

This permit conveys no right to occupy any street, alley or sidewalk, or any part thereof, either temporarily or permanently except that specifically provided for in the building code. It is the owner's responsibility to determine that the proposed construction does not violate existing private covenants and/or subdivision restrictions applicable to subject property or violate any State, Municipal or County zoning, subdivision or health department law, rule, ordinance, or regulations.

Approved plans MUST be retained on job and this card KEPT POSTED until final inspection has been made.
Such building SHALL NOT BE OCCUPIED until FINAL INSPECTION has been made and approved.

POST THIS CARD

BUILDING INSPECTION APPROVALS	PLUMBING INSPECTION APPROVALS	ELECTRICAL INSPECTION APPROVALS

FOOTING OR FOUNDATIONS
DATE _____
INSPECTOR _____
SLAB
DATE _____
INSPECTOR _____
FRAMING–PRIOR TO
SHEETROCK OR LATH
DATE _____
INSPECTOR _____
FINAL INSPECTION
DATE _____
INSPECTOR _____
	1. PLUMBING IN SLAB
	   DATE _____ INSP. _____
	2. ROUGH DRAINAGE AND VENTS
	   DATE _____ INSP. _____
	3. ROUGH WATER IN BUILDING
	   DATE _____ INSP. _____
	4. BUILDING SEWER
	   DATE _____ INSP. _____
	5. FINAL INSPECTION
	   DATE _____ INSP. _____
		1. ELECTRICAL TEMPORARY
		   DATE _____ INSP. _____
		2. ROUGH ELECTRICAL
		   DATE _____ INSP. _____
		3. FINAL INSPECTION
		   DATE _____ INSP. _____

MARKS		**GAS INSPECTION APPROVALS**
		1. ROUGH GAS PIPING
		   DATE _____ INSP. _____
		2. FINAL GAS PIPING
		   DATE _____ INSP. _____

	WORK SHALL NOT PROCEED UNTIL EACH DIVISION HAS APPROVED THE VARIOUS STAGES OF CONSTRUCTION.	**MECHANICAL INSPECTION APPROVALS**
		1. GAS VENT
		   DATE _____ INSP. _____
		2. FINAL MECHANICAL
		   DATE _____ INSP. _____

EXCAVATORS SHOULD CALL
LINE LOCATION CENTER
1-800-666-3000
48 HOURS BEFORE DIGGING

THE INSPECTION BY THE CITY OR COUNTY OF THIS BUILDING CONSTITUTES NO REPRESENTATION EXPRESS OR IMPLIED BY THE CITY OR COUNTY OR ITS EMPLOYEES AS TO THE COMPLIANCE OF THIS BUILDING WITH THE REQUIRED SET-BACK LINES OR THE QUALITY OF THE WORKMANSHIP OR MATERIALS USED THEREIN.

UNLAWFUL TO REMOVE OR DEFACE THIS CARD
UNTIL CONSTRUCTION IS COMPLETE

FIGURE 1.7 A building permit form must be posted at the job site during the period of construction.

electrical equipment have been installed. This inspection allows the inspector to verify that the service, branch circuits, and outlet locations comply with *National Electrical Code* requirements.

Finish (Final) Inspection. The second inspection takes place when the electrical installation is complete. This inspection allows the inspector to verify that the electrical system operates safely, that GFCI and AFCI protection are in place, and that all other applicable *Code* requirements have been met.

What Is "Red Tagging"? When an inspector finds *Code* violations on a job in progress, he or she typically leaves behind a list of deficiencies that must be corrected before approval will be granted. Often this form is posted at the service, or next to the electrical permit, and is typically called a "red tag" in the field (although it may not actually be red). Normally, the local jurisdiction will grant an occupancy permit, and the local utility will turn on power, only when all deficiencies have been corrected and the AHJ has approved the job.

Licensing

Many jurisdictions require that only licensed contractors and electricians perform work within the scope of the *NEC*. Some that require licensing provide an exception that allows owners and/or their employees (such as maintenance personnel and building engineers) to perform electrical work on the owner's own property. It is important for installers to know and comply with the applicable licensing regulations in the jurisdiction(s) where they are working.

"Qualified Person." The *National Electrical Code* doesn't require that work be performed by licensed electricians. The *Code* describes what must be done to ensure electrical safety, but not who must do the work. Some types of work covered by *NEC* rules are typically performed by engineers, technicians, or other specialized installers and maintenance personnel who, while not electricians, are technically competent to perform the work safely. Lastly, not all states and municipalities require licensing of electrical installers.

Rather than addressing licensing or training, the *Code* often refers to work being done by a "qualified person." Article 100 defines *qualified person* as "one who has skills and knowledge related to the construction and operation of the electrical equipment and installations and has received safety training on the hazards involved."

Working Safely is the First Priority

The purpose of the *National Electrical Code*, as stated in 90.1(A), is "the practical safeguarding of persons and property from hazards arising from the use of electricity." In other words, the intent of the *NEC* is to ensure that electrical distribution systems in homes and other buildings are safe for people to use. However, safety for workers on the job site while constructing buildings is equally important.

Construction work is inherently hazardous, and electrical construction even more so as electricity is dangerous. The hazards of working around electricity include shock and electrocution, fire, and arc-flash. Arc-flash (also called *arc-blast*) is a high energy "explosion" that can occur when something happens such as accidental shorting across the bus bars in a residential panelboard by, for example, dropping a pair of pliers.

NFPA 70E-2004, *Standard for Electrical Safety in the Workplace*, is the governing standard for protection against electrical hazards on construction sites (which are also workplaces). The electrical protection techniques described in this section are based on the safety rules of NFPA 70E (Figure 1.8).

In addition to electrical hazards, construction work also involves other dangers such as falling from roofs and ladders, injuries from dropped tools and materials, and accidents with power tools. Entire books have been written about construction safety. This section summarizes essential safety precautions when performing electrical construction work in one- and two-family dwellings.

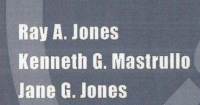

FIGURE 1.8 Because construction sites are workplaces, the electrical safety requirements of NFPA 70E apply.

Only Qualified Persons. *National Electrical Code* Article 100 defines a *qualified person* as "one who has skills and knowledge related to the construction and operation of the electrical equipment and installations and has received safety training on the hazards involved." NFPA 70E uses the same definition.

To help prevent accidents and injuries, only qualified persons meeting this definition should perform electrical construction and maintenance work. Untrained, unqualified persons should never be allowed to do electrical work.

Wear the Right Personal Protective Equipment (PPE). PPE is the first line of defense against injury. The minimum PPE for safe electrical construction work includes the following (Figure 1.9):

- Long-sleeved shirt and pants of natural fibers, such as cotton or wool. Don't wear synthetic fabrics such as polyester or nylon since they can melt and catch fire in case of an electrical arc–blast.
- Steel-toed boots.
- Hard hat. Only plastic hard hats can be worn for electrical work.
- Safety glasses or goggles.
- Work gloves.

In addition, workers should never wear metal jewelry such as rings, wristwatches, and bracelets when working around electrical circuits and equipment. Gold and silver are excellent conductors of electricity.

Shock Protection. Article 590 contains requirements for temporary power on construction sites. Among the most important safety rules is one requiring that all 125-volt, single-phase, 15-, 20-, and 30-ampere receptacles used for construction purposes such as supplying power tools and portable work lights have ground-fault circuit-interrupter (GFCI) protection for personnel (Figure 1.10).

This requirement applies whether the receptacles being used to supply construction equipment are themselves temporary or part of the permanent wiring of the

FIGURE 1.9 Hard hats and safety glasses or goggles are important items of personal protective equipment (PPE).

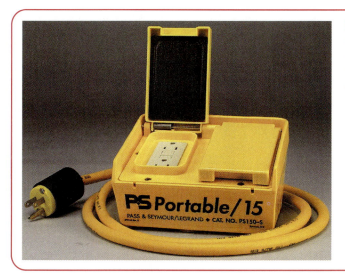

FIGURE 1.10 Portable plug-in GFCI units are often used on construction sites to comply with *NEC* 590.6(A).

dwelling [590.6(A)]. That means that when repairs or renovations in an existing residence are being done, *all* receptacles used for construction purposes must be GFCI-protected.

Use Power Tools Safely. Follow these precautions when using power tools:

- Always plug 120-volt power tools into GFCI-protected receptacles.
- When using 120-volt tools, be sure that extension cords are in good condition.
- Battery-powered tools are safer on construction sites. They reduce the risk of electric shock and also minimize tripping hazards from extension cords.
- Always wear safety glasses or goggles when using power tools.

Don't Work "Live." One of the most critical safety techniques is to work only on circuits and equipment that have been turned off or disconnected from the power supply, whenever possible. During new construction, circuits may not be live because the utility normally doesn't turn on power until the home has passed inspection by the AHJ. But in the latter stages of construction, circuits will often be live, and final testing and check-out can only be performed on energized systems. Workers must be extremely careful not to contact live parts during such testing.

Always test electrical circuits and parts before working on them to be sure the power is off. Common testing devices used in residential work include multimeters (Figure 1.11) and outlet testers. When work must be performed "live," such as testing and checking energized electrical systems, be sure to wear proper PPE and not contact the live parts during the testing operations.

Working around live panelboards with their covers off is particularly hazardous. A short circuit or faulty circuit breaker in an energized panelboard could result in an arc-blast, causing severe burns and other injuries to the workers involved. NFPA 70E requires the following additional PPE when performing "switching operations" on live panelboards (Figure 1.12):

- Fire-resistant (FR) clothing
- FR flash jackets or suits with hoods over the FR clothing

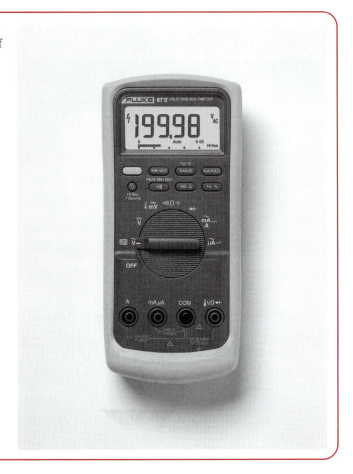

FIGURE 1.11 A digital multimeter typical of those used to test for voltage. (Courtesy of Fluke)

- Arc-rated face shield
- Hearing protection
- Voltage-rated gloves
- Voltage-rated tools

The subject of PPE against electrical hazards is complex, and the correct PPE needed depends upon the type of work being done. For complete information about this subject, see NFPA 70E-2004, *Standard for Electrical Safety in the Workplace.*

PPE isn't needed when there are no electrical hazards to protect against. So, the simplest safety rule for residential electrical construction is—*DON'T WORK LIVE!*

Lockout/Tagout. When functioning residential electrical systems are de-energized to perform work safely (for example, by turning off circuit breakers or removing fuses), precautions must be taken to ensure that circuits are not accidentally turned back on while work is going on.

Lockout/tagout is the preferred method of controlling electrical energy sources to minimize hazards to personnel (Figure 1.13). The details are complex and beyond the scope of this book. But every electrical contractor should have a company lockout/tagout procedure that should always be followed when electrical circuits are de-energized during construction or maintenance work. For more information, refer to NFPA 70E, Annex G, "Sample Lockout /Tagout Procedure."

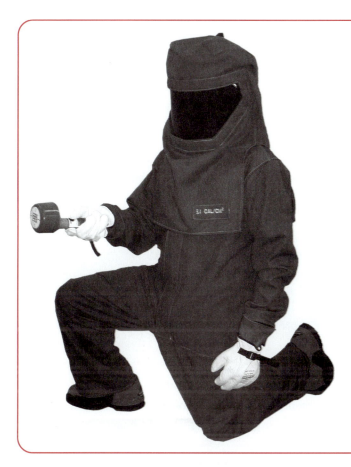

FIGURE 1.12 This worker is wearing complete PPE for working on an energized panelboard. (Courtesy of Salisbury)

FIGURE 1.13 Lockout devices are used to ensure that electrical circuits remain de-energized while work is being performed. (Courtesy of Salisbury)

Watch Out for Overhead Power Lines. Contacting overhead power lines (such as service drop conductors) with objects such as ladders, lengths of metal conduit, radio antennas, bucket trucks, and backhoes is a *major cause* of electrical injuries and deaths on construction sites (Figure 1.14). Take the following precautions to minimize this hazard:

• Determine the location of all overhead lines (if any) before beginning work.
• Where possible, barricade the areas under overhead lines or post warning signs.

FIGURE 1.14 Construction equipment must be used safely around overhead power lines.

- Keep all material and equipment at least 10 feet away from overhead lines.
- Use only nonconductive (fiberglass) ladders for electrical construction work.

Fall Protection. Falls are another major cause of construction accidents. Equipment and systems designed to prevent workers from falling include guardrails, plastic netting, personal fall restraint systems, and controlled access zones. As a general rule of thumb, fall protection should be used whenever there is a danger of falling six feet or more.

Normally, fall protection systems such as guardrails and plastic netting are provided by the general contractor to protect all trades on a construction job. Individual protection such as fall restraint systems (safety harnesses) are normally provided by each contractor to be used as needed for specific tasks.

Use Ladders Safely. Unsafe use of ladders causes many falling incidents on construction sites. Follow these precautions when using ladders.

- As previously mentioned, use only nonconductive (fiberglass) ladders for electrical construction work.
- Use a ladder long enough for the job. When climbing from one level of a structure to the next, at least three feet of the ladder should project above the top level.
- Place the legs of a ladder on a firm, level surface. Don't shim or prop up a ladder by using boards, bricks, or construction debris.
- Place ladders against walls using the "one-to-four rule." The base of the ladder should be one foot away from the wall or other vertical surface for every four feet of height to the point of support (Figure 1.15).
- When possible, secure the top of the ladder to the structure.

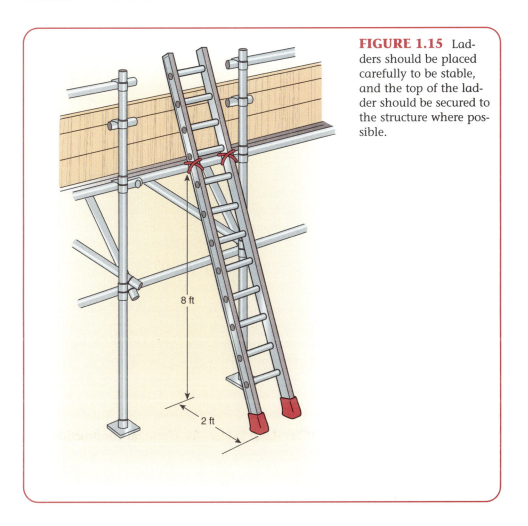

FIGURE 1.15 Ladders should be placed carefully to be stable, and the top of the ladder should be secured to the structure where possible.

- Never use ladders closer than 10 feet to overhead power lines such as service drop conductors.
- Always open stepladders to their fullest extent and lock the braces.
- Don't leave tools or materials on the shelf or top of a stepladder

Always use the "three-point contact" method when going up or down ladders. In other words, both feet and one hand, or both hands and one foot, should always be on the ladder simultaneously. This means that workers must face the ladder at all times and should not try to carry tools and materials in their hands when climbing ladders. Instead wear a tool belt and hand or lift up other materials after the worker has climbed the ladder.

When working on a ladder, the worker should hold a rung or rail with one hand at all times. If a task requires two hands, the worker should wear a safety belt or fall restraint harness.

SYMBOLS FOR RESIDENTIAL WIRING

NECA 100-1999, *Symbols for Electrical Construction Drawings* (ANSI), is the official U.S. standard for electrical symbols. Many symbols apply broadly to all types of electrical installations. Others are primarily related to commercial, industrial, and occupational construction and have little to do with residential wiring. Drawing symbols that apply to electrical systems in dwellings are shown in Figures 1.16 through 1.20.

BASICS OF RESIDENTIAL ELECTRICAL SYSTEMS

This book doesn't teach basic electrical theory. It assumes that users have a basic grounding in electricity and understand such concepts as current, voltage, resistance, and Ohm's Law. However, a few fundamentals about wiring systems installed in dwellings are worth mentioning before turning to a discussion of wiring methods in the next chapter.

Series and Parallel Circuits

The two basic types of electrical circuits are *series* and *parallel*. Parallel circuits are much more practical for nearly all real-life applications in ac power distribution. An example of an ac series circuit is old-fashioned, 2-wire strings of Christmas tree lighting from the 1960s. The filament of each lamp was a part of the series circuit, so if one burned out, it interrupted the circuit and the entire string went dark (Figure 1.21).

By contrast, modern decorative lighting strings connect lamps in parallel, so a single lamp can fail without affecting any of the others. The same is true of branch circuits supplying luminaires (lighting fixtures), receptacles, and other electrical equipment in houses.

Electrical plans illustrate circuits in a simplified form that makes it look as though devices and utilization equipment are wired in series. However, the conductors are actually connected in parallel, as shown in Figures 1.22 and 1.23.

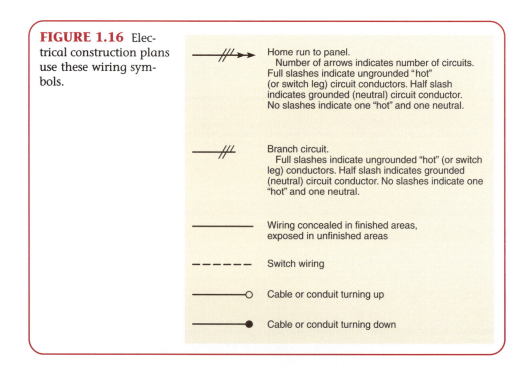

FIGURE 1.16 Electrical construction plans use these wiring symbols.

Home run to panel.
   Number of arrows indicates number of circuits. Full slashes indicate ungrounded "hot" (or switch leg) circuit conductors. Half slash indicates grounded (neutral) circuit conductor. No slashes indicate one "hot" and one neutral.

Branch circuit.
   Full slashes indicate ungrounded "hot" (or switch leg) conductors. Half slash indicates grounded (neutral) circuit conductor. No slashes indicate one "hot" and one neutral.

Wiring concealed in finished areas, exposed in unfinished areas

Switch wiring

Cable or conduit turning up

Cable or conduit turning down

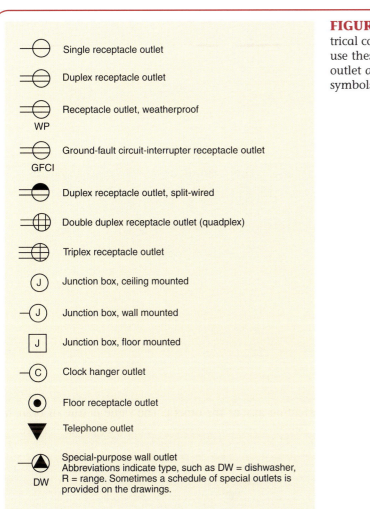

FIGURE 1.17 Electrical construction plans use these receptacle outlet and junction box symbols.

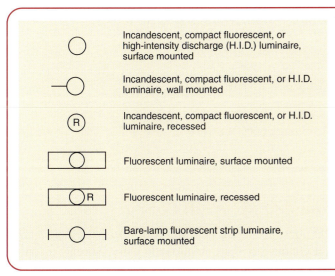

FIGURE 1.18 Electrical construction plans use these symbols for lighting outlets and luminaires.

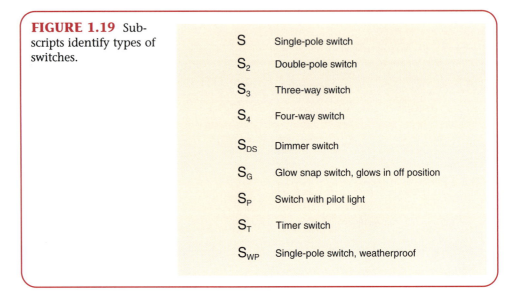

FIGURE 1.19 Subscripts identify types of switches.

S	Single-pole switch
S₂	Double-pole switch
S₃	Three-way switch
S₄	Four-way switch
S_DS	Dimmer switch
S_G	Glow snap switch, glows in off position
S_P	Switch with pilot light
S_T	Timer switch
S_WP	Single-pole switch, weatherproof

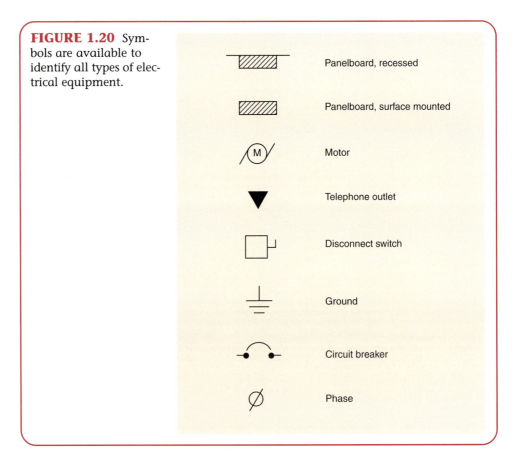

FIGURE 1.20 Symbols are available to identify all types of electrical equipment.

	Panelboard, recessed
	Panelboard, surface mounted
M	Motor
▼	Telephone outlet
	Disconnect switch
	Ground
	Circuit breaker
Ø	Phase

Hot (Phase) and Grounded (Neutral) Conductors

The *National Electrical Code* doesn't contain definitions of *hot*, *phase*, or *neutral* conductors. It refers to circuit conductors that intentionally carry current to supply outlets or utilization equipment as *ungrounded conductors* and to circuit conductors that are intentionally connected to ground (usually at the service equipment) as

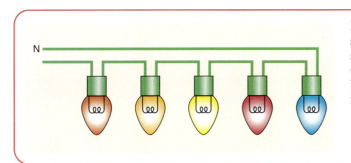

FIGURE 1.21 In a series circuit, such as a string of old-style Christmas lights, if one bulb burns out, there is no current to the others.

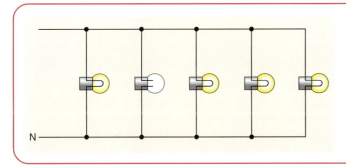

FIGURE 1.22 When lamps are connected in parallel, one can burn out without affecting others on the same circuit.

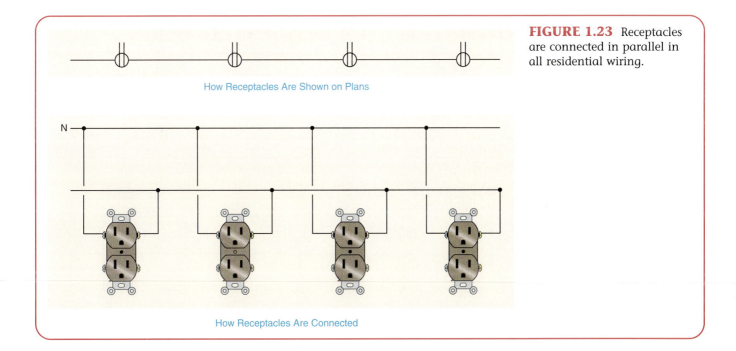

How Receptacles Are Shown on Plans

How Receptacles Are Connected

FIGURE 1.23 Receptacles are connected in parallel in all residential wiring.

grounded conductors, a term that also is defined in Article 100. Grounded conductors normally have white or gray insulation, and are covered by the rules of Article 200.

For everyday purposes, electricians refer to *ungrounded conductors* as "phase" or "hot" wires. They refer to all intentionally *grounded conductors* as "neutrals," although, strictly speaking, only the non–current-carrying conductor of a balanced,

3-phase, 4-wire, wye-connected circuit is neutral. Grounded conductors of 2-wire circuits (Figure 1.24) carry current under normal operating conditions and, for this reason, cannot be considered "neutral."

One reason electricians don't normally use the official *Code* terminology for these circuit conductors is to avoid confusion with *grounding conductors*, which are used to connect equipment and grounded conductors to grounding electrodes. For clarity, this book refers to *grounded conductors* as "grounded (neutral) conductors." This subject is covered in greater detail in Chapter 3, Services, Service Equipment, and Grounding.

Using Standard *Code* Terminology

In every other respect, this book uses standard *NEC* terminology as defined in Article 100. For example, it uses the term *panelboard* consistently, rather than *panel* or *load center*. Similarly, this book refers to cable types and other products by their official *Code* names rather than by the alternate terms by which they are also known.

Many electrical products used in residential wiring systems have slang or "field" names. Sometimes these names are based on the common brand name of a product. The origins of other field names are unknown.

Whatever the source, field names for electrical products aren't the names used in the *NEC*. Using field names leads to confusion and misunderstanding. Some common field names to be avoided are listed in Table 1.1.

DEFINITIONS FOR RESIDENTIAL WIRING

Article 100 of the *National Electrical Code* contains definitions. Many apply broadly to all types of electrical installations. Others are primarily related to commercial,

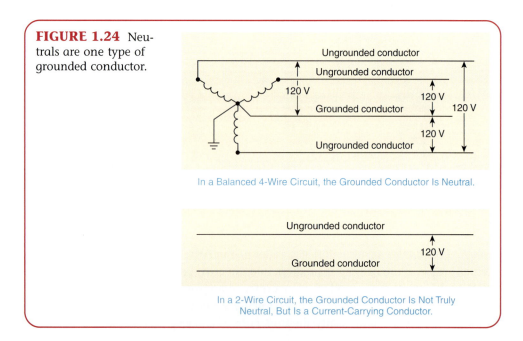

FIGURE 1.24 Neutrals are one type of grounded conductor.

In a Balanced 4-Wire Circuit, the Grounded Conductor Is Neutral.

In a 2-Wire Circuit, the Grounded Conductor Is Not Truly Neutral, But Is a Current-Carrying Conductor.

TABLE 1.1 Field Names for Electrical Products

Field Name	*Code* Name	Article
BX®	Armored cable, Type AC	320
Condulet	Conduit body	314
Flex	Flexible metal conduit, Type FMC	348
GFI	Ground-fault circuit interrupter (GFCI)	210, 680, and others
Greenfield	Flexible metal conduit, Type FMC	348
Green ground	Grounding electrode conductor	250 and others
Guts	Panelboard	408
Load center	Panelboard	408
Plug	Receptacle	210, 406, and others
Plugmold	Multioutlet assembly	380
Red head, red devil	Anti-short bushing for AC cable	320
Romex®	Nonmetallic-sheathed cable, Type NM	334
Rope	Nonmetallic-sheathed cable, Type NM	334
Sealtight, sealtite	Liquidtight flexible metal conduit, Type LFMC	350
Smurf tube	Electrical nonmetallic tubing, Type ENT	362
Sub-panel	Panelboard	408
Toggle switch	Snap switch, wall switch	210, 404, and others
Thin-wall	Electrical metallic tubing, Type EMT	358
Wiremold	Surface metal raceway	386
Wiremold	Surface nonmetallic raceway	388

industrial, and occupational construction and have little to do with residential wiring. Official *NEC* definitions that apply to electrical systems in dwellings are listed in the "*NEC* Definitions" section at the back of this book. A list entitled "Important *NEC* Terms" is given at the beginning of each chapter for the reader's convenience and contains terms related to that chapter.

1 CHAPTER REVIEW

MULTIPLE CHOICE

1. The *National Electrical Code* chapter that covers special occupancies is

A. Article 90
B. Chapter 1
C. Chapter 5
D. Chapter 745

2. The first four chapters of the *NEC*

A. Apply generally to all electrical installations.
B. Cover communications systems including telephone and cable television wiring used in homes.
C. Contain tables relating to conductors installed in raceways.
D. Both A and B

3. Which of the following meets the *Code* definition of *dwelling unit*?

A. Penthouse condominium
B. Townhouse
C. Extended-stay hotel room with full kitchen
D. All of the above

4. The following *NEC* article defines wiring practices for temporary power on construction sites:

A. Article 250
B. Article 305
C. Article 590
D. Article 840

5. The National Electrical Installation Standards (NEIS) are

A. Building codes
B. Annexes to the *National Electrical Code*
C. Quality standards for electrical construction
D. Both B and C

6. *BX*, *greenfield*, *guts*, and *toggle switch* are all examples of

A. Electrical products used in dwellings
B. Wiring methods
C. Field names
D. None of the above

7. The *Comprehensive Consensus Codes*® (C3®) are

A. Product listing agencies
B. Wiring rules for dwellings
C. Training materials
D. A series of building codes that includes the *NEC*

8. When an AHJ conducts more than one inspection of a dwelling under construction, these are typically called

A. Rough
B. Final
C. Red tag
D. Both A and B

9. Listed electrical products are

A. Tested for safety and compliance with *NEC* requirements
B. Marked or labeled in some way
C. Included in a published directory
D. All of the above

10. A *qualified person*, as defined in Article 100, is required to have

A. An electrician license
B. Graduated from an apprenticeship program
C. Skills, knowledge, and safety training
D. *NEC* certification

FILL IN THE BLANKS

1. The title of NFPA 70 is _____.

2. A housekeeping unit with space for eating, living, and sleeping, and permanent provisions for cooking and sanitation is known as a _____.

3. Products can only be approved by _____.

4. _____ is the official *Code* name for what is commonly called a "neutral."

5. A _____ is one who has skills and knowledge related to the construction and operation of the electrical equipment and has received safety training on the hazards involved.

6. The quality and workmanship standards published by the National Electrical Contractors Association (NECA) are known as _____.

7. Subjects covered by Chapter 2 of the *National Electrical Code* are _____.

8. Romex® is a common brand name for a wiring method whose proper *Code* name is _____.

9. The purpose of the *National Electrical Code*, as stated in _____, is "the practical safeguarding of persons and property from hazards arising from the use of electricity."

10. Article 100 defines *dwelling unit* as "a single unit, providing complete and independent living facilities for one or more persons, including permanent provisions for _____."

TRUE OR FALSE

1. The *NEC* requires a minimum of 12 AWG conductors for branch-circuit wiring.

    ____ True   ____ False

2. The *NEC* is a building code.

    ____ True   ____ False

3. The *NEC* refers to product testing and listing agencies as Nationally Recognized Testing Laboratories (NRTLs).

    ____ True   ____ False

4. "Listed" means the same as "labeled."

    ____ True   ____ False

5. All electricians are required to be licensed.

    ____ True   ____ False

6. The *National Electrical Code* automatically supersedes (is enforced in place of) local electrical codes.

    ____ True   ____ False

7. Other NFPA codes besides the *NEC* have electrical requirements that apply to dwelling units.

    ____ True   ____ False

8. Organizations such as Underwriters Laboratories Inc. approve electrical products for use according to applicable *Code* rules.

    ____ True   ____ False

9. The AHJ is responsible for approving equipment, materials, installations, and procedures.

    ____ True   ____ False

10. Article 590 requires that all 125-volt, single-phase, 15-, 20-, and 30-ampere receptacles used for construction purposes have GFCI protection for personnel.

    ____ True   ____ False

CHALLENGE QUESTIONS

1. Why isn't the commonly used "*NEC* approved" or "approved by the *NEC*" accurate?

2. *Code* users frequently confuse the concepts of Article and Section. Explain the difference.

3. Discuss why the *National Electrical Code* uses the concept of "qualified person" rather than simply requiring that all work be performed by licensed electricians. What are the two key requirements for a "qualified person"?

4. Why can't the *grounded* conductor of a two-wire branch circuit be considered a neutral? And what is an *ungrounded* conductor?

5. Is the typical hotel/motel room considered a *dwelling unit*? How about an extended-stay type hotel room that includes a kitchenette? Explain your answer.

6. What is the difference between *listed* and *labeled* electrical equipment?

Planning the Installation
Required Branch Circuits and Load Calculations

2

OBJECTIVES

After completing this chapter, the student will be able to understand the following:

- Required branch circuits in dwellings
- Branch circuit ratings
- Load calculations
- Connected loads vs. demand loads
- Special rules for two-family dwellings

INTRODUCTION

This chapter explains planning and design considerations for residential electrical construction, according to the *National Electrical Code®*. The following major topics are included:

- Residential branch circuits
- Planning the installation
- Determining the number of branch circuits required
- Calculating the service load
- Computing floor area for service load calculations
- Special rules for two-family dwellings

IMPORTANT NEC TERMS

Branch Circuit
Branch Circuit, Appliance
Branch Circuit, General-Purpose
Branch Circuit, Individual
Building
Circuit Breaker
Continuous Load
Demand Factor
Dwelling Unit
Dwelling, One-Family
Dwelling, Two-Family

RESIDENTIAL BRANCH CIRCUITS

Although not intended as a design specification or instruction manual [90.1(C)], the *National Electrical Code*® provides a great deal of guidance in planning the circuits of a residential installation. Article 100 defines *branch circuit* as follows.

> **Branch Circuit:** The circuit conductors between the final overcurrent device protecting the circuit and the outlet(s).

An *individual branch circuit* supplies only one utilization equipment, sometimes through a single receptacle outlet (Figure 2.1). A *general-purpose branch circuit* supplies two or more receptacles or outlets for lighting and appliances.

Branch-Circuit Ampere Ratings and Overcurrent Protection

The rating of a branch circuit is not determined by conductor size, but by the rating of its overcurrent device. A 15-ampere branch circuit is protected by a 15-ampere circuit breaker or fuse. If a 30-ampere overcurrent device is used, then it's a 30-ampere branch circuit [210.3].

Branch circuits supplying multiple outlets are permitted to be rated 15, 20, 30, 40, and 50 amperes [210.3]. Most branch circuits used in homes to supply receptacle and lighting outlets are rated either 15 or 20 amperes. Individual branch circuits used to supply large loads—such as electric laundry and cooking equipment, water heaters, furnaces, and air-conditioning units—may have higher ratings, such as 40, 50, 60, and 100 amperes.

Summary of Branch-Circuit Requirements. *NEC* Table 210.24, shown in Exhibit 2.1, provides a convenient summary of branch-circuit requirements.

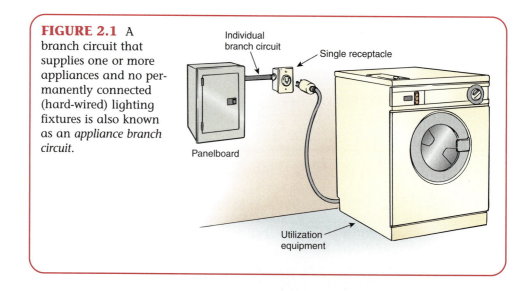

FIGURE 2.1 A branch circuit that supplies one or more appliances and no permanently connected (hard-wired) lighting fixtures is also known as an *appliance branch circuit*.

Individual branch circuit

Single receptacle

Panelboard

Utilization equipment

EXHIBIT 2.1

NEC TABLE 210.24 Summary of Branch-Circuit Requirements

Circuit Rating	15 A	20 A	30 A	40 A	50 A
Conductors (min. size):					
Circuit wires[1]	14	12	10	8	6
Taps	14	14	14	12	12
Fixture wires and cords (See 240.5)					
Overcurrent Protection	**15 A**	**20 A**	**30 A**	**40 A**	**50 A**
Outlet devices:					
Lampholders permitted	Any type	Any type	Heavy duty	Heavy duty	Heavy duty
Receptacle rating[2]	15 max. A	15 or 20 A	30 A	40 or 50 A	50 A
Maximum Load	**15 A**	**20 A**	**30 A**	**40 A**	**50 A**
Permissible load	See 210.23(A)	See 210.23(A)	See 210.23(B)	See 210.23(C)	See 210.23(C)

[1]These gauges are for copper conductors.
[2]For receptacle rating of cord-connected electric-discharge luminaires (lighting fixtures), see 410.30(C).

Branch-Circuit Loading

NEC Article 100 defines *continuous load* as follows.

Continuous Load: A load where the maximum current is expected to continue for 3 hours or more.

Lighting and ceiling paddle fans are good examples of continuous loads (Figure 2.2). Kitchen food waste disposers (which typically operate for no more than 30 seconds at a time) and clothes washers (30 minutes at a time) are *noncontinuous* loads. Many other electric loads in dwelling units—water heaters and cooking, heating, and air-conditioning equipment—are also treated as *noncontinuous* loads because they are controlled by thermostats that prevent them from operating without a break for 3 hours or more.

Where a branch circuit supplies a continuous load or any combination of continuous and noncontinuous loads, the rating of the overcurrent device is required to be

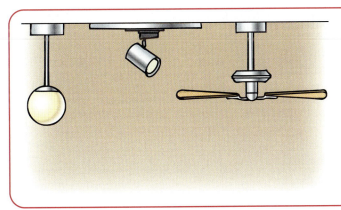

FIGURE 2.2 Luminaires, lighting track, and ceiling-suspended paddle fans are among those loads that must be regarded as *continuous*.

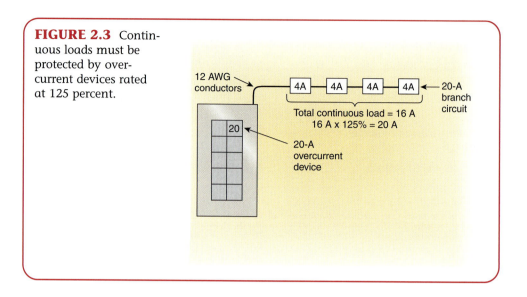

FIGURE 2.3 Continuous loads must be protected by overcurrent devices rated at 125 percent.

at least the noncontinuous load plus 125 percent of the continuous load [210.20(A)] (Figure 2.3).

Circuit Breaker Ratings. In addition, molded-case circuit breakers for residential applications are not listed for use at more than 80 percent of their ratings. Thus, a 20-ampere circuit breaker can actually support only 16 amperes of load ($20 \times 0.80 = 16$), and a 15-ampere circuit breaker can support only 12 amperes of load ($15 \times 0.80 = 12$).

In practical terms, these two facts mean that branch circuits in dwellings are never loaded beyond 80 percent of their overcurrent device ratings.

Minimum Branch-Circuit Ratings

The minimum circuit conductor size permitted by the *NEC* is 14 AWG, protected by a 15-ampere breaker or fuse. Certain branch circuits in dwellings are required to use 12 AWG with 20-ampere overcurrent protection. This requirement is primarily due to increasing loads for household appliances plugged into convenience receptacles.

A 15-ampere branch circuit can supply a maximum continuous load of 1440 volt-amperes (120 volts $\times$ 15 amperes $\times$ 0.80 = 1440 volt-amperes), and a 20-ampere branch circuit can supply a maximum continuous load of 1920 volt-amperes (120 volts $\times$ 20 amperes $\times$ 0.80 = 1920 volt-amperes).

But listed plug-in appliances such as hair dryers and coffee makers are available in ratings up to 1800 watts, which exceeds the continuous duty rating of a 15-ampere circuit breaker. For this reason, the *Code* requires bathroom branch circuits and small-appliance branch circuits used in kitchens and dining rooms to be rated 20 amperes. If exposed to loads exceeding 1440 volt-amperes for an extended period, 15-ampere circuit breakers would trip.

Branch-Circuit Voltages

Most homes have services rated 120/240 volts, single-phase, 3-wire. Branch circuits supplied by these services are rated either 120 volts or 240 volts. Some larger, all-electric homes have 208Y/120-volt, 3-phase, 4-wire services similar to those in commercial buildings. Branch circuits supplied by these services are rated either 120 volts or 208 volts.

Services rated 480Y/277 volt, 3-phase, 4-wire are not used for one- and two-family dwellings. Some high-rise apartment and condominium buildings do have 480Y/277-volt services, with the voltage stepped down to 120/240 volts for individual dwelling units. However, multifamily dwellings are outside the scope of this book.

Equipment Voltage Ratings. Motors and heating equipment intended for use on 240-volt residential branch circuits must be rated 240 volts. If "commercial" electrical appliances rated 208 volts were used, motors would run slower and the output of heating equipment would be lower. And from a safety standpoint, lower voltage means higher amperes, which increases the risk of overloads. For this reason, *NEC* 110.4 requires that "the voltage rating of electrical equipment shall not be less than the nominal voltage of a circuit to which it is connected."

Voltage Drop Considerations

The *National Electrical Code* does not have requirements for voltage drop. However, for reasonable efficiency of operation, Fine Print Notes (FPNs) to 210.19(A)(1) and 215.2(A)(3) recommend that the maximum total voltage drop on both feeders and branch circuits should be 5 percent.

Excessive voltage drop degrades the operating characteristics of electrical equipment. Motors run hotter and produce less torque. Heating appliances run hotter and have a shorter useful life. Incandescent light output is reduced. Fluorescent and high-intensity discharge (HID) luminaires may experience starting problems and reduced light output.

Calculation Example

Use the following basic formula to determine the voltage drop in a 2-wire ac circuit:

$$E_{VD} = IR$$

where:

E_{VD} = voltage drop

I = current in amperes

R = resistance in ohms

Use the formula to determine the voltage drop over two 12 AWG solid copper conductors, each 85 ft long, serving a 20-ampere load. The total circuit length is 170 ft.

NEC Table 8, Conductor Properties, gives different resistance values for coated and uncoated wire. Since branch-circuit wiring is insulated, use the value of 1.93 ohms per thousand feet:

$$I = 20 \text{ amperes}$$

$$R = 1.93 \text{ ohms per 1000 ft}$$

$$\frac{85 \times 2}{1000} = 0.170 \times 1.93$$

$$= 0.3281 \text{ ohms}$$

Substituting gives

$$E_{VD} = IR$$
$$= 20 \text{ A} \times 0.3281 \text{ } \Omega$$
$$= 6.562 \text{ V}$$

Using Larger Conductors to Minimize Voltage Drop. Few residential installers perform voltage-drop calculations for feeders and branch circuits. However, there is a trend toward using 12 AWG wiring for more branch circuits in dwellings, even where the practice is not specifically required by the *Code*. Using 20-ampere branch circuits with larger wires reduces voltage drop, wastes less energy through overheated wires, and minimizes tripping of circuit breakers due to overloads.

Another approach is to install 12 AWG wire for all 15-ampere homeruns or branch circuits longer than 50 ft. This method is an easy, "rule-of-thumb" approach that helps keep total voltage drop between the service equipment and the final outlet on a branch circuit within the 5 percent maximum recommended by the *Code*.

The additional cost to install 12 AWG conductors rather than 14 AWG during original construction is very low.

NOTE: Some jurisdictions don't permit the use of 14 AWG wire for branch circuits, requiring minimum 12 AWG instead. This requirement, in turn, means that the minimum overcurrent device rating used will be 20 amperes, rather than 15 amperes.

PLANNING THE INSTALLATION

It isn't necessary to design or lay out the entire electrical system in advance in order to calculate a residential service size or select the service panelboard. However, it is necessary to select the cooking equipment; water heater; clothes dryer; and heating, ventilating, and air-conditioning (HVAC) equipment beforehand.

The use of high-load electrical equipment rated 240 volts increases both the service size and the number of pole spaces in the panelboard. By contrast, fossil-fuel heating appliances (natural gas or oil) are small electrical loads that use 120-volt power only to operate equipment such as controls, clocks, and gas igniters. These appliances in turn reduce the number of pole spaces required. The three major steps in planning for a typical residential electrical installation follow.

Step 1. *Select major appliances.* This step focuses on the wide variety of electrical appliances available today (Figure 2.4). Normally this step is done by the builder, architect, or property owner, who furnishes information on the electrical characteristics to the electrical designer. Often this electrical designer is the contractor who will wire the house, although some residences have electrical systems designed in advance by consulting engineers.

Step 2. *Determine the number of branch circuits (and panelboard pole spaces) required.* The electrical designer determines these numbers based on the rules of *NEC* Article 210, Branch Circuits, and the voltage ratings of the major appliances. A 120-volt branch circuit requires one pole space, while a 240-volt branch circuit requires two pole spaces.

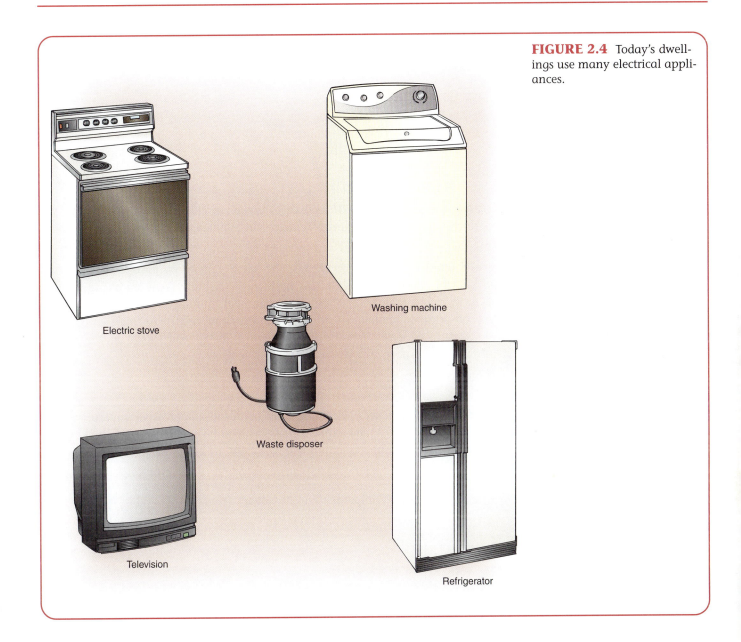

FIGURE 2.4 Today's dwellings use many electrical appliances.

Electric stove

Washing machine

Waste disposer

Television

Refrigerator

Step 3. *Calculate the service load.* The electrical designer calculates the service load based on the rules of *NEC* Article 220, Branch Circuit, Feeder, and Service Calculations.

DETERMINING THE REQUIRED NUMBER OF BRANCH CIRCUITS

The number of branch circuits required is based on the square footage (usable floor area) of the residence, specific branch circuits required in all dwelling units, and the major appliances selected. Article 210 contains the rules for branch circuits, which are explained in the following sections.

General-Purpose Branch Circuits

The terms *lighting load* and *lighting branch circuits* are actually not quite accurate when applied to houses. Receptacle outlets and permanently installed lighting are generally on different branch circuits in commercial, industrial, and institutional construction (they may even operate at different voltages). In residential wiring, however, receptacle outlets for lighting and permanently installed lighting normally share the same branch circuit—a *general-purpose branch circuit*—which is defined in Article 100 as follows.

> **Branch Circuit, General-Purpose:** A branch circuit that supplies two or more receptacles or outlets for lighting and appliances.

General Lighting Load. In dwellings, the same general-purpose branch circuits supply not only permanently installed luminaires (lighting fixtures) but receptacle outlets for floor and table lamps, home entertainment equipment, personal computers, clocks, and portable appliances, all of which consume relatively small amounts of power. Thus, in the case of homes, the general lighting load shown in *NEC* Table 220.12 is actually more of a general electrical load.

Calculation Example

NEC Table 220.12 specifies a lighting load of 3 volt-amperes per square foot for dwelling units, and 210.11(B) requires that this load be evenly proportioned among multioutlet branch circuits. Assuming a house floor area of 2200 sq ft, the minimum number of lighting branch circuits needed is calculated using the following formula:

$$\frac{3 \text{ volt-amperes (VA)} \times \text{square feet}}{120 \text{ volts}} = \text{amperes}$$

which gives

$$\frac{3 \text{ VA} \times 2200 \text{ sq ft}}{120 \text{ V}} = 55 \text{ A}$$

Using 15-ampere branch circuits,

$$\frac{55 \text{ A}}{15 \text{ A}} = 3.67 \text{ branch circuits (round up to 4 circuits)}$$

This result works out to one 15-ampere lighting branch circuit per each 550 sq ft (2200 ÷ 4 = 550).

Using 20-ampere branch circuits,

$$\frac{55 \text{ A}}{20 \text{ A}} = 2.75 \text{ branch circuits (round up to 3 circuits)}$$

This result works out to one 20-ampere lighting branch circuit per each 733 sq ft (2200 ÷ 3 = 733).

Code **Minimums and Design Considerations.** In both calculations shown in the previous example, the "rounded up" answer is only slightly larger than the calculated answer. But flexibility and convenience are often increased by exceeding *Code* safety minimums.

The last sentence of 210.11(B) states that "branch-circuit overcurrent devices and circuits shall only be required to be installed to serve the connected load." However, many electricians feel it is a good rule of thumb to provide one 15-ampere lighting circuit for every 500 sq ft of living space. This rule gives 3.6 volt-amperes per square foot:

$$120 \text{ volts} \times 15 \text{ amperes} = 1800 \text{ volt-amperes (VA)}$$
$$1800 \text{ VA} \div 500 \text{ sq ft} = 3.6 \text{ VA/sq ft}$$

Where 20-ampere branch circuits are used, installing one for every 650 sq ft of living space provides a nearly equal power density of 3.69 volt-amperes per square foot:

$$120 \text{ V} \times 20 \text{ A} = 2400 \text{ VA}$$
$$2400 \text{ VA} \div 500 \text{ sq ft} = 3.69 \text{ VA/sq ft}$$

No Additional Loads for General-Use Outlets. The general lighting load of 3 volt-amperes per square foot covers all general-use lighting and receptacle outlets in a dwelling. No additional load calculations are required for these outlets, according to 220.14(J). [This rule is different from the *Code* rule for commercial construction: 220.14(I) requires that a load of 180 volt-amperes be calculated for each receptacle outlet. However, this rule doesn't apply to dwellings.]

The *NEC* doesn't limit the number of receptacle outlets that can be installed on a single 15- or 20-ampere branch circuit. The reason is that receptacles in homes tend to be lightly loaded (compared to commercial buildings) for at least two reasons:

1. Most home appliances, other than electric cooking equipment, personal grooming appliances, and power tools, don't draw much power.
2. Receptacle outlets located behind furniture may never have anything plugged into them.

CAUTION: Some jurisdictions have their own rules that limit the number of receptacle outlets that may be installed on a single circuit in residential construction to the same maximums permitted in commercial construction (Figure 2.5).

Other Required Branch Circuits

Small-Appliance Branch-Circuit Loads. Section 210.11(C)(1) requires a minimum of two 20-ampere branch circuits to supply receptacle outlets in the kitchen, dining room, pantry, and breakfast room of a dwelling. Some dwellings have more than two small-appliance branch circuits, even though only two are required.

Laundry Branch Circuit. At least one 20-ampere branch circuit is required to supply a receptacle outlet for laundry equipment, as shown in Figure 2.6. This circuit isn't permitted to have other outlets [210.11(C)(2)].

Bathroom Branch Circuit. At least one 20-ampere branch circuit is required to supply a receptacle outlet(s) in bathrooms. This circuit is not permitted to have other

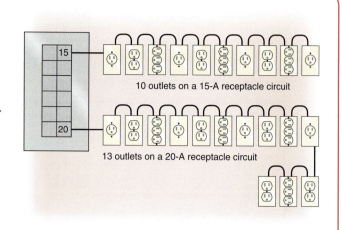

FIGURE 2.5 For occupancies other than dwellings, *NEC* 220.14(I) specifies the maximum number of receptacle outlets permitted on 15- and 20-ampere branch circuits.

10 outlets on a 15-A receptacle circuit

13 outlets on a 20-A receptacle circuit

FIGURE 2.6 A receptacle outlet supplied by a 20-ampere branch circuit is required for laundry equipment.

outlets, unless it supplies only a single bathroom [210.11(C)(3) Exception]. In other words, a branch circuit that serves only a single bathroom may supply all the outlets in that bathroom: lighting, receptacles, exhaust fan, wall heater, and so on.

Appliance Branch Circuits

NEC Article 100 defines an *appliance branch circuit* as follows.

Branch Circuit, Appliance: A branch circuit that supplies energy to one or more outlets to which appliances are to be connected and that has no permanently connected luminaires (lighting fixtures) that are not a part of an appliance.

TABLE 2.1   Types of Appliance Branch Circuits

Type	Voltage	Pole Spaces
Range (electric)	240	2
Range (gas)	*	*
Oven (electric)	240	2
Oven (gas)	*	*
Refrigerator[1]	120	1
Water heater (electric)	240	2
Water heater (gas)	120	1
Furnace (electric)	240	2
Furnace (gas, oil)	120	1
Dishwasher/disposer	120	1
Clothes dryer (electric)	240	2
Clothes dryer (gas)	**	**

*Supplied by small-appliance branch circuits [210.52(B)(2), Exception No. 2].
**Supplied by laundry branch circuits [210.11(C)(2)].
[1]The receptacle outlet for a refrigerator may be supplied by the small-appliance branch circuits [210.52(B)(1)] or by a separate circuit [210.52(B)(1), Exception No. 2].

The *NEC* has no rule specifying minimum numbers of appliance branch circuits. The number varies from one dwelling to another, depending on the types of appliances installed. Some appliances for which separate branch circuits are often provided are listed in Table 2.1.

Additional Branch Circuits

In addition to the required branch circuits and appliance branch circuits listed previously, many dwellings have additional branch circuits to supply items such as the following:

- Workshops with power tools and multi-outlet assemblies
- Outdoor lighting
- Swimming pools, fountains, and spas
- Snow-melting equipment
- Home theater rooms

Panelboard Configurations

The total pole spaces required to serve the circuits in a dwelling can be determined based on the factors described previously. Following the three steps is an important part of advance planning for a residential electrical installation.

Figure 2.7 shows a typical circuit directory for a residential panelboard. Note that circuits 1-3 for the central air conditioner and circuits 2-4 for the electric range are actually 2-pole, 240-volt circuit breakers. All other circuits in this panelboard have single-pole, 120-volt breakers. Circuits 12, 13, and 19, serving lighting and receptacle outlets in bedrooms, are all protected by arc-fault circuit-interrupter (AFCI) circuit breakers, as required by *NEC* 210.12(B). This subject is covered in Chapter 9, The Bedroom.

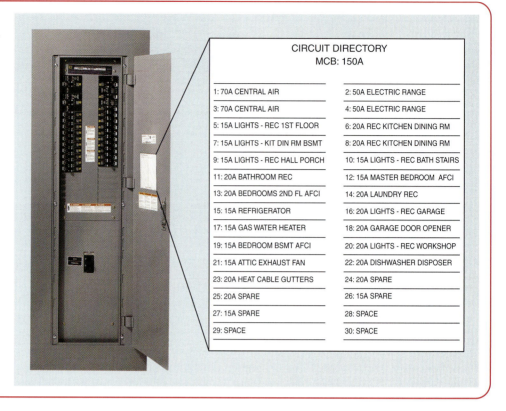

FIGURE 2.7 A typical panelboard directory is filled in by the installer, identifying each circuit. (Photo portion courtesy of Schneider Electric)

CIRCUIT DIRECTORY
MCB: 150A

1: 70A CENTRAL AIR	2: 50A ELECTRIC RANGE
3: 70A CENTRAL AIR	4: 50A ELECTRIC RANGE
5: 15A LIGHTS - REC 1ST FLOOR	6: 20A REC KITCHEN DINING RM
7: 15A LIGHTS - KIT DIN RM BSMT	8: 20A REC KITCHEN DINING RM
9: 15A LIGHTS - REC HALL PORCH	10: 15A LIGHTS - REC BATH STAIRS
11: 20A BATHROOM REC	12: 15A MASTER BEDROOM AFCI
13: 20A BEDROOMS 2ND FL AFCI	14: 20A LAUNDRY REC
15: 15A REFRIGERATOR	16: 20A LIGHTS - REC GARAGE
17: 15A GAS WATER HEATER	18: 20A GARAGE DOOR OPENER
19: 15A BEDROOM BSMT AFCI	20: 20A LIGHTS - REC WORKSHOP
21: 15A ATTIC EXHAUST FAN	22: 20A DISHWASHER DISPOSER
23: 20A HEAT CABLE GUTTERS	24: 20A SPARE
25: 20A SPARE	26: 15A SPARE
27: 15A SPARE	28: SPACE
29: SPACE	30: SPACE

The rating of the service and main overcurrent device (in this case, a main circuit breaker, or MCB) are determined as described in the next section.

CALCULATING THE SERVICE LOAD

Article 220 contains rules for branch-circuit, feeder, and service calculations. All dwellings have branch circuits. Few have feeders (typically only those with subpanels for pool equipment, accessory buildings, and the like).

Every dwelling has a service. The size of the service and the rating of the main overcurrent protective device are based on the calculated load. This load is calculated according to the rules of Article 220, based on the loads of the major appliances and other equipment that will be installed in the house and the numbers and types of branch circuits specified in Article 210. (Required branch circuits are discussed in the previous section of this chapter.)

Examples of residential load calculations appear in Annex D of the *National Electrical Code* and are shown in Exhibit 2.4 later in this chapter. Explanations of each part of the sample calculations follow.

Calculated Load and Demand Factors

Calculated load consists of the general lighting load specified by 220.12, certain required branch circuits, cooking loads, laundry loads, and heating/air-conditioning loads. Parts of the connected load are subject to *demand factors*. These factors recognize that the demand load of electrical circuits (actual power draw) is often less than the connected load (based on equipment ratings).

In other words, all the lights in a dwelling will rarely be on at the same time, and all four burners of an electric range will rarely be turned to high at the same moment. For this reason, it makes sense to permit the calculated service load to be based on a lower figure.

A discussion of the general load for residential service calculations follows. Also see "Computing Floor Area for Load Calculations" at the end of this section.

Lighting Load. *NEC* Table 220.12 specifies the general lighting load for dwelling units as 3 volt-amperes per square foot. This lighting load doesn't depend on the number of general-purpose branch circuits installed to serve lighting and receptacle outlets. Instead, it is calculated based on the usable floor area of a dwelling unit. Demand factors for lighting load in dwelling units are provided in *NEC* Table 220.42, shown in Exhibit 2.2.

Small-Appliance Branch Circuits. The standard calculation for residential service size includes 1500 volt-amperes for each of two 20-ampere small-appliance branch circuits (3000 volt-amperes total). But if more than two small-appliance branch circuits are actually installed in a house, 1500 volt-amperes for each circuit must be included for calculation purposes [220.52(A)]. The demand factors of Table 220.42 can also be applied to this portion of the service load.

Laundry Branch Circuit. The calculation for service size includes 1500 volt-amperes for the 20-ampere laundry circuit. If more than one laundry branch circuit is actually installed in a house, 1500 volt-amperes for each circuit must be included

EXHIBIT 2.2

NEC TABLE 220.42 Lighting Load Demand Factors

Type of Occupancy	Portion of Lighting Load to Which Demand Factor Applies (Volt-Amperes)	Demand Factor (Percent)
Dwelling units	First 3000 or less at	100
	From 3001 to 120,000 at	35
	Remainder over 120,000 at	25
Hospitals*	First 50,000 or less at	40
	Remainder over 50,000 at	20
Hotels and motels, including apartment houses without provision for cooking by tenants*	First 20,000 or less at	50
	From 20,001 to 100,000 at	40
	Remainder over 100,000 at	30
Warehouses (storage)	First 12,500 or less at	100
	Remainder over 12,500 at	50
All others	Total volt-amperes	100

*The demand factors of this table shall not apply to the calculated load of feeders or services supplying areas in hospitals, hotels, and motels where the entire lighting is likely to be used at one time, as in operating rooms, ballrooms, or dining rooms.

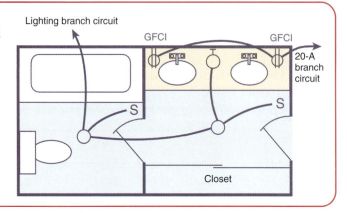

FIGURE 2.8 No load calculation is performed for the required 20-ampere branch circuit supplying bathroom receptacles.

for calculation purposes [220.42(B)]. The demand factors of Table 220.42 can also be applied to this portion of the service load.

Bathroom Branch Circuit. No load is included for the 20-ampere bathroom branch circuit required by 210.11(C)(3) and shown in Figure 2.8. This circuit is considered part of the general lighting load of 3 volt-amperes per square foot.

Electric Heat. Fixed electric space-heating equipment is included at 100 percent of total connected load. This equipment includes both electric furnaces and other permanently installed units such as wall heaters and baseboard heaters. The Exception to 220.51 allows the authority having jurisdiction (AHJ) to permit smaller service conductors under certain circumstances.

Miscellaneous Appliances. Section 220.53 permits applying a demand factor of 75 percent to the loads of four or more appliances fastened in place. This factor doesn't apply to cooking, laundry, space heating, or air-conditioning equipment. These types of appliances all have their own *Code* rules governing how loads are computed for dwelling service calculations.

Calculation Example

Determine the 120/240-volt service load needed for the following fastened-in-place appliances in a dwelling unit:

Appliance	Rating	Load
Water heater	4500 W, 240 V	4500 VA
Kitchen waste disposer	¼ hp, 120 V	696 VA
Dishwasher	1250 W, 120 V	1250 VA
Furnace motor	¼ hp, 120 V	696 VA
Whole-house fan	½ hp, 120 V	1176 VA

First, calculate the total of the five fastened-in-place appliances:

Total load = 4500 VA + 696 VA + 1250 VA + 696 VA + 1176 VA
= 8318 VA

Because the load is for more than four appliances, apply a demand factor of 75 percent:

$$8318 \text{ VA} \times 0.75 = 6239 \text{ VA}$$

Electric Dryer. The service load for a household electric clothes dryer shall be either 5000 watts (volt-amperes) or the nameplate rating, whichever is higher [220.54]. No load is included for a gas clothes dryer, which is plugged into a receptacle outlet on the laundry branch circuit. This appliance uses only a small amount of electricity for controls and the gas igniter.

Electric Range and Other Cooking Appliance(s). The service load for household electric ranges, wall-mounted ovens, counter-mounted cooking units (commonly called *cooktops*), and other permanently installed cooking appliances rated more than 1¾ kW (1750 volt-amperes) can be calculated in accordance with *NEC* Table 220.55 (Exhibit 2.3). This table specifies demand factors to be used with different numbers and ratings of electric cooking appliances.

Note 4 to Table 220.55 allows a demand factor to be applied even on the branch-circuit load for a single electric range. The reason is that all burners plus the oven will rarely be turned on at the same time. However, this same Note 4 specifies that the load for one electric wall oven or one electric cooktop shall be the nameplate rating of the appliance.

Gas Appliances. No load is included in the service calculation for gas cooking appliances, which are typically plugged into receptacle outlets on the small-appliance branch circuit(s). These appliances use only a small amount of electricity for items such as controls, clocks, lights, and gas igniters.

Cord-and-Plug-Connected Appliances. No load is included in the service calculation for cord-and-plug-connected electric cooking appliances such as microwave ovens, toasters, and coffee makers. These are plugged into receptacle outlets on the small-appliance branch circuit(s).

Noncoincident Loads — Electric Heat and Air Conditioning. Noncoincident loads are those not likely to be used at the same time, such as electric heat and air conditioning. Section 220.60 permits only the largest noncoincident load to be used when calculating service load; the smaller one can be ignored. In an all-electric dwelling, the heating load is normally larger than the air-conditioning load, which means that only the electric heating load must be considered when calculating service capacity.

Nonconincendent Loads — Gas or Oil Heating. Fossil fuel heating systems typically have a 120-volt, 15-ampere branch circuit to supply electric controls and a gas igniter. In these cases, the air-conditioning load is normally larger than the heating load and is used to calculate service capacity.

EXHIBIT 2.3

NEC **TABLE 220.55 Demand Factors and Loads for Household Electric Ranges, Wall-Mounted Ovens, Counter-Mounted Cooking Units, and Other Household Cooking Appliances over 1¾ kW Rating (Column C to be used in all cases except as otherwise permitted in Note 3.)**

| | Demand Factor (Percent) (See Notes) | | Column C |
Number of Appliances	Column A (Less than 3½ kW Rating)	Column B (3½ kW to 8¾ kW Rating)	Maximum Demand (kW) (See Notes) (Not over 12 kW Rating)
1	80	80	8
2	75	65	11
3	70	55	14
4	66	50	17
5	62	45	20
6	59	43	21
7	56	40	22
8	53	36	23
9	51	35	24
10	49	34	25
11	47	32	26
12	45	32	27
13	43	32	28
14	41	32	29
15	40	32	30
16	39	28	31
17	38	28	32
18	37	28	33
19	36	28	34
20	35	28	35
21	34	26	36
22	33	26	37
23	32	26	38
24	31	26	39
25	30	26	40
26–30	30	24	15 kW + 1 kW for each range
31–40	30	22	
41–50	30	20	25 kW + ¾ kW for each range
51–60	30	18	
61 and over	30	16	

1. Over 12 kW through 27 kW ranges all of same rating. For ranges individually rated more than 12 kW but not more than 27 kW, the maximum demand in Column C shall be increased 5 percent for each additional kilowatt of rating or major fraction thereof by which the rating of individual ranges exceeds 12 kW.
2. Over 8¾ kW through 27 kW ranges of unequal ratings. For ranges individually rated more than 8¾ kW and of different ratings, but none exceeding 27 kW, an average value of rating shall be calculated by adding together the ratings of all ranges to obtain the total connected load (using 12 kW for any range rated less than 12 kW) and dividing by the total number of ranges. Then the maximum demand in Column C shall be increased 5 percent for each kilowatt or major fraction thereof by which this average value exceeds 12 kW.
3. Over 1¾ kW through 8¾ kW. In lieu of the method provided in Column C, it shall be permissible to add the nameplate ratings of all household cooking appliances rated more than 1¾ kW but not more than 8¾ kW and multiply the sum by the demand factors specified in Column A or B for the given number of appliances. Where the rating of cooking appliances falls under both Column A and Column B, the demand factors for each column shall be applied to the appliances for that column, and the results added together.
4. Branch-Circuit Load. It shall be permissible to calculate the branch-circuit load for one range in accordance with Table 220.55. The branch-circuit load for one wall-mounted oven or one counter-mounted cooking unit shall be the nameplate rating of the appliance. The branch-circuit load for a counter-mounted cooking unit and not more than two wall-mounted ovens, all supplied from a single branch circuit and located in the same room, shall be calculated by adding the nameplate rating of the individual appliances and treating this total as equivalent to one range.
5. This table also applies to household cooking appliances rated over 1¾ kW and used in instructional programs.

COMPUTING DWELLING UNIT FLOOR AREA FOR SERVICE LOAD CALCULATIONS

Residential load calculations are based in part on floor area. Section 220.12 requires a general lighting load of 3 volt-amperes per square foot for dwelling units. The floor area is computed from the *outside dimensions* of the building or apartment-condominium unit. Open porches, garages, or other unused or unfinished spaces, if they are not adaptable for future use, can be omitted from the total floor area used to calculate general lighting load.

However, basements (or portions of them) are often finished into living space, such as spare bedrooms or recreation rooms, after original construction. Garages are often turned into home offices or family rooms, porches are sometimes enclosed, and attic spaces with adequate headroom may be converted into usable rooms later. In order to calculate a service size that meets a homeowner's long-term needs, it is often more realistic to consider part or all of the square footage of these "unfinished spaces" as usable floor area when performing load calculations.

Figures 2.9 and 2.10 show the same house plans with the floor area to be used in computing service size figured two ways: Figure 2.9 omits the basement and garage, while Figure 2.10 includes the garage floor area plus half the basement. (The other half of the basement is a laundry room and mechanical equipment area.)

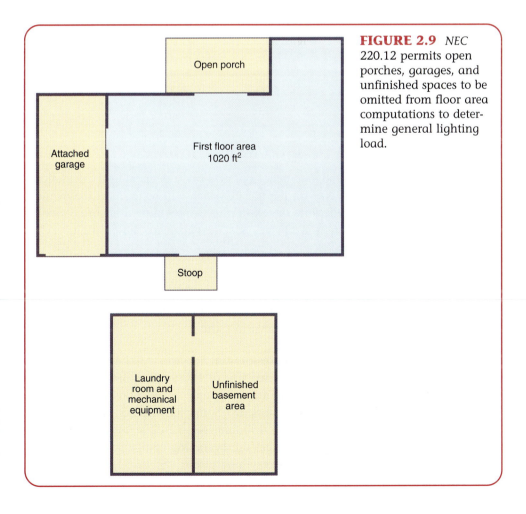

FIGURE 2.9 *NEC* 220.12 permits open porches, garages, and unfinished spaces to be omitted from floor area computations to determine general lighting load.

Open porch

First floor area
1020 ft^2

Attached garage

Stoop

Laundry room and mechanical equipment

Unfinished basement area

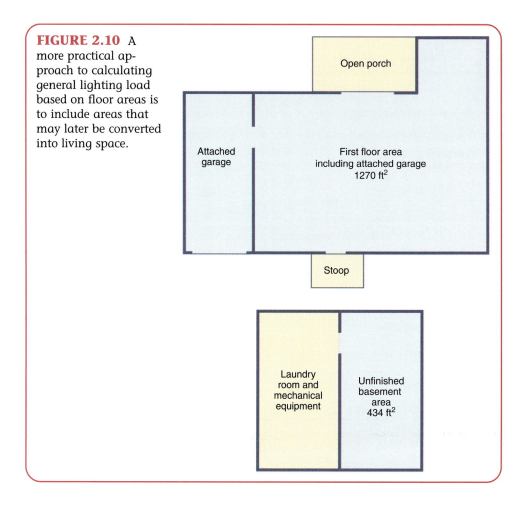

FIGURE 2.10 A more practical approach to calculating general lighting load based on floor areas is to include areas that may later be converted into living space.

Open porch

Attached garage

First floor area including attached garage 1270 ft²

Stoop

Laundry room and mechanical equipment

Unfinished basement area 434 ft²

Code Minimums and Design Considerations

Including parts of the garage and basement floor area in the demand load calculations may result in a larger service size for the house, which will better meet the owning family's long-term needs. And while the extra cost of providing a larger service at the time of original construction is fairly low, "heavying up" an existing service later on is much more expensive.

Service Calculation Examples

Sample service calculations for single-family dwellings are given in Examples D1(a) through D2(c) of *NEC* Annex D and shown in Exhibit 2.4.

Importance of Installing an Adequate Residential Service. Although the *National Electrical Code* requires a minimum service size of 100 ampere, 3-wire, for a one-family dwelling [230.79(C)], all but one of the sample calculations in Exhibit 2.4 result in a service size exceeding 100 amperes. Some jurisdictions now require a minimum service size for new houses of 150 amperes, 120/240 volt, single-phase, 3-wire; and 200-ampere residential services are becoming common.

Article 220 load calculations, like everything else in the *NEC*, represent minimums required for safety. However, the service installed in any residence should

EXHIBIT 2.4

NEC Annex D Examples

Example D1(a) One-Family Dwelling

The dwelling has a floor area of 1500 ft^2, exclusive of an unfinished cellar not adaptable for future use, unfinished attic, and open porches. Appliances are a 12-kW range and a 5.5-kW, 240-V dryer. Assume range and dryer kW ratings equivalent to kVA ratings in accordance with 220.54 and 220.55. **Calculated Load** *[see 220.40]*

General Lighting Load: 1500 ft^2 at 3 VA per ft^2 = 4500 VA

Minimum Number of Branch Circuits Required *[see 210.11(A)]*

General Lighting Load: 4500 VA ÷ 120 V = 37.5 A
  This requires three 15-A, 2-wire or two 20-A, 2-wire circuits.
  Small Appliance Load: Two 2-wire, 20-A circuits *[see 210.11(C)(1)]*
  Laundry Load: One 2-wire, 20-A circuit *[see 210.11(C)(2)]*
  Bathroom Branch Circuit: One 2-wire, 20-A circuit (no additional load calculation is required for this circuit) *[see 210.11(C)(3)]*

Minimum Size Feeder Required *[see 220.40]*

General Lighting		4,500 VA
Small Appliance		3,000 VA
Laundry		1,500 VA
	Total	9,000 VA
3000 VA at 100%		3,000 VA
9000 VA − 3000 VA = 6000 VA at 35%		2,100 VA
	Net Load	5,100 VA
Range *(see Table 220.55)*		8,000 VA
Dryer Load *(see Table 220.54)*		5,500 VA
	Net Calculated Load	18,600 VA

Net Calculated Load for 120/240-V, 3-wire, single-phase service or feeder

18,600 VA ÷ 240 V = 77.5 A

Sections 230.42(B) and 230.79 require service conductors and disconnecting means rated not less than 100 amperes.

Calculation for Neutral for Feeder and Service

Lighting and Small Appliance Load		5,100 VA
Range: 8000 VA at 70% *(see 220.61)*		5,600 VA
Dryer: 5500 VA at 70% *(see 220.61)*		3,850 VA
	Total	14,550 VA

Calculated Load for Neutral

14,550 VA ÷ 240 V = 60.6 A

Example D1(b) One-Family Dwelling

Assume same conditions as Example No. D1(a), plus addition of one 6-A, 230-V, room air-conditioning unit and one 12-A, 115-V, room air-conditioning unit,* one 8-A, 115-V, rated waste disposer, and one 10-A, 120-V, rated dishwasher. See Article 430 for general motors and Article 440, Part VII, for air-conditioning equipment. Motors have nameplate ratings of 115 V and 230 V for use on 120-V and 240-V nominal voltage systems.
  *(For feeder neutral, use larger of the two appliances for unbalance.)
  From Example D1(a), feeder current is 78 A (3-wire, 240 V).

	Line A	Neutral	Line B
Amperes from Example D1(a)	78	61	78
One 230-V air conditioner	6	—	6
One 115-V air conditioner and 120-V dishwasher	12	12	10
One 115-V disposer	—	8	8
25% of largest motor *(see 430.24)*	3	3	2
Total amperes per line	99	84	104

Therefore, the service would be rated 110 A.

Example D2(a) Optional Calculation for One-Family Dwelling, Heating Larger Than Air Conditioning
[see 220.82]

The dwelling has a floor area of 1500 ft^2, exclusive of an unfinished cellar not adaptable for future use, unfinished attic, and open porches. It has a 12-kW range, a 2.5-kW water heater, a 1.2-kW dishwasher, 9 kW of electric space heating installed in five rooms, a 5-kW clothes dryer, and a 6-A, 230-V, room air-conditioning unit. Assume range, water heater, dishwasher, space heating, and clothes dryer kW ratings equivalent to kVA.

Air Conditioner kVA Calculation

6 A × 230 V ÷ 1000 = 1.38 kVA

This 1.38 kVA [item 1 from 220.82(C)] is less than 40% of 9 kVA of separately controlled electric heat [item 6 from 220.82(C)], so the 1.38 kVA need not be included in the service calculation.

General Load

1500 ft^2 at 3 VA		4,500 VA
Two 20-A appliance outlet circuits at 1500 VA each		3,000 VA
Laundry circuit		1,500 VA
Range (at nameplate rating)		12,000 VA
Water heater		2,500 VA
Dishwasher		1,200 VA
Clothes dryer		5,000 VA
	Total	29,700 VA

Application of Demand Factor *[see 220.82(B)]*

First 10 kVA of general load at 100%		10,000 VA
Remainder of general load at 40% (19.7 kVA × 0.4)		7,880 VA
	Total of general load	17,880 VA
9 kVA of heat at 40% (9000 VA × 0.4) =		3,600 VA
	Total	21,480 VA

Calculated Load for Service Size

21.48 kVA = 21,480 VA

21,480 VA ÷ 240 V = 89.5 A

Therefore, the minimum service rating would be 100 A in accordance with 230.42 and 230.79.

Feeder Neutral Load, per 220.61

1500 ft^2 at 3 VA		4,500 VA
Three 20-A circuits at 1500 VA		4,500 VA
	Total	9,000 VA
3000 VA at 100%		3,000 VA
9000 VA − 3000 VA = 6000 VA at 35%		2,100 VA
	Subtotal	5,100 VA
Range: 8 kVA at 70%		5,600 VA
Clothes dryer: 5 kVA at 70%		3,500 VA
Dishwasher		1,200 VA
	Total	15,400 VA

(Continues)

EXHIBIT 2.4

NEC Annex D Examples *(Continued)*

Calculated Load for Neutral

15,400 VA ÷ 240 V = 64.2 A

Example D2(b) Optional Calculation for One-Family Dwelling, Air Conditioning Larger Than Heating *[see 220.82(A) and 220.82(C)]*

The dwelling has a floor area of 1500 ft², exclusive of an unfinished cellar not adaptable for future use, unfinished attic, and open porches. It has two 20-A small appliance circuits, one 20-A laundry circuit, two 4-kW wall-mounted ovens, one 5.1-kW counter-mounted cooking unit, a 4.5-kW water heater, a 1.2-kW dishwasher, a 5-kW combination clothes washer and dryer, six 7-A, 230-V room air-conditioning units, and a 1.5-kW permanently installed bathroom space heater. Assume wall-mounted ovens, counter-mounted cooking unit, water heater, dishwasher, and combination clothes washer and dryer kW ratings equivalent to kVA.

Air Conditioning kVA Calculation

Total amperes = 6 units × 7 A = 42 A

42 A × 240 V ÷ 1000 = 10.08 kVA (assume PF = 1.0)

Load Included at 100%

Air Conditioning: Included below *[see item 1 in 220.82(C)]*
Space Heater: Omit *[see item 5 in 220.82(C)]*

General Load

1500 ft² at 3 VA	4,500 VA
Two 20-A small appliance circuits at 1500 VA each	3,000 VA
Laundry circuit	1,500 VA
Two ovens	8,000 VA
One cooking unit	5,100 VA
Water heater	4,500 VA
Dishwasher	1,200 VA
Washer/dryer	5,000 VA
Total general load	32,800 VA
First 10 kVA at 100%	10,000 VA
Remainder at 40% (22.8 kVA × 0.4 × 1000)	9,120 VA
Subtotal general load	19,120 VA
Air conditioning	10,080 VA
Total	29,200 VA

Calculated Load for Service

29,200 VA ÷ 240 V = 122 A (service rating)

Feeder Neutral Load, per 220.61

Assume that the two 4-kVA wall-mounted ovens are supplied by one branch circuit, the 5.1-kVA counter-mounted cooking unit by a separate circuit.

1500 ft² at 3 VA		4,500 VA
Three 20-A circuits at 1500 VA		4,500 VA
	Subtotal	9,000 VA
3000 VA at 100%		3,000 VA
9000 VA − 3000 VA = 6000 VA at 35%		2,100 VA
	Subtotal	5,100 VA

Two 4-kVA ovens plus one 5.1-kVA cooking unit = 13.1 kVA. Table 220.55 permits 55% demand factor or 13.1 kVA × 0.55 = 7.2 kVA feeder capacity.

Subtotal from above	5,100 VA
Ovens and cooking unit: 7200 VA × 70% for neutral load	5,040 VA
Clothes washer/dryer: 5 kVA × 70% for neutral load	3,500 VA
Dishwasher	1,200 VA
Total	14,840 VA

Calculated Load for Neutral

14,840 VA ÷ 240 V = 61.83 A (use 62 A)

Example D2(c) Optional Calculation for One-Family Dwelling with Heat Pump (Single-Phase, 240/120-Volt Service) *(see 220.82)*

The dwelling has a floor area of 2000 ft², exclusive of an unfinished cellar not adaptable for future use, unfinished attic, and open porches. It has a 12-kW range, a 4.5-kW water heater, a 1.2-kW dishwasher, a 5-kW clothes dryer, and a 2½-ton (24-A) heat pump with 15 kW of backup heat.

Heat Pump kVA Calculation

24 A × 240 V ÷ 1000 = 5.76 kVA

This 5.76 kVA is less than 15 kVA of the backup heat; therefore, the heat pump load need not be included in the service calculation *[see 220.82(C)]*.

General Load

2000 ft² at 3 VA	6,000 VA
Two 20-A appliance outlet circuits at 1500 VA each	3,000 VA
Laundry circuit	1,500 VA
Range (at nameplate rating)	12,000 VA
Water heater	4,500 VA
Dishwasher	1,200 VA
Clothes dryer	5,000 VA
Subtotal general load	33,200 VA
First 10 kVA at 100%	10,000 VA
Remainder of general load at 40% (23,200 VA × 0.4)	9,280 VA
Total net general load	19,280 VA

Heat Pump and Supplementary Heat*

240 V × 24 A = 5760 VA

15 kW Electric Heat:

5760 VA + (15,000 VA × 65%) = 5.76 kVA
+ 9.75 kVA = 15.51 kVA

*If supplementary heat is not on at same time as heat pump, heat pump kVA need not be added to total.

Totals

Net general load	19,280 VA
Heat pump and supplementary heat	15,510 VA
Total	34,790 VA

Calculated Load for Service

34.79 kVA × 1000 ÷ 240 V = 144.96 A

Therefore, this dwelling unit would be permitted to be served by a 150-A service.

also provide spare capacity to serve future loads and additions without overloading the service conductors or requiring a later "heavy up." Owner convenience and long-term satisfaction often are improved by exceeding *NEC* minimums; see 90.8(A) and the FPN to 90.1(B).

> ### *General Guidelines for Planning a Service*
> - An apartment-condominium unit should have a minimum service size of 100 ampere, 3-wire.
> - A townhouse should have a minimum service size of 150 amperes, 3-wire.
> - A detached house should have a minimum service size of 200 amperes, 3-wire.
> - The service rating should exceed the calculated demand load by 25 to 50 percent. [Example D2(c), shown in Exhibit 2.4, recommends a 150-ampere service to serve a calculated demand load of 144.96 amperes, but there is very little spare capacity for future increases in electricity use. In this case, 200 amperes might be a better choice for a service rating.]
> - The service panel should provide at least 25 percent spare pole spaces for future overcurrent protective devices. (So if a total of 32 pole spaces are needed to serve branch circuits at the time of original construction, installing a 40- or 42-pole panel would provide ample spare capacity for future expansion—for example, adding more branch circuits to serve garages, basements, and attics converted into living spaces.)

Considerations for 3-Phase Services. Very large custom homes sometimes require 208Y/120-volt, 3-phase, 4-wire services, particularly if they are all-electric or have large loads such as guest houses, swimming pool heaters, or extensive snow-melting equipment.

1. This type of service requires special planning and advance consultation with the local utility, since 3-phase service often isn't available in residential areas.
2. Houses with 3-phase services must use heating equipment and motors rated for operation at 208 volts rather than 240 volts. The *Code* requires that the voltage rating of electrical equipment not be less than the nominal voltage of a circuit to which it is connected [110.4].

TWO-FAMILY DWELLINGS

Each dwelling unit in a two-family or multifamily dwelling follows essentially the same *Code* rules as a single-family dwelling, with some minor exceptions. For example, the panelboard in each unit isn't required to be listed as service equipment. Instead, one set of service equipment for the whole building supplies each individual unit by means of a feeder.

The *NEC* also has some additional special rules for two-family dwellings. These rules are summarized and explained in more detail at the end of each chapter in this book.

What Is a Two-Family Dwelling?

NEC Article 100 defines a dwelling units as follows.

> **Dwelling Unit:** A single unit, providing complete and independent living facilities for one or more persons, including permanent provisions for living, sleeping, cooking, and sanitation.

Article 100 also provides the following definitions.

> **Building:** A structure that stands alone or that is cut off from adjoining structures by fire walls with all openings therein protected by approved fire doors.
>
> **Dwelling, One-Family:** A building that consists solely of one dwelling unit.
>
> **Dwelling, Two-Family:** A building that consists solely of two dwelling units.

According to these definitions, typical duplex houses are not two-family dwelling units (Figure 2.11). Instead, they are two separate buildings standing side-by-side and sharing a common fire wall. In other words, typical duplex houses are actually pairs of one-family dwellings. They are wired in accordance with all the standard *Code* rules that apply to one-family dwellings.

A two-family dwelling is a building that contains two dwelling units not separated by fire walls. "Two-plexes" or "two-deckers" consisting of two apartments located one above the other are the most common type of two-family dwelling, though this arrangement isn't the only design.

Each unit in a two-family dwelling must comply with all the standard wiring rules for dwelling units, as well as the special *Code* rules that apply to two-family dwellings.

FIGURE 2.11 Duplex houses separated by a fire wall are actually pairs of one-family dwellings.

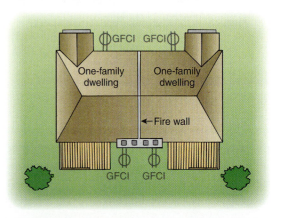

SPECIAL RULES FOR TWO-FAMILY DWELLINGS

- ***Common Area Branch Circuits Prohibited.*** Branch circuits in a dwelling unit are permitted to supply only loads located within, or associated with, that dwelling unit. Section 210.25 prohibits branch circuits from feeding more than one dwelling unit and prohibits the sharing of systems such as fire alarms, telephones, laundry room circuits, and lighting for public and common areas.

  All such common equipment is required to be supplied from a separate "house loads" panelboard. This method permits access to the branch-circuit disconnecting means without the need to enter any individual tenant's space. It also prevents a tenant from turning off important circuits that might affect other tenants.

  Not every two-family dwelling has common loads or a house panel. Some two-family dwellings are designed in such a way that all loads can be supplied from individual tenant panels. For example, each unit may have its own washer and dryer, rather than sharing laundry equipment.

- ***Meters and Service Equipment.*** Two-family dwellings often have three meter housings, one for each living unit and the third for the "house panel," which serves stair or hall lighting, outdoor lighting, basements, storage areas, and laundry rooms (Figure 2.12). Each of the three panelboards is required to be marked as being suitable for use as service equipment [230.66]. Additional details of designing and installing services for two-family dwellings are discussed in Chapter 3, Services, Service Equipment, and Grounding.

- ***Outdoor Receptacles.*** Each unit of a two-family dwelling that is located at grade level must have at least one GFCI-protected receptacle outlet installed at the front and back of the dwelling [210.8(A)(3), 210.52(E)]. Figure 2.13 illustrates a situation in which, because of the sloping grade, the lower unit of a two-family vacation house has only a single GFCI-protected receptacle outlet. There is no "back" side to the bottom dwelling unit. The receptacle on the deck of the upper unit isn't required by the *Code*. However, if an outdoor receptacle outlet is provided, it must have GFCI protection [210.8(A)(3)].

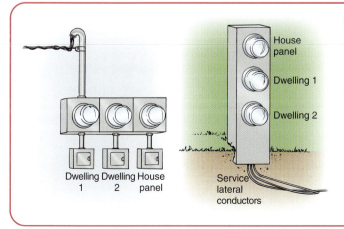

FIGURE 2.12 These overhead and underground services for two-family dwellings include meters for "house load" panelboards

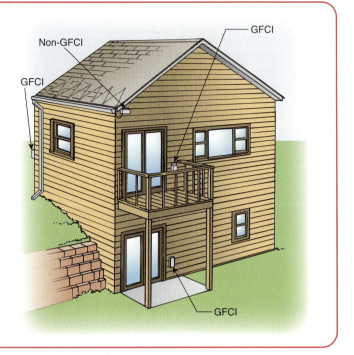

FIGURE 2.13 This illustration shows the GFCI-protected receptacle outlets required for a two-family dwelling built into a hillside.

MULTIPLE CHOICE

1. The minimum wire size permitted by the *NEC* for branch circuits is
   - A. 14 AWG
   - B. 12 AWG
   - C. 12 AWG for small-appliance and bathroom branch circuits in dwelling units
   - D. Both A and C

2. A dwelling unit, as defined in the *NEC*, must have
   - A. Permanent provisions for living, sleeping, cooking, and sanitation
   - B. Space for living and provisions for cooking
   - C. Ground-fault circuit-interrupter protection
   - D. Both A and C

3. Branch-circuit ratings depend on
   - A. The voltage of the circuit
   - B. The size of the ungrounded conductor(s) in the branch circuit
   - C. The rating or setting of the overcurrent protective device
   - D. Section 220.17(A)

4. The most common service voltage for one- and two-family dwellings is
   - A. 208Y/120 volts, 3-phase, 4-wire
   - B. 480Y/277 volts, 3-phase, 4-wire
   - C. 120/240 volts, single-phase, 3-wire
   - D. 240/120 volts, 3-phase, 4-wire (delta)

5. The maximum number of receptacle outlets permitted to be served by a 15-ampere branch circuit in a dwelling unit is
   - A. 10
   - B. 13
   - C. 16
   - D. None of the above

6. The *Code* specifies a general lighting load for dwellings of
   - A. 3 volt-amperes per square foot
   - B. 3 watts per square foot
   - C. 33 volt-amperes per square foot
   - D. 33 watts per square foot

7. The following branch circuit(s) is (are) required in every dwelling unit
   - A. Small-appliance branch circuits
   - B. Bathroom branch circuit
   - C. Sump pump branch circuit
   - D. Both A and B

8. As a general rule of thumb for planning residential services, it is a good idea to allow this (these) minimum(s)
   - A. 100 ampere, 3-wire, for an apartment-condominium unit
   - B. 150 ampere, 3-wire, for a townhouse
   - C. 200 ampere, 3-wire, for a detached house
   - D. All of the above

9. General-purpose branch circuits of 120 volts are rated
   - A. 15 amperes
   - B. 20 amperes
   - C. 25 amperes
   - D. Both A and B

10. A 20-ampere branch circuit can supply a maximum continuous load of
    - A. 1440 volt-amperes
    - B. 1800 volt-amperes
    - C. 1920 volt-amperes
    - D. 1200 volt-amperes

FILL IN THE BLANKS

1. A *continuous load* is a load in which the maximum current is expected to continue for _____ hours or more.

2. The minimum 3-wire size service permitted by the *NEC* for a one-family dwelling is _____ amperes.

3. Branch circuits supplying multiple outlets are permitted to have the following ampere ratings: _____.

4. When computing usable floor area for branch circuit, feeder, and service calculations according to Article 220, name at least one type of area that is permitted to be excluded if "not adaptable for future use": _____.

5. The demand loads for electric ranges and cooking appliances in dwellings are determined according to Table _____ in the *NEC*.

6. Each 240-volt branch circuit requires _____ pole space(s).

7. Small-appliance branch circuits and laundry branch circuits are required to be rated _____ amperes.

8. Section 220.53 permits applying a demand factor of _____ percent to the loads of four or more appliances fastened in place. (This rule doesn't apply to cooking, laundry, space heating, or air-conditioning equipment, which have their own *Code* rules governing how loads are computed.)

9. Large custom homes sometimes require services rated _____, particularly if they are all-electric or have large loads such as guest houses and snow-melting equipment.

10. The *NEC* recommends a maximum total voltage drop of _____ on both feeders and branch circuits.

TRUE OR FALSE

1. A branch circuit that supplies energy only for outlets to which appliances and permanently installed luminaires (lighting fixtures) are connected is called an *appliance branch circuit.*

_____ True   _____ False

2. A building that consists of two apartments, one above the other, is a two-family dwelling.

_____ True   _____ False

3. A one-family dwelling with two laundry rooms or areas is required to have two laundry branch circuits.

_____ True   _____ False

4. Voltage drop causes electrical equipment to run cooler and last longer.

_____ True   _____ False

5. When calculating connected load, outdoor lighting is required to be added to the interior lighting load of 3 volt-amperes per square foot before the demand factor is applied.

_____ True   _____ False

6. In a two-family dwelling, the common area loads are divided between the two dwelling unit panelboards.

_____ True   _____ False

7. Townhouses divided from the units on either side by firewalls are considered to be individual buildings (i.e., one-family dwellings).

_____ True   _____ False

8. The general lighting load for dwellings includes loading for receptacle outlets.

_____ True   _____ False

9. When gas cooking appliances will be installed in a residence, a load of 180 volt-amperes must be included in the service calculation to cover such items as an integral clock, convenience receptacle, and control circuitry.

_____ True   _____ False

10. The *National Electrical Code* requires a minimum of three branch circuits to supply appliances.

_____ True   _____ False

CHALLENGE QUESTIONS

1. Where a branch circuit supplies a continuous load or a combination of continuous and noncontinuous loads, the rating of the circuit breaker or fuse is required to be at least the noncontinuous load plus 125 percent of the continuous load. Discuss the reasons for this.

2. Most branch circuits in dwelling units are rated 15 amperes but the *Code* requires that bathroom, small-appliance, and laundry branch circuits be rated 20 amperes. What is the reason for this requirement?

3. Discuss how voltage drop causes performance problems for appliances, luminaires, and other utilization equipment found in residences.

4. Explain why the general lighting load of Table 220.12 also covers dwelling-unit receptacle outlets, while in other occupancies, receptacle outlets are assigned a load of 180-volt amperes for calculation purposes.

5. What is the general concept behind the use of demand factors in Article 220 Branch-Circuit, Feeder, and Service Calculations?

6. Why are electric heat and air conditioning considered to be "noncoincident loads?"

7. Discuss why *NEC* 220.12 permits open porches, garages, and unfinished spaces to be omitted from floor area computations to determine general lighting load.

8. Why do most newly installed residential services exceed the minimum *Code* rating of 100 amperes for a 120/240 volt, single-phase, 3-wire system?

Services, Service Equipment, and Grounding

3

OBJECTIVES

After completing this chapter, the student will be able to understand the following:

- Service equipment
- Overhead services, including required clearances
- Underground services
- Working space around electrical equipment
- Grounding and bonding requirements for dwellings

INTRODUCTION

This chapter covers services for one- and two-family dwellings, including the important subject of grounding. The following major topics are included:

- Selecting service equipment
- Services
- Service point
- Sizing service-entrance conductors and raceways
- Grounding
- Special rules for two-family dwellings

IMPORTANT NEC TERMS

Bonding Jumper, Main
Ground
Grounded
Grounded, Effectively
Grounded Conductor
Grounding Conductor
Grounding Conductor, Equipment
Grounding Electrode Conductor
Interrupting Rating
Panelboard
Service
Service Cable
Service Drop
Service-Entrance Conductors, Underground System
Service-Entrance Conductors, Overhead System
Service Equipment
Service Lateral
Service Point

SELECTING SERVICE EQUIPMENT

Residential service equipment typically consists of a panelboard, which must be listed for use as service equipment. Circuit-breaker overcurrent protection is used for almost all new construction, though fuse-type panelboards are very common in older, existing houses.

Panelboards used in dwellings are usually constructed with each pole space a half-inch high, while commercial panelboards more often have inch-high pole spaces to accept circuit breakers that are physically larger.

Residential panelboards that use smaller circuit breakers are often known as *load centers,* and most manufacturers catalog them this way (see Figure 3.1). However, the term *load center* doesn't appear in the *National Electrical Code®* or in UL 67, *Safety Standard for Panelboards*, which contains requirements for listing all panelboards. In other words, load centers are actually panelboards and are subject to all the requirements of Article 408, Switchboards and Panelboards.

Listed Service Equipment

The panelboard main circuit breaker (MCB) frequently serves as the service disconnecting means required by 230.70. When this is the case, the panelboard is required to be listed for use as service equipment (Figure 3.2). Panelboards listed for use as service equipment come equipped with a main bonding jumper (MBJ), which connects the grounding busbar to the panelboard enclosure [250.24(B) and 408.3(C)]. Typically this MBJ is a green-colored metal screw [250.28(B)] or a metal strap. Panelboards used at locations other than services aren't permitted to have their grounding and neutral buses connected. This type of connection is only allowed at the service.

Sometimes structural considerations or other circumstances make it impossible to locate the panelboard near the point of entrance of the service conductors. In such

FIGURE 3.1 These are examples of panelboards of the type marketed as load centers. (Courtesy of EATON'S Cutler-Hammer business line)

FIGURE 3.2 This panelboard with main circuit-breaker disconnect is rated for use as service equipment. (Courtesy of Schneider Electric)

cases, a separate service disconnecting means must be provided [230.70(A)(1)], as shown in Figure 3.3. When a disconnect switch is used for this purpose, it must be listed as service equipment and come furnished with an MBJ. The panelboard protected by the disconnect switch can then be a lighting and appliance branch-circuit panelboard.

Panelboard Types and Construction

Nearly all distribution equipment installed in dwellings are lighting and appliance branch-circuit panelboards, listed for use as service equipment. A lighting and appliance branch-circuit panelboard is one that has at least 10 percent of its overcurrent devices protecting lighting and appliance branch circuits, which are defined as circuits rated 30 amperes or less and having a grounded conductor—in other words, 120-volt circuits protected by single-pole breakers.

Split-Bus Panelboards. Section 408.36(A) permits lighting and appliance branch-circuit panelboards to have up to two main circuit breakers. These split-bus panelboards typically supply general lighting and receptacle branch circuits from one half of the panelboard, and major loads such as cooking and HVAC equipment from the other half. Both main circuit breakers must be turned off to disconnect the entire installation (Figure 3.4). Some jurisdictions do not allow the use of split-bus panelboards.

An older type of split-bus panelboard had six 2-pole circuit breakers or fuses in the upper portion. Five of the circuit breakers served major 240-volt loads such as the furnace, central air-conditioner, water heater, range, and wall oven. The sixth

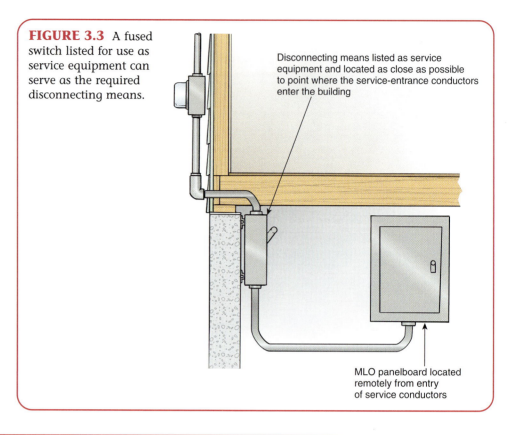

FIGURE 3.3 A fused switch listed for use as service equipment can serve as the required disconnecting means.

Disconnecting means listed as service equipment and located as close as possible to point where the service-entrance conductors enter the building

MLO panelboard located remotely from entry of service conductors

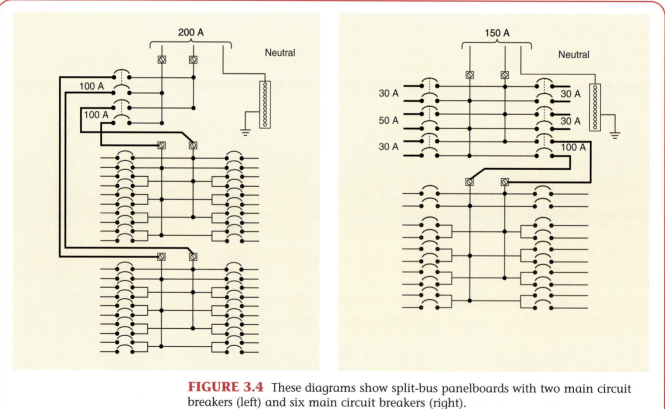

FIGURE 3.4 These diagrams show split-bus panelboards with two main circuit breakers (left) and six main circuit breakers (right).

2-pole overcurrent device protected the lower portion, which contained general lighting and receptacle branch circuits operating at 120 volts.

Maximum Number of Poles. Panelboards of any type are permitted to have a maximum of 42 pole spaces for the installation of branch-circuit overcurrent devices [408.35]. Each 120-volt circuit is protected by a single-pole circuit breaker. Each 240-volt circuit is protected by a 2-pole circuit breaker. There are no 3-pole circuit breakers in a residential panelboard rated 120/240 volt, single-phase, 3-wire.

In most homes, 42 pole spaces are sufficient to serve the loads. Some very large and/or all-electric homes need so many circuits that they use a commercial-style, double-section panelboard providing up to 84 pole spaces. An 84-pole panelboard is actually two 42-pole panelboards mounted side by side, configured so that one has a main circuit breaker and subfeeds the other. In other cases, sub-panels are installed in other locations away from the service entrance. (See "Sub-Panels," the next section of this chapter.)

Typically, panelboards with the following main circuit breaker ratings have the numbers of pole spaces shown:

Rating	Pole Spaces
100 A	16 poles
125 A	24 poles
150 A	30 poles
200 A	40 poles
225 A	42 poles

Sub-Panels

A sub-panel is a secondary panelboard supplied from the service equipment. (Note that the term *sub-panel* doesn't appear in the *Code* or in panelboard listing standards, but is used here for convenience and ease of understanding.) Sub-panels are typically used for two reasons:

1. The total number of branch circuits in a dwelling exceeds 42 pole spaces.
2. Parts of the home located away from the service equipment have a heavy concentration of loads or a number of branch circuits, such as a workshop in a basement or garage, a pool equipment building, or an accessory apartment ("granny flat") with a separate kitchen.

Sub-panels normally don't have main overcurrent devices and are known as *main lugs only (MLO)* panelboards. They are protected by a 2-pole, 240-volt breaker in the service panelboard (Figure 3.5).

Locating Sub-Panels. Sub-panels must comply with all standard *NEC* requirements for locating panelboards. They can't be installed in clothes closets because of the presence of easily ignitible materials [240.24(D)], or in bathrooms because of the presence of moisture [240.24(E)]. In addition, the same clear space, working space, and illumination requirements that apply to service equipment apply to sub-panels as well. (See "Locating and Mounting Considerations" later in this chapter.)

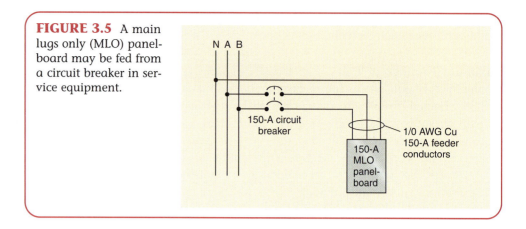

FIGURE 3.5 A main lugs only (MLO) panelboard may be fed from a circuit breaker in service equipment.

Locating and Mounting Considerations

Height. Panelboards and disconnects must be mounted so that the center of the operating handle of the main switch or circuit breaker is not more than 6 ft 7 in. above floor level in its highest position [404.8].

Working Space. Section 110.26 provides detailed rules for providing safe working space around electrical equipment. The rules that apply to residences can be summarized as follows:

- *Working space* not less than 3 ft deep, 6½ ft high, and 30 in. wide (or the width of the equipment, whichever is greater) must be provided in front of panelboards and service disconnects.
- *Panelboards* may not be mounted above permanently installed workbenches, shelves, freezers, laundry equipment, and such. Doing so violates the working space requirement.
- *Other associated equipment* (such as a wireway above or below the panelboard) is permitted to extend no more than 6 in. beyond the front of the electrical equipment within the working space.
- *Dedicated electrical space* must be equal to the width and depth of the service equipment and extend from the floor to the structural ceiling (or a height of 6 ft above the equipment). No piping, ducts, storage shelves, or "other equipment foreign to the electrical installation" are permitted within this zone.
- *Hinged panelboard doors* must be able to open a full 90 degrees (Figure 3.6).

Adequate Lighting. Section 110.26(D) requires that illumination be provided for working spaces around service equipment installed indoors. It does not specify a minimum footcandle level or require dedicated lighting outlets for service equipment.

Mounting Methods. Electrical equipment is required to be firmly secured to the surface on which it is mounted. Typical acceptable mounting methods include the use of concrete anchors or metal strut-type channels, or a plywood backboard can be installed for mounting electrical equipment. Securing electrical equipment to wooden plugs driven into holes in masonry, concrete, plaster, and similar materials is *prohibited* by 110.13(A).

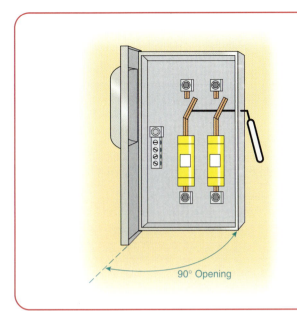

FIGURE 3.6 Equipment doors are required to open a full 90 degrees to ensure a safe working space.

90° Opening

Damp or Wet Locations. Residential service equipment is sometimes mounted outdoors in wet or damp locations. Similarly, some basements can be categorized as damp locations. Electrical equipment installed in such areas is required to have a ¼-in. air space between the enclosure and the wall surface [312.2(A)]. Most panelboards and disconnect switches are manufactured with raised or "dimpled" mounting holes that satisfy this *Code* requirement.

Prohibited Locations. Service equipment can't be installed in clothes closets because of the presence of easily ignitible materials [240.24(D)] or in bathrooms because of the presence of moisture [230.70(A)(2), 240.24(E)].

Interrupting Ratings

Service equipment must have an interrupting rating sufficient to interrupt the fault current, also called "short-circuit current," available at its terminals [110.9]. Because one- and two-family dwellings are typically served by pole- or pad-mounted transformers of relatively low kVA ratings, the fault currents available at residential service equipment are also normally low. However, it is important to contact the electric utility to obtain the actual fault current prior to installing service equipment.

Circuit Breaker A.I.R. Ratings. Molded-case circuit breakers with ½-in. frames, of the type most often used in residential load center panelboards, are classified as having a 10,000 amperes interrupting rating (A.I.R.). Main circuit breakers come in both 10,000 and 22,000 A.I.R. Panelboards can be upgraded in the field to withstand higher-than-normal available fault currents simply by replacing the standard 10,000 A.I.R. main circuit breaker with a higher-rated 22,000 A.I.R. model (Figure 3.7). Circuit breakers and fuses with higher ratings such as 42,000 A.I.R. are also available for use where factors such as low-impedance utility transformers result in high available fault currents.

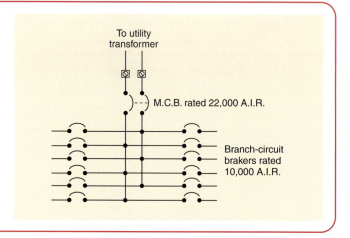

FIGURE 3.7 A panelboard of this sort is considered "series rated" under UL Standard 67; a panelboard in which all devices have the same A.I.R. rating is considered "fully rated."

Determining Available Fault (Short-Circuit) Current. Normally it isn't necessary to calculate the fault current available at a dwelling. This information is provided by the serving utility.

However, fault current for a single-phase transformer can be calculated using the following formula:

Short-circuit current = Full-load current × Multiplier

The full-load current and multiplier are calculated as follows:

$$\text{Full-load current} = I = \frac{\text{kVA} \times 1000}{E}$$

$$\text{Multiplier} = \frac{100}{Z}$$

where:

I = transformer full-load current in amperes

kVA = transformer rating in kilovolt-amperes

E = transformer secondary line-to-line voltage

Z = transformer-rated impedance (percent)

Calculation Example

For a pole-mounted transformer rated 75 kVA, 120/240 volts, single-phase, with a nameplate impedance of 1.5 percent Z, the full-load current is

$$I = \text{kVA} \times \frac{1000}{E}$$

$$= 75 \times \frac{1000}{240}$$

$$= 312.5 \text{ A}$$

The multiplier is found using the formula

$$\text{Multiplier} = \frac{100}{Z}$$

$$= \frac{100}{1.5}$$

$$= 66.67$$

The short-circuit current is found using the formula

$$\text{Short-circuit current} = \text{Full-load current} \times \text{Multiplier}$$

$$= 312.5 \times 66.67$$

$$= 20{,}834 \text{ A}$$

A main circuit breaker rated 22,000 A.I.R. can protect against this available fault current.

Circuit Directories and Marking

Panelboards are required to have a directory of circuits located on or inside the panel door [408.4]. All new panelboards come from the factory furnished with a blank directory. It's also permissible to use a different typewritten card, a computer-printed form, or another substitute for the directory provided by the manufacturer.

Disconnect switches must also be marked to indicate their purpose "unless located and arranged so the purpose is evident" [110.22]. This requirement is sometimes a matter for discussion with the authority having jurisdiction (AHJ). While a non-fused disconnect located outdoors near an air-conditioning unit might not require identification, a service disconnect installed other than in the room in which the house panelboard is located might need a sign or label to indicate its purpose.

The purpose of all circuit directories and marking is safety, enabling the correct circuits to be identified and turned off in an emergency or during maintenance.

SERVICES

Article 100 in the *National Electrical Code* defines *service* as follows.

> **Service:** The conductors and equipment for delivering electric energy from the serving utility to the wiring system of the premises served.

The power company normally determines whether the service will be overhead or underground and where the meter will be located. The meter socket is normally, but not always, provided by the utility and installed by the electrical contractor. In cases where the contractor is responsible for purchasing the meter socket, it must be a brand and model specified by the utility.

Overhead Services

Figure 3.8 illustrates a common design of overhead service. The conductors from the utility pole to the house are called *service-drop conductors* and are normally installed by the utility. Typically, there are two phase conductors and one grounded (neutral) conductor. At the house, the service-drop conductors (installed and owned by the utility) are spliced to the *service-entrance conductors* (installed by the electrical contractor and owned by the customer).

In most installations, this location represents the *service point*, which is discussed later in this chapter. Overhead services can be installed at houses with the service drop attached to the side of the building or with the service drop attached to a mast.

General Guidelines for Installing a Service Drop

Service Drop Attached to Side of Building

- The service-drop conductors should extend from the utility pole to a raintight service head (also called a *weatherhead*) located on an exterior wall or above the roof. They may be wrapped around a messenger cable for support, and the messenger serves as the grounded (neutral) conductor in most installations. The messenger is connected to a bracket screwed to the side of the house or is attached to the raceway leading from the weatherhead to the meter enclosure. The conductors are formed in a "drip loop" to enter the weatherhead from below.

- Sometimes the three service-drop conductors are run in free air and are not supported by a messenger cable. In such cases, the conductors may be attached to an insulator rack mounted on the exterior wall below the level of the service head (Figure 3.9), to serve the same purpose as a drip loop—that is, to prevent rainwater and condensation from migrating into the meter enclosure by following the conductors.

- Service-drop conductors should be spliced to the service-entrance conductors at the weatherhead. The service-entrance conductors are either Type SE cable or individual conductors installed in conduit. The cable or conduit is attached to the wall with the appropriate type of straps.

- Type SE cable enters the meter socket through a fitting with a watertight seal. Conduit is attached to the meter socket using the appropriate fitting for the type of conduit used.

- The conductors leading from the meter socket enclosure to the service equipment are sometimes called *service-entrance conductors*. Typically, the service equipment is located indoors, as close as practical to the meter location and often on the opposite side of the same wall, as shown in Figure 3.3.

- When Type SE cable is used, a fitting called a *sill plate* is used to protect the cable where the conduit enters the building. The sill plate is packed with sealing compound to keep water out.

- As shown in Figure 3.10, when individual conductors in conduit are used and where different temperatures inside and outside the building may cause

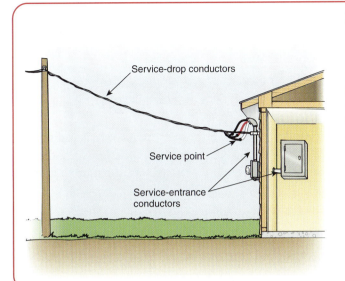

FIGURE 3.8 An overhead system utilizes a service drop from a utility pole to an attachment point on a house and service-entrance conductors from the point of attachment (spliced to service-drop conductors), down the side of the house, through the meter socket, and terminating in the service equipment.

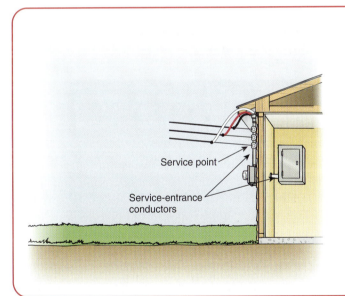

FIGURE 3.9 An overhead system may also have service drop conductors attached to a wall-mounted insulator rack.

condensation to form, the service raceway must be sealed with an approved material [300.7(A)].

Service Drop Attached to Mast

- The three *service-drop conductors* extend from the utility pole to a service head (also called a *weatherhead*) installed on a service mast. They are formed in a drip loop to enter the service head from below.

(Continues)

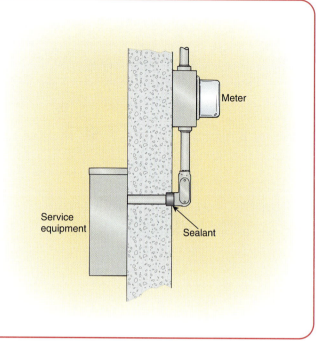

FIGURE 3.10 Raceways passing from the exterior to the interior of a building must be filled with an approved material to prevent the formation of condensation.

Meter

Service equipment

Sealant

General Guidelines for Installing a Service Drop (Continued)

NOTE: Instead of a service head (weatherhead), a gooseneck design can be used. In this case, the mast is sometimes formed in a curve so that the service conductors enter from below. The end of this "gooseneck" is then taped and painted to provide additional water resistance (Figure 3.11).

- The service mast is typically rigid metal conduit (Type RMC) or intermediate metal conduit (Type IMC), either steel or aluminum, in trade size 2. Some utilities have their own design guidelines that specify the precise type and size of metal conduit that can be used.

- Depending on the house construction, the service mast may be attached to the exterior wall or may pass through the roof. When the service mast penetrates a roof, metal flashing with a neoprene seal is used to prevent the entry of water (Figure 3.12).

- The RMC or IMC is attached to the meter socket using the appropriate fitting for the type of conduit used. The service-entrance conductors continue from the meter socket to the service equipment in conduit.

Clearances for Overhead Services

Service conductors are required to be kept at specified distances from roofs, windows, porches, driveways, and parking areas to prevent people and vehicles from accidentally contacting them. Sections 230.9 and 230.24 provide detailed clearance rules

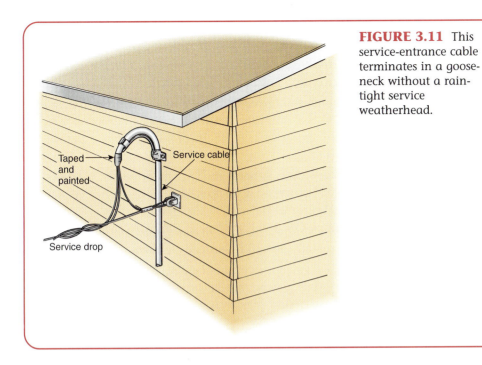

FIGURE 3.11 This service-entrance cable terminates in a gooseneck without a raintight service weatherhead.

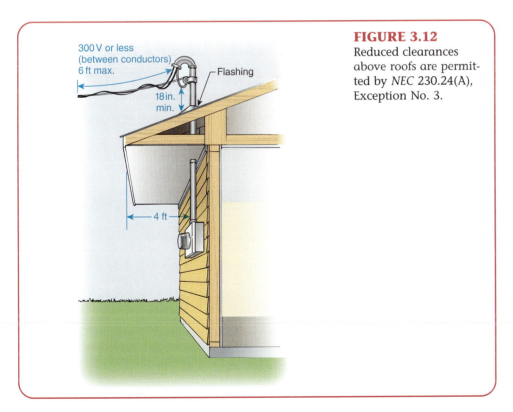

FIGURE 3.12 Reduced clearances above roofs are permitted by *NEC* 230.24(A), Exception No. 3.

for services. The minimum clearances required for residential services operating at not more 240 volts to ground can be summarized as follows.

- *Above Grade.* Distance of at least 10 ft to the lowest point of the drip loop and/or point of attachment is required.

• *Above Residential Property Including Driveways.* Distance of at least 12 ft to service-drop conductors is required.

• *Above Streets, Alleys, Roads, and Other Areas Subject to Truck Traffic.* Distance of at least 18 ft to service-drop conductors is required.

• *Above Roof.* Service conductors passing above a roof with a slope of at least 4 in. in 12 in. must be at least 3 ft from the nearest point of the roof. For flatter roofs, the service conductors must be at least 18 in. from the nearest part of the roof, as shown in Figure 3.12.

• *Windows.* Distance of at least 3 ft horizontally from or below windows designed to be opened is required. There is no restriction on running service conductors above an openable window (Figure 3.13).

• *Doors, Porches, Balconies, Ladders, Fire Escapes, and Similar Locations.* Distances of at least 3 ft horizontally and 10 ft vertically are required (Figure 3.14).

• *Other Openings.* Service conductors cannot be installed where they obstruct an opening through which materials may be moved, such as in commercial and farm buildings, and they cannot be installed beneath such openings.

• *Vegetation.* There is no *Code* rule requiring a minimum distance from trees. However, trees and other vegetation cannot be used to support overhead service conductors [230.10].

• *Communications Conductors.* There is no rule requiring a minimum clearance between electric service conductors and other services such as telephone, cable TV, and network-powered broadband communications systems. However, only power service-drop conductors are permitted to be attached to a service mast [230.28]. The service mast *cannot* also be used for incoming telephone wires, cable TV, or other systems.

Underground Services

Underground services are increasingly common, particularly in new suburban housing developments. Typically, the developers of new subdivisions want to avoid the visual clutter associated with utility poles, transformers, and overhead wiring of all kinds. Local jurisdictions may also have ordinances that require underground power distribution for new construction.

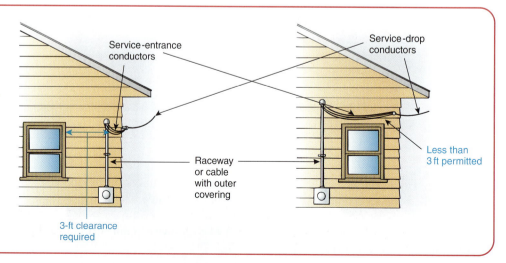

FIGURE 3.13 Minimum clearances are required for service conductors located alongside or below a window (left) but not above the top level of a window designed to be opened (right).

Service-entrance conductors

Service-drop conductors

Raceway or cable with outer covering

3-ft clearance required

Less than 3 ft permitted

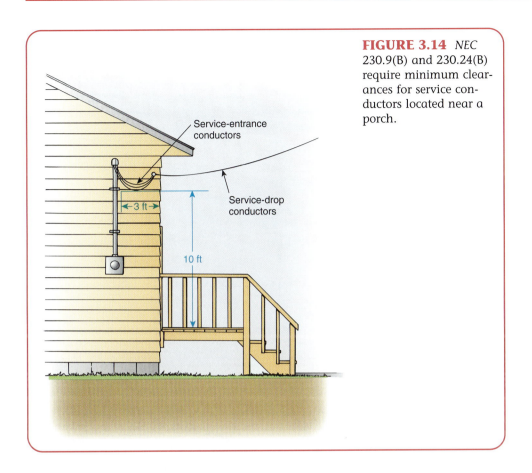

Service-entrance
conductors

3 ft

10 ft

Service-drop
conductors

FIGURE 3.14 *NEC* 230.9(B) and 230.24(B) require minimum clearances for service conductors located near a porch.

Underground distribution systems may be fed either from tank-type transformers mounted on poles or from pad-mounted transformers at grade. The buried conductors from the transformer to the house are called *service-lateral conductors* and are normally installed by the utility. There may be three single Type USE conductors (two phase conductors and one grounded conductor) or Type USE cable with a concentric neutral (Figure 3.15). In many areas, service-lateral and communications cables (telephone and cable TV) are installed together in a common trench during new construction.

Figures 3.16 and 3.17 illustrate underground services fed from pole-mounted and pad-mounted transformers:

- The service-lateral conductors extend from the pole- or pad-mounted transformer to the meter enclosure. Two different types of meter installations are shown at the end of this chapter, in Figures 3.29 and 3.30.

- In cases where a conventional meter and enclosure are used, direct-buried cable or conductors must be physically protected by being run in a raceway where they emerge from grade (Figure 3.16).

- When the service lateral is buried at least 18 in. below grade and not encased in concrete, a warning ribbon must be placed in the trench at least 12 in. above the conductors [300.5(D)(3)]. The warning ribbon is typically buried about 6 in. deep to identify the location of the service conductors for people digging in the yard later on.

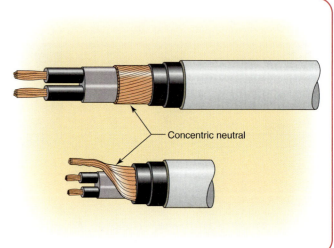

FIGURE 3.15 Some service-entrance cable has a so-called concentric neutral. The neutral conductor is not insulated but is made up of bare strands wrapped around the other, insulated conductors. When using Type SE and Type USE cable of this type, the separate strands are twisted together into one big, bare conductor at termination points. See 338.10(B)(2).

Concentric neutral

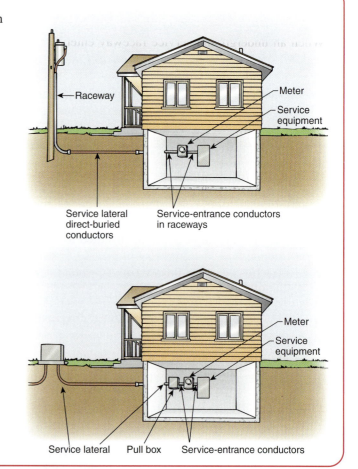

FIGURE 3.16 Underground systems can be served from either pole- or pad-mounted transformers.

Raceway

Meter

Service equipment

Service lateral direct-buried conductors

Service-entrance conductors in raceways

Meter

Service equipment

Service lateral   Pull box   Service-entrance conductors

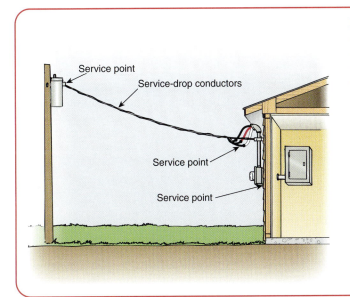

FIGURE 3.17 There are several possible locations of the "service point" in an overhead system.

- Service-entrance conductors run from the meter enclosure to the service equipment, which typically is located indoors. The service-entrance conductors are either Type USE cable or individual conductors installed in conduit.
- When an underground service raceway enters a building, the raceway must be sealed to prevent the entry of moisture [230.8, 300.5(G)]. This is particularly important for service equipment located physically downhill from the transformer or other utility supply equipment.

SERVICE POINT

NEC Article 100 provides the following definition.

Service Point: The point of connection between the facilities of the serving utility and the premises wiring.

Everything upstream of the service point (or on the "line side") belongs to the utility and is governed by the utility's own design and installation practices. Often such practices are based on ANSI/IEEE C2-2002, *National Electrical Safety Code*, which covers distribution systems for both power and communications. Normally, utility wiring and equipment are not subject to approval by the local AHJ.

Everything downstream of the service point (or on the "load side") is premises wiring. Whether located inside or outside of the residence, premises wiring is subject to all the requirements of the *National Electrical Code* and to approval by the AHJ.

Service Point Location and Responsibility

Although the definition of *service point* seems simple and straightforward, determining where the service point is actually located on a given installation can be complex. In most cases, the service point location depends on the practices and regulations of the serving utility. Figures 3.17 and 3.18 illustrate several possible service point locations for the same installation, depending on the practices of the local utility.

For *overhead services,* the service point could be at the following locations:

- Load side of pole-mounted transformer
- Splice (point of attachment) at weatherhead, gooseneck, or wall-mounted rack
- Customer side of meter

For *underground services,* the service could be at the following locations:

- Load side of pad-mounted transformer
- Pull box between transformer and meter
- Customer side of meter

It's important to understand that the location of the service point doesn't depend on who does the work of installing a particular piece of wiring or equipment. In some parts of the country, the utility installs service laterals. Elsewhere, this work is the electrical contractor's responsibility.

A meter enclosure may be installed by either a utility or a contractor, but the meter itself is always owned by the utility. In some places, the contractor installs the service mast or underground raceway, and then the utility installs conductors as far as the meter. In other areas, the serving utility hires the contractor to install utility-owned equipment such as pad-mounted transformers and service laterals.

According to the Article 100 definition of *service point,* the dividing line between "facilities of the serving utility" and "premises wiring" is not who *performs* the work, but who *owns and controls* the finished installation. The following distinction determines whether *NEC* rules apply to an installation:

- Wiring and equipment *downstream* of the service point are premises wiring, which must be installed according to *NEC* rules and approved by the AHJ.
- Wiring and equipment *upstream* of the service point are part of the utility's facilities, and aren't subject to the *NEC* or the AHJ.

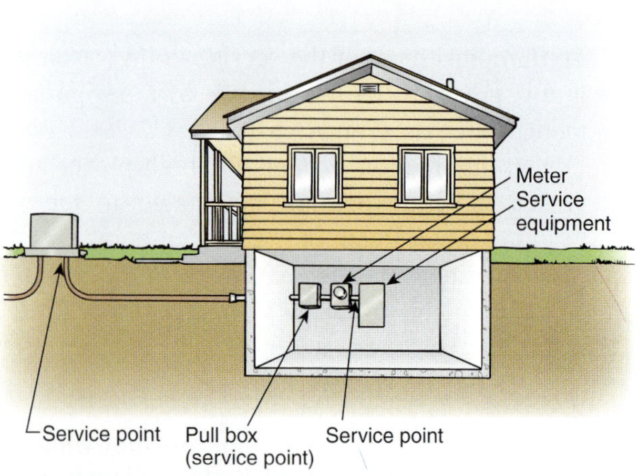

FIGURE 3.18 There are several possible locations of the *service point* in an underground system.

Meter
Service equipment

Service point   Pull box   Service point
(service point)

Service-Drop and Service-Lateral Conductor Sizes

As we have seen, service-drop and service-lateral conductors are owned by the utility company. Frequently, messenger-type overhead conductors or direct-burial cables are used. Less often, utility service conductors are installed in raceways. The sizes of these service-drop conductors, service-lateral conductors, and raceways (when used) are determined according to the rules of ANSI/IEEE C2-2002, *National Electrical Safety Code*, which is the wiring code for electric utilities.

Frequently, sizes of service conductors and raceways selected according to NESC rules are smaller than those of the service-entrance conductors (selected under *NEC* rules) to which they are spliced.

The reason for the size difference is similar to the reason behind the less-stringent *Code* rules for overcurrent protection, overload protection, and physical protection of conductors in supervised installations, such as those at industrial plants that have their own on-site maintenance and engineering personnel [240.90].

Because electric utilities have maintenance crews on duty around the clock to service their own systems, they are permitted to exercise greater control over what types of conductors, raceways, and equipment they install.

Importance of the Service Point. As this discussion shows, it is very important for an electrical contractor helping to build a new house to contact the serving utility in advance to obtain its rules, regulations, and design guidelines. For example, if the service point is considered to be the load-side terminals of a pole- or pad-mounted transformer, then the conductors between the transformer and house are *service-entrance conductors* that must comply with the *Code* and be approved by the AHJ. But if the service point is considered to be at the meter on the exterior of the house, then the conductors from the meter to the transformer are *service-lateral conductors* that do not have to comply with the *Code* and are exempt from inspection.

SIZING SERVICE-ENTRANCE CONDUCTORS AND RACEWAYS

Service-entrance conductors are those that run between the terminals of the service equipment and the point of connection to the service drop or service lateral. Sometimes Type SE or Type USE cable is used for the entire run from the point of connection to the meter enclosure and from the meter enclosure inside to the service equipment. In other installations, individual conductors of types such as THW, THHN, and THWN are installed in conduit.

Permitted wiring methods for service-entrance conductors are specified in 230.43 of the *NEC*, which is shown in Exhibit 3.1. Some of these—Type IGS cable, wireways, busways, auxiliary gutters, cablebus, and Type MI (mineral-insulated, metal-sheathed) cable—aren't normally used for services to one- and two-family dwellings.

Conductor Sizes. The *NEC* has a special table for selecting conductor sizes for 120/240-volt, 3-wire, single-phase dwelling services. Table 310.15(B)(6) gives conductor sizes for services ranging from 100 amperes [the minimum permitted by 230.79(C)] to 400 amperes (the maximum rating generally considered practical for a single-phase service).

EXHIBIT 3.1

NEC 230.43 Wiring Methods for 600 Volts, Nominal, or Less

Service-entrance conductors shall be installed in accordance with the applicable requirements of this *Code* covering the type of wiring method used and shall be limited to the following methods:

(1) Open wiring on insulators
(2) Type IGS cable
(3) Rigid metal conduit
(4) Intermediate metal conduit
(5) Electrical metallic tubing
(6) Electrical nonmetallic tubing (ENT)
(7) Service-entrance cables
(8) Wireways
(9) Busways
(10) Auxiliary gutters
(11) Rigid nonmetallic conduit
(12) Cablebus
(13) Type MC cable
(14) Mineral-insulated, metal-sheathed cable
(15) Flexible metal conduit not over 1.8 m (6 ft) long or liquidtight flexible metal conduit not over 1.8 m (6 ft) long between raceways, or between raceway and service equipment, with equipment bonding jumper routed with the flexible metal conduit or the liquidtight flexible metal conduit according to the provisions of 250.102(A), (B), (C), and (E)
(16) Liquidtight flexible nonmetallic conduit

The conductor sizes in *NEC* Table 310.15(B)(6) are smaller, in some cases, than those required by Table 310.16 for the same loads. The reason for this is that, as 310.15(B)(6) states, "the feeder conductors to a dwelling unit shall not be required to have an allowable ampacity rating greater than their service-entrance conductors." See the preceding discussion, "Service-Drop and Service-Lateral Conductor Sizes."

NEC Table 310.15(B)(6), shown in Exhibit 3.2, applies only to 3-wire residential services. Conductor sizes for 120/208-volt, 3-phase, 4-wire services are selected from *NEC* Table 310.16.

Raceway Sizes. Raceway sizes are determined using the tables in Annex C when all conductors in a raceway are the same size (the most common situation in residential wiring). When conductors of different sizes are mixed together, raceway size is determined using *NEC* Table 5 in Chapter 9. For example, a service panelboard with a 200-ampere MCB requires 2/0 AWG service-entrance conductors, according to Table 310.15(B)(6). Three 2/0 AWG conductors of types THW, THHW, or THWN require the following raceway sizes:

NEC TABLE 310.15(B)(6) *NEC* **Conductor Types and Sizes for 120/240-Volt, 3-Wire, Single-Phase Dwelling Services and Feeders. Conductor Types RHH, RHW, RHW-2, THHN, THHW, THW, THW-2, THWN, THWN-2, XHHW, XHHW-2, SE, USE, USE-2**   **EXHIBIT 3.2**

Conductor (AWG or kcmil)		Service or Feeder Rating (Amperes)
Copper	Aluminum or Copper-Clad Aluminum	
4	2	100
3	1	110
2	1/0	125
1	2/0	150
1/0	3/0	175
2/0	4/0	200
3/0	250	225
4/0	300	250
250	350	300
350	500	350
400	600	400

Raceway Type	Trade Size	Table
Electrical metallic tubing (EMT)	1½	C1
Rigid metal conduit (RMC)	1½	C8
Rigid PVC Conduit, Schedule 80	2	C9

Some utilities have requirements for minimum service-entrance raceway sizes that differ from *NEC* requirements. For example, trade size 2 rigid metal conduit (RMC) or intermediate metal conduit (IMC) is a common size for raceways used as service masts. Contractors should always contact the local utility for information before installing service-entrance conductors and raceways.

GROUNDING

Grounding is one of the most important safety concepts in the *National Electrical Code*. Proper grounding practices protect homeowners from electric shock and ensure the correct operation of overcurrent protective devices.

NEC Article 100 defines *grounded* and *effectively grounded* as follows.

Grounded: Connected to earth or to some conducting body that serves in place of the earth.

Grounded, Effectively: Intentionally connected to earth through a ground connection or connections of sufficiently low impedance and having sufficient current-carrying capacity to prevent the buildup of voltages that may result in undue hazards to connected equipment or to persons.

The fault-current path from the last outlet on a branch circuit, back to the service entrance, must be of sufficiently low impedance to allow fault current to rise quickly. This rise causes the circuit breaker or fuse to open the affected circuit quickly, minimizing damage to the electrical distribution system and helping prevent secondary hazards such as shock and fire damage. Safe grounding of residential services involves connections to ground on both the utility and customer sides.

Grounded and Ungrounded Conductors

Grounded Conductor. A *grounded conductor* is a system or circuit conductor intentionally connected to ground. Often, grounded conductors are called *neutrals*, though technically they are only non–current-carrying when the load is balanced on the ungrounded conductors in the same feeder or branch circuit. In new construction, grounded conductors normally have insulation that is white or gray in color [200.7(A)].

Ungrounded Conductor. An *ungrounded conductor* is a system or circuit conductor *not* intentionally connected to ground. Usually, ungrounded conductors are called *hot* or *phase* conductors. Ungrounded conductors typically have black, blue, or red insulation. However, other colors are also permitted—except for white, gray, and green.

Types of Grounding Conductors

A *grounding conductor* is used to connect equipment, or the grounded conductor of a wiring system, to a grounding electrode or electrodes. There are two types of grounding conductors: *equipment grounding conductors* and *grounding electrode conductors*, which are defined as follows in Article 100.

> **Grounding Conductor, Equipment:** The conductor used to connect the non–current-carrying metal parts of equipment, raceways, and other enclosures to the system grounded conductor, the grounding electrode conductor, or both, at the service equipment or at the source of a separately derived system.
>
> **Grounding Electrode Conductor:** The conductor used to connect the grounding electrode(s) to the equipment grounding conductor, to the grounded conductor, or to both, at the service, at each building or structure where supplied by a feeder(s) or branch circuit(s), or at the source of a separately derived system.

Grounding conductors are typically bare or have insulation that is predominantly green in color.

Utility Transformer and Service Equipment

The utility transformer is grounded. The service drop or service lateral from that transformer consists of two ungrounded (hot or phase) conductors and one grounded conductor. At the service equipment, the grounded (white) and equipment grounding (green) systems are connected together and to ground (earth), as follows:

- The grounded conductors are connected to screw terminals on the neutral busbar. The equipment grounding conductors are connected to screw terminals on the grounding busbar.
- Panelboards listed as service equipment come from the factory furnished with a main bonding jumper (MBJ). The MBJ may be a copper strap that connects the neutral busbar to the equipment grounding busbar, or simply a green-colored metal screw that connects the neutral bar to the grounded panelboard enclosure. (Normally the neutral busbar is insulated from contact with the metal enclosure.)
- The service panelboard is then connected to the grounding electrode with a grounding electrode conductor (Figure 3.19).

Branch Circuits

Code rules for grounded and grounding conductors in 120- and 240-volt branch circuits can be summarized as follows:

- Every 120-volt branch circuit consists of one grounded conductor, which has white insulation as required by 200.6(A), and one ungrounded (hot or phase) conductor, which is usually black though it may also be red or blue.
- Every 240-volt branch circuit consists of two ungrounded (hot or phase) conductors, which are usually black, red, or blue [240-volt circuits don't have grounded (white) conductors].
- All 120-volt and 240-volt branch circuits have equipment grounding conductors. Sometimes these are called "green grounds" in the field, because many equip-

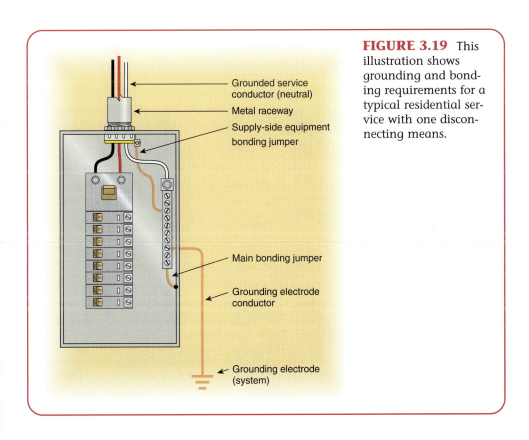

Grounded service conductor (neutral)

Metal raceway

Supply-side equipment bonding jumper

Main bonding jumper

Grounding electrode conductor

Grounding electrode (system)

FIGURE 3.19 This illustration shows grounding and bonding requirements for a typical residential service with one disconnecting means.

ment grounding conductors have an outer covering that's predominantly green in color, as required by 250.119. However, 250.118 also permits bare conductors, metallic raceways, and the metal sheaths of some cables to be used for equipment grounding purposes.

- The grounded (white) and equipment grounding (green) conductors are not permitted to be connected together at any location except the service equipment.

- At panelboards other than service equipment, the equipment grounding conductor busbar is bonded to the metal enclosure or cabinet. But the neutral busbar is insulated to prevent electrical contact between it and the grounded metal panelboard enclosure.

Grounding Electrode Systems

Sections 250.50 and 250.52 specify types of metal equipment that are required to be bonded together to form a grounding electrode system (Figure 3.20). Of these, only metal underground water pipe and ground rods are commonly available at dwellings. Service equipment is grounded to metal water piping systems as follows.

Step 1. *Determine size of grounding electrode conductor and bonding jumpers using NEC 250.66 and Table 250.66. For example, NEC Table 310.15(B)(6) specifies 2/0 AWG aluminum or copper-clad aluminum conductors for a dwelling service rated 150 amperes. NEC Table 250.66 states that for this size of service-entrance conductor, the required grounding electrode conductor is 6 AWG copper or 4 AWG aluminum or copper-clad aluminum.*

Step 2. *Connect the grounding electrode conductor to the metal water piping system within 5 ft of the place where it enters the building [250.52(A)(1)] and ahead of the water meter [250.53(C)(1)] (Figure 3.21). The grounding electrode conductor is connected to the metal water pipe using a pipe clamp listed for the purpose (Figure 3.22). Since interior water piping systems frequently*

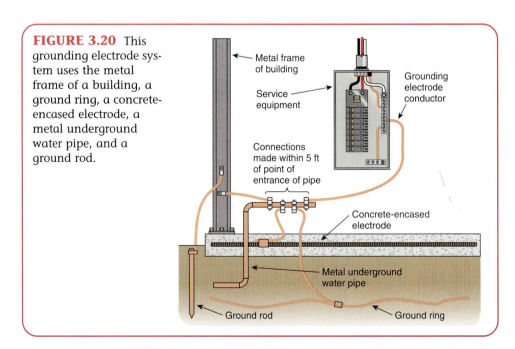

FIGURE 3.20 This grounding electrode system uses the metal frame of a building, a ground ring, a concrete-encased electrode, a metal underground water pipe, and a ground rod.

Metal frame of building

Service equipment

Grounding electrode conductor

Connections made within 5 ft of point of entrance of pipe

Concrete-encased electrode

Metal underground water pipe

Ground rod

Ground ring

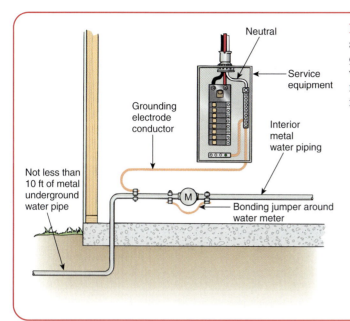

FIGURE 3.21 This service equipment is grounded to metal water piping within 5 ft of entry into a dwelling.

Neutral

Service equipment

Grounding electrode conductor

Interior metal water piping

Not less than 10 ft of metal underground water pipe

Bonding jumper around water meter

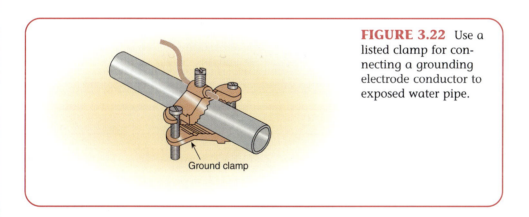

FIGURE 3.22 Use a listed clamp for connecting a grounding electrode conductor to exposed water pipe.

Ground clamp

include nonconductive plastic pipe or fittings, service grounding connections are not permitted at other locations within the house.

Step 3. *Install bonding jumpers around the water meter, water heater, filtration devices, and other equipment* (Figure 3.23). Continuity of the grounding path and the bonding connection to interior water piping is not permitted to rely on water meters and similar equipment [250.53(D)(1)].

Step 4. *Install a supplemental grounding electrode if needed.* When underground metal water piping systems are used as grounding electrodes, the *Code* requires that a supplemental grounding electrode be installed as well [250.53(D)(2)]. Typically, a supplemental grounding electrode is a copper or copper-plated steel ground rod a minimum of ⅝ in. in diameter and at least 8 ft long. Section 250.53(G) provides detailed guidance for installing rod and pipe electrodes (Figure 3.24).

The supplemental grounding electrode is permitted to be connected to the grounding system, in one of several ways. It can be bonded to the

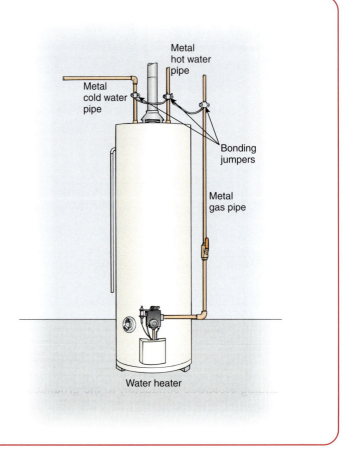

FIGURE 3.23 Bonding jumpers at a gas-fired water heater must be long enough to permit removal of the equipment without losing the integrity of the grounding path.

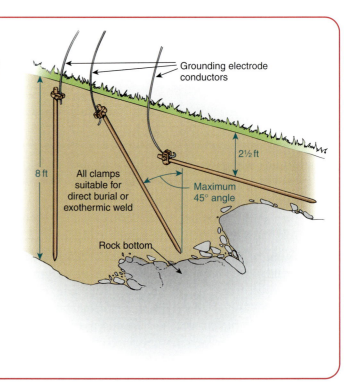

FIGURE 3.24 The installation requirements for rod and pipe electrodes are specified in *NEC* 250.53(G).

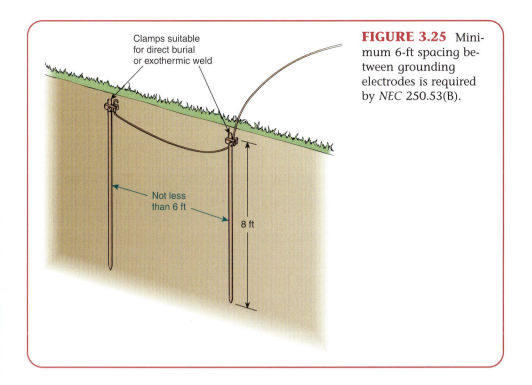

Clamps suitable
for direct burial
or exothermic weld

Not less
than 6 ft

8 ft

FIGURE 3.25 Minimum 6-ft spacing between grounding electrodes is required by *NEC* 250.53(B).

grounding electrode conductor, to the grounded service-entrance conductor, to the grounded metal service raceway (but not if the raceway is a flexible type), or to the grounded service enclosure. The supplemental electrode bonding conductor is not required to be larger than 6 AWG copper or 4 AWG aluminum [250.53(E)].

Step 5. Install multiple supplemental grounding electrodes if necessary. Where a single grounding electrode doesn't have a resistance to ground of 25 ohms or less, it must be augmented by one additional electrode located not closer than 6 ft away [250.53(B), 250.56] (Figure 3.25). Achieving low ground resistance is typically a problem in dry, rocky soil and easier in moist, loamy soil. Only one extra electrode is required by the *Code*—it isn't necessary to keep installing additional electrodes until 25 ohms to ground is achieved, which may not be possible in certain locations.

Concrete-Encased Electrodes. Where the concrete footing or foundation for a one- or two-family dwelling contains steel reinforcing bars (rebar) at least ½ in. in diameter, or at least 20 ft of bare copper conductor 4 AWG or larger, they are required to be connected to the grounding electrode system, as shown in Figure 3.26 [250.52(A)(3)]. Lightweight steel reinforcing mesh of the type often used for building houses doesn't have to be bonded.

Physical Protection of Grounding Electrode Conductors. A 6 AWG grounding electrode conductor is permitted to run along the surface of the building. However, small conductors are required to be protected by conduits, or raceways, as are larger ones if exposed to severe physical damage [250.64(B)]. Metal raceways or cable armor must be electrically continuous from the service equipment to the grounding electrode or bonded to the grounding electrode conductor as shown in Figure 3.27.

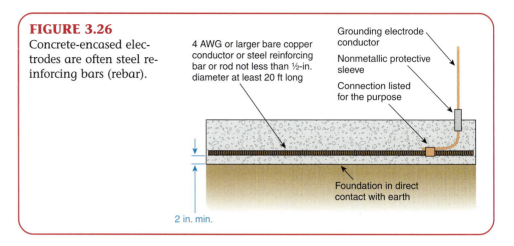

FIGURE 3.26
Concrete-encased electrodes are often steel reinforcing bars (rebar).

4 AWG or larger bare copper conductor or steel reinforcing bar or rod not less than ½-in. diameter at least 20 ft long

Grounding electrode conductor

Nonmetallic protective sleeve

Connection listed for the purpose

Foundation in direct contact with earth

2 in. min.

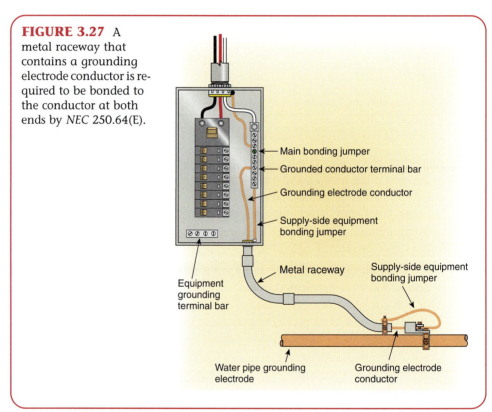

FIGURE 3.27 A metal raceway that contains a grounding electrode conductor is required to be bonded to the conductor at both ends by *NEC* 250.64(E).

Main bonding jumper

Grounded conductor terminal bar

Grounding electrode conductor

Supply-side equipment bonding jumper

Equipment grounding terminal bar

Metal raceway

Supply-side equipment bonding jumper

Water pipe grounding electrode

Grounding electrode conductor

SPECIAL RULES FOR TWO-FAMILY DWELLINGS

- *Meters and Service Equipment.* Two-family dwellings commonly have either two or three meter housings (or a two- or three-gang meter housing). The third meter is for the "house loads panel" serving common area facilities such as hall or stairway lighting, outdoor security lighting, or lighting for basements, storage areas, and laundry rooms. Figure 3.28 shows a typical multi-meter installation used with an overhead service.

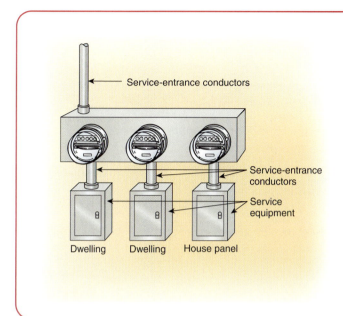

FIGURE 3.28 This illustration shows meters and service equipment for a two-family dwelling with an overhead service.

Figures 3.29 and 3.30 show two different types of meter installations for underground services. Each of the three panelboards or main disconnects is required to be marked as being suitable for use as service equipment [230.66].

- *Grounding.* Two-family dwellings with more than one service equipment must comply with all the same rules for service grounding discussed in this chapter. However, it's important to understand that a two-family dwelling has one utility service drop or lateral. This circumstance means that only a single grounding electrode conductor is required to the grounding electrode (most commonly, metal underground water piping). Each of the two or three service equipments *does not require* its own separate connection to the grounding electrode.

  Figure 3.31 shows one arrangement for connecting three service disconnecting means (in this case, fused switches). An alternate design would be to connect the grounding electrode conductor to the first disconnect located on the left. Bonding jumpers would then be run through the raceway nipples and wireway from one equipment grounding busbar to another.

- *Conductor Sizing.* Equipment bonding jumpers on the supply side of the service are based on the size of the service-entrance conductors. The conductors in the riser conduit determine the size of the bonding jumpers for the riser and wireway enclosure. The bonding jumper for each nipple is determined by the size of the service conductors they each contain. All supply-side bonding jumpers are sized using *NEC* Table 250.66.

  For example, *NEC* Table 310.15(B)(6) specifies 3/0 AWG copper conductors for a dwelling service rated 225 amperes. Table 250.66 states that for this size of service-entrance conductor, the required grounding electrode conductor is 4 AWG copper or 2 AWG aluminum or copper-clad aluminum.

FIGURE 3.29 This illustration shows meters and panelboards for a two-family dwelling with an underground service; the panelboards are not required to be listed as service equipment.

Service-entrance conductors

Service disconnecting means

Feeder conductors

Service-lateral conductors

Panelboards (not listed as service equipment)

Dwelling Dwelling House panel

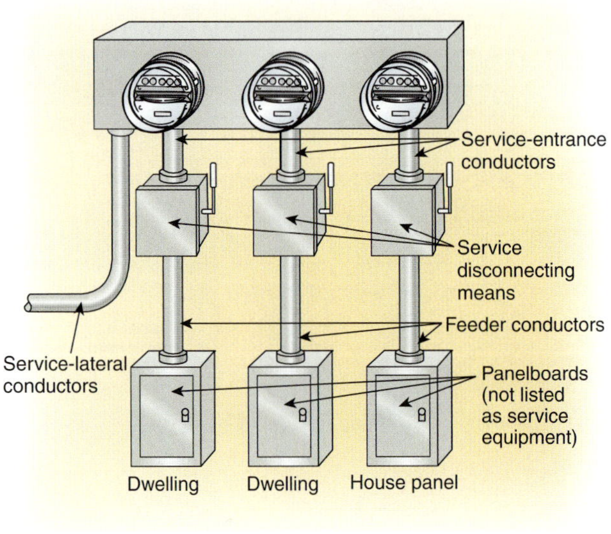

FIGURE 3.30 A meter pedestal can be used with underground service for a two-family dwelling. Indoor service equipment is on the other side of the wall.

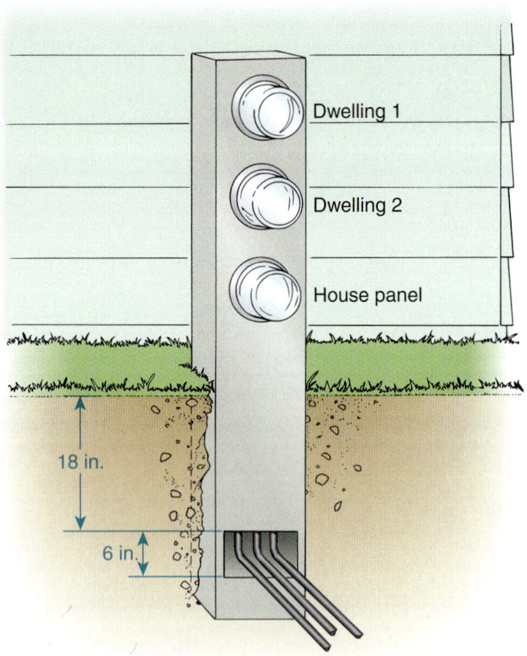

Dwelling 1

Dwelling 2

House panel

18 in.

6 in.

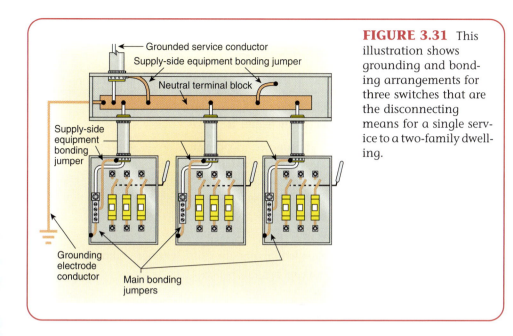

FIGURE 3.31 This illustration shows grounding and bonding arrangements for three switches that are the disconnecting means for a single service to a two-family dwelling.

MULTIPLE CHOICE

1. Lighting and appliance panelboards have
   A. At least 10 percent of their overcurrent devices protecting lighting and appliance branch circuits
   B. At least 10 percent of their circuit breakers rated 30 amperes or less, single-pole
   C. A main bonding jumper (MBJ)
   D. Both A and B

2. Grounded and equipment grounding conductors are connected together at
   A. Each outlet
   B. Each panelboard
   C. Service equipment only
   D. Service point

3. Service equipment is required to withstand
   A. Seismic forces
   B. Available fault current
   C. Rough handling
   D. None of the above

4. Split-bus panelboards are permitted to have
   A. A maximum of 42 branch-circuit pole spaces
   B. Six main overcurrent devices
   C. Fused overcurrent devices only
   D. None of the above

5. Meter pedestals are used with
   A. Overhead services
   B. Underground services
   C. Indoor meters
   D. Digital watt-hour meters

6. Where a single grounding electrode doesn't have a resistance to ground of 25 ohms or less, it must be augmented by the following number of additional electrodes, each located not closer than 6 ft away:
   A. One
   B. Two, if the first two do not achieve a maximum resistance of 25 ohms
   C. The number required to achieve a minimum resistance of 50 ohms
   D. Additional grounding electrodes are not required

7. Watt-hour meters are installed by the
   A. Electrical contractor
   B. Serving utility
   C. Authority having jurisdiction
   D. Both A and B

8. Pad-mounted transformers are used with the following type of residential service:
   A. Overhead
   B. Underground
   C. Lateral
   D. Aerial

9. Residential panelboards without main overcurrent devices are known as
   A. Fuse boxes
   B. Load centers
   C. MLO
   D. None of the above

10. The conductor used to connect non–current-carrying metal parts of equipment, raceways, and other enclosures to the system grounded conductor is called the
    A. Grounded conductor
    B. Grounding electrode conductor
    C. Equipment grounding conductor
    D. Redundant grounding conductor

FILL IN THE BLANKS

1. The conductor that connects the service equipment to the grounding electrode is known as a(n) _____.

2. Grounded conductors in 120/240-volt, single-phase, 3-wire systems typically have insulation that is _____ in color.

3. The most common grounding electrode used in residential construction is _____.

4. Service-entrance conductors are those that connect the _____ or service-lateral conductors with the service equipment.

5. A(n) _____ is used to ensure the continuity of the grounding path around a water meter or water heater.

6. It isn't usually necessary to calculate the fault current available at a dwelling, because this information is normally provided by _____.

7. Panelboards and disconnect switches listed for use as service equipment come equipped with

a(n) _____, which connects the grounding busbar to the panelboard enclosure.

8. Wiring and equipment downstream of the service point are called _____, which must be installed according to *NEC* rules and are subject to approval by the AHJ.

9. Grounding electrode conductors smaller than _____ and run exposed are required to be protected by metallic conduit or raceway armor.

10. Service equipment must have an interrupting rating sufficient to interrupt the _____ available at its terminals.

TRUE OR FALSE

1. Equipment grounding conductors are required to be copper.

   _____ True   _____ False

2. The minimum size conductor permitted for connecting a supplemental grounding electrode is 6 AWG copper or 4 AWG aluminum or copper-clad aluminum.

   _____ True   _____ False

3. Some large all-electric houses use 480Y/277-volt, 3-phase, 4-wire services.

   _____ True   _____ False

4. The service grounding electrode conductor is required to be connected to the metal underground water piping system within the first 10 ft after it enters the house, unless a bonding jumper is installed around the water heater.

   _____ True   _____ False

5. Rigid metal conduit (RMC) and intermediate metal conduit (IMC) are commonly used for service masts.

   _____ True   _____ False

6. A 120/240-volt, single-phase, 3-wire residential service consists of two ungrounded (hot) conductors and one grounded conductor.

   _____ True   _____ False

7. The *NEC* has a special table for selecting conductor sizes for 120/240-volt, 3-wire, single-phase services for dwellings.

   _____ True   _____ False

8. The sizes of service conductors and raceways selected according to *National Electrical Safety Code* (NESC) rules sometimes may be smaller than those of the service entrance conductors (selected under *NEC* rules) to which they are spliced.

   _____ True   _____ False

9. Grounding electrode conductors and bonding jumpers are required to be copper or copper-clad aluminum conductors only.

   _____ True   _____ False

10. When a two-family dwelling has common area facilities such as stairway lighting, storage areas, and laundry rooms, branch circuits serving these areas must originate at service equipment protected by fuses.

   _____ True   _____ False

CHALLENGE QUESTIONS

1. Why aren't panelboards other than those listed as service equipment furnished with main bonding jumpers (MBJ)?

2. What is the reason for requiring that all switches and circuit breakers used as switches be located so that the operating handle is not more than 6 ft 7 in. above the floor or working platform?

3. Electrical closets and other spaces where panelboards are located are frequently used for storage. What are the *Code* implications of this practice?

4. Discuss the difference between *service-drop, service-lateral,* and *service-entrance conductors.* Assume that a utility-owned transformer supplies a one-family house with a watt-hour meter mounted on the exterior wall and a service panelboard inside the house.

5. Metal water piping is frequently used as all or part of the grounding electrode systems for dwellings. Why must bonding jumpers be used around items such as meters, filtration devices, and water heaters that are connected in line with the metal water piping?

6. *NEC* 110.9 requires that service equipment have an interrupting rating sufficient for the current available at the line terminals of the equipment. Explain why residential load-center panelboards that must withstand current levels higher than 10,000 A.I.R. are typically series-rated rather than fully rated.

Wiring Methods

4

OBJECTIVES

After completing this chapter, the student will be able to understand the following:

- General wiring rules in the *National Electrical Code*®
- Wiring methods used in dwellings
- The differences between rules for installing Type NM and Type AC cable
- Installation rules for electrical boxes
- How to calculate box fill and determine what size boxes are needed
- Special rules for two-family dwellings

INTRODUCTION

This chapter covers wiring methods used in dwellings, including installation rules and practices for boxes, both metal and nonmetallic. The following major topics are included:

- General wiring rules for dwellings
- Nonmetallic-sheathed cable (Type NM)
- Armored cable (Type AC)
- Box fill and selection
- General requirements for installing boxes
- Tubular raceways
- Surface raceways
- Installing conductors in raceways
- Using oversized conductors to minimize voltage drop
- Special rules for two-family dwellings

IMPORTANT NEC TERMS

Accessible (as applied to wiring methods)
Armored Cable, Type AC
Bonding (Bonded)
Concealed
Conductor, Bare
Conductor, Covered
Conductor, Insulated
Conduit Body
Copper-Clad Aluminum Conductors
Device
Electrical Metallic Tubing (EMT)
Electrical Nonmetallic Tubing (ENT)
Exposed (as applied to wiring methods)
Fitting
Flexible Metal Conduit
Grounded
Grounded Conductor
Grounding Conductor
Lighting Outlet
Liquidtight Flexible Metal Conduit (LFMC)
Liquidtight Flexible Nonmetallic Conduit (LFNC)
Nonmetallic-Sheathed Cable, Type NM
Outlet
Raceway
Receptacle Outlet
Rigid Nonmetallic Conduit (RNC)

GENERAL WIRING RULES FOR DWELLINGS

The most common wiring methods used in new one- and two-family dwellings are nonmetallic-sheathed cable (Type NM) and armored cable (Type AC). Cable wiring methods are less expensive to install than raceway systems and require less training for the installer. They offer the advantage of low first cost, but changes after initial construction are fairly difficult and expensive to make.

Individual conductors installed in tubular raceways are much less common in residential construction. Raceway wiring systems are more expensive to install and require more training. They have the advantage that wires can be pulled out and replaced with new ones after original construction. Metal raceways—chiefly electrical metallic tubing (EMT)—are widely used in the Chicago metropolitan area, where local codes prohibit most cable wiring methods.

Certain general rules and considerations apply to all wiring methods, including the components and equipment used in the installation.

Solid and Stranded Conductors

Nonmetallic-sheathed cable (Type NM) and armored cable (Type AC) are manufactured with solid conductors in 14, 12, and 10 AWG sizes. The *Code* requires that conductors installed in raceways must be stranded in sizes 8 AWG and larger [310.3]. However, many installers prefer to use stranded 10 AWG conductors as well, because they are much easier to pull.

Conductor Terminations and Splices

Equipment Terminals. Circuit breakers, disconnect switches, many wiring devices, and other electrical equipment used in dwellings are manufactured with screw terminals. Some wiring devices have holes that 14 AWG solid conductors can be pushed into without using tools. Wiring device listing requirements don't allow these push-in connectors to be used with 12 AWG conductors.

Splice and Tap Connections. Conductors are only permitted to be spliced or tapped within enclosures. Splices cannot be made inside raceways [300.13(A)] or left exposed [300.15]. Twist-on, solderless connectors are the most common method used to join conductors in outlet boxes. They are frequently known in the field by trade names such as Wirenut® and Scotchlok®.

Connector Marking. Equipment terminals and wire connectors without special markings are intended for use with copper conductors only. Terminals and connectors marked AL/CU are suitable for use with copper, aluminum, or copper-clad aluminum conductors. Terminals and connectors marked AL are suitable for use only with aluminum conductors. In the case of small parts such as twist-on connectors, the marking may be on the package or instructions rather than on the connector itself.

Box. A box, conduit body, or fitting must be installed at each outlet, switch location, or conductor splice point [300.15]. Removable covers must be accessible after installation in order to provide access to the wires inside [300.15(A)] (Figure 4.1).

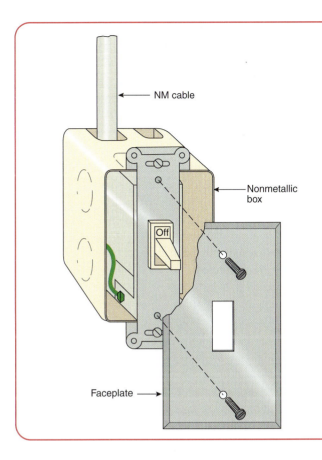

FIGURE 4.1 A typical residential wall switch installation includes a nonmetallic box, Type NM cable, and a faceplate (removable cover).

NM cable

Nonmetallic box

Off

Faceplate

Cables and Raceways

Securing and Supporting. Raceways, cables, boxes, cabinets, and fittings must be securely fastened in place and adequately supported. Section 300.11(A) states that "cables and raceways shall not be supported by ceiling grids." Raceways can sometimes be used to support other raceways, cables, or equipment as described in *NEC* 300.11(B) (Figure 4.2). But cables are never permitted to support other items [300.11(C)].

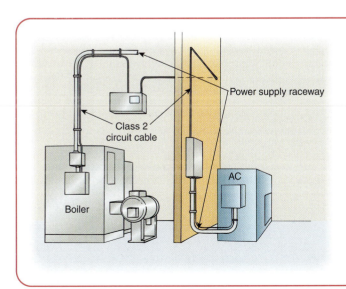

FIGURE 4.2 Raceways are permitted to support Class 2 thermostat cables, per *NEC* 300.11(B).

Power supply raceway

Class 2 circuit cable

Boiler

AC

Continuous Between Boxes. Raceways and cable armors or sheaths must be continuous between boxes, cabinets, or other enclosures [300.12]. Raceway systems consisting of conduit or tubing and boxes must be completely installed before conductors are pulled into them.

Section 300.18(A) permits a raceway system with conductors to be installed up to the point of utilization without a final box where such an installation is required to connect to motorized equipment, appliances, or luminaires (lighting fixtures) in dropped ceilings using fixture whips (Figure 4.3).

The Exception to 300.18(A) allows short sections of raceways to be installed—as sleeves through floors, for example—in order to protect conductors or cables from physical damage.

Bonding and Grounding. Metal cable armor, raceways, and enclosures must be metallically joined together (bonded) so that they are electrically continuous [300.10]. Metal raceway systems are required to be grounded.

Underground Installations. There are many *Code* rules for protecting underground cables and raceways from damage. They are covered in more detail in Chapter 12, Outdoor Areas.

NONMETALLIC-SHEATHED CABLE (TYPE NM)

Type NM cable consists of two or three insulated conductors enclosed in a nonmetallic outer jacket or cover. It typically has a separate bare copper equipment grounding conductor. Conductor sizes are 14 AWG through 2 AWG for copper, and 12 AWG through 2 AWG for aluminum or copper-clad aluminum.

NOTE: Nonmetallic-sheathed cable with larger AWG sizes of aluminum or copper-clad aluminum conductors is sometimes used to supply high-load electric appliances such as ranges and ovens, water heaters, and clothes dryers.

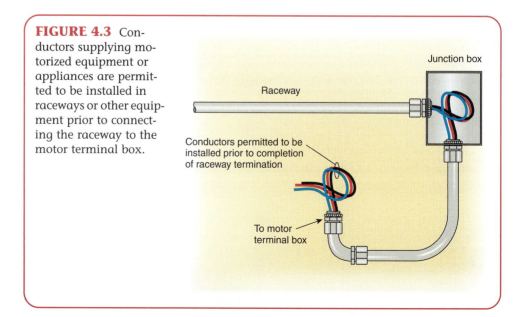

FIGURE 4.3 Conductors supplying motorized equipment or appliances are permitted to be installed in raceways or other equipment prior to connecting the raceway to the motor terminal box.

Junction box

Raceway

Conductors permitted to be installed prior to completion of raceway termination

To motor terminal box

Types of NM Cable

Type NM cable is manufactured in three varieties:

1. *Type NM* is permitted in normally dry locations such as the air voids in masonry block or tile walls.

2. *Type NMC* is also permitted in damp or corrosive locations.

3. *Type NMS* (also known as ''Smart House cable''), is a special type of Type NM cable with integral communications and signaling conductors under the same jacket as power conductors.

NOTE: Type NMS cable is used only for closed-loop and programmed power systems covered by Article 780. Houses are no longer being constructed using this special technology, and Type NMS cable is used only for repairing and maintaining existing ''Smart Houses.''

What Are Types NM-B and NMC-B? All nonmetallic-sheathed cable sold and used today is marked *NM-B* or *NMC-B* because, when originally developed, Type NM cable conductors had insulation rated 60°C (140°F). However, the insulation often became brittle due to overheating when NM cable was connected to ceiling-mounted luminaires.

Beginning in 1984, all nonmetallic-sheathed cable was required to have conductor insulation rated 90°C (194°F) [334.112 FPN]. The new variety was marked with the suffix -*B* to distinguish it from the older cable rated 60°C (140°F). All NM cable available today is NM-B or NMC-B (Figure 4.4).

NOTE: The conductor ampacity for nonmetallic-sheathed cable is based on the 60°C (140°F) column of NEC Table 310.16, shown in Exhibit 3.3 in Chapter 3 of this book. The ampacity given in the 90°C (194°F) column is only permitted to be used for derating purposes, such as might be necessary when NM cable is installed in attics buried under thermal insulation.

Type NM Cable with Equipment Grounding Conductor. Types NM and NMC cables are manufactured in different versions, both with and without an equipment grounding conductor (either bare uninsulated, or with a green cover). The type with ground wire must be used for all residential wiring purposes, since all receptacles,

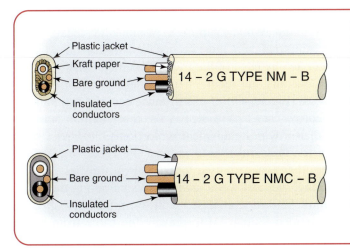

FIGURE 4.4 Nonmetallic-sheathed cable, Type NM, consists of two or more insulated circuit conductors with a bare equipment grounding conductor. Type NM-B (top) is permitted in normally dry locations; Type NMC-B (bottom) can be used in either dry or damp locations.

wall switches, metal faceplates, and exposed parts of metal luminaires (lighting fixtures) are required to be grounded.

Installing Nonmetallic-Sheathed Cable

The *NEC* provides a number of basic rules for installing nonmetallic-sheathed cables.

Securing. Type NM cable must be secured in place at intervals not greater than 4½ ft and within 12 in. from every box, cabinet, or termination. Staples or cable ties are most commonly used. The *Code* permits running two flat NM cables together, one on top of the other, and securing both using the same staple or tie. However, cables must be stapled flat, not on edge [334.30].

When NM cable is run horizontally through holes or notches in framing members spaced not more than 4½ ft apart, it is considered to be adequately supported. Ties are not required where the horizontal cable passes through these members. However, the cable must still be secured within 12 in. of each box [334.30(A)].

Type NM cable can be used unsupported where fished, when installed in panels for prefabricated houses, and in whips up to 4½ ft long for connecting to luminaires (lighting fixtures) or equipment installed within accessible ceilings [334.30(B)].

Plastic Device Boxes. When NM cable is used with single-gang nonmetallic boxes, it doesn't have to be secured to the box if the cable is fastened within 8 in. of the nonmetallic box. The *Code* permits more than one NM cable to enter through a single cable knockout opening [314.17(C), Exception].

NM Cable in Dropped or Suspended Ceilings. Nonmetallic-sheathed cables are permitted to be installed above dropped or suspended ceilings *only in one- and two-family dwellings.* This wiring practice is prohibited in other types of buildings [334.12(A)(2)]. Type NM cable runs in dropped ceiling spaces must be properly secured and supported as required by 300.11(A).

Bending Radius. The radius of the curved inner edge of a NM cable bend shall not be less than five times the diameter of the cable. The diameters of 14/2, 14/3, 12/2, and 12/3 cables (the sizes most often used for branch circuits in dwellings) range from about ½ to ⅝ in. So in practical terms, the minimum radius to the *inner edge of the cable* can't be less than 2½ in. for smaller NM cables and 3⅛ in. for larger NM cables [334.24].

Protection Against Physical Damage (Wood Framing). Nonmetallic-sheathed cables that are run parallel to framing members must be installed so that the edge of the cable is no closer than 1¼ in. to any edge of the framing member where screws or nails are likely to penetrate. Generally, this requirement applies to *vertical* framing members such as studs [300.4(D)] (Figure 4.5). If this distance can't be maintained, the NM cable must be protected by a steel plate, sleeve, or other guard at least ¹⁄₁₆ in. thick (Figure 4.6), or one that provides equivalent protection [300.4(A)(1) and Exception No. 2]. The reason for these rules is to prevent the cable from being damaged by drywall screws or other fasteners used to attach wall finishes.

When holes are drilled through vertical wood studs or other framing members for running AC cables, they must be drilled at least 1¼ in. from the nearest edge, or a steel protector must be provided [300.4(A)] to prevent the cable from being damaged.

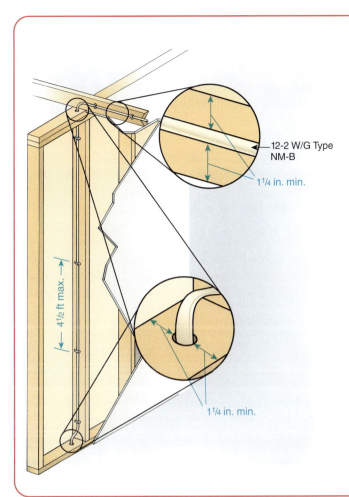

FIGURE 4.5 This Type NM cable is installed parallel to framing members in accordance with *NEC* 300.4(D).

12-2 W/G Type NM-B

1¼ in. min.

4½ ft max.

1¼ in. min.

FIGURE 4.6 Steel plates are used to protect a nonmetallic-sheathed cable installed within 1¼ in. of the edge of a wood stud. (Courtesy of RACO Interior Products Inc.)

When drilling through floor joist members, the edge of a bored hole should be kept at least 2 in. from the bottom chord of the joist. Prefabricated trusses cannot be drilled.

Protection Against Physical Damage (Metal Framing). When nonmetallic-sheathed cables pass through openings in steel studs and framing members, they must be protected by listed bushings or listed grommets that cover all metal edges [300.4(B)(1)]. Factory-made openings in steel studs are located to keep cables more than 1¼ in. from the "front" edge of the stud. When holes are field-cut, they must comply with the same rules as holes drilled through wood framing members.

Protection Against Physical Damage (Through Floors). When exposed Type NM cable passes through a floor, it must be enclosed in rigid metal conduit, intermediate metal conduit, electrical metallic tubing, Schedule 80 PVC rigid nonmetallic conduit, or other approved means extending at least 6 in. above the floor [334.15(B)]. This rule doesn't apply to NM cables that pass through floor plates in the void spaces of walls, but such cables must also be kept at least 1¼ in. back from the edge of framing members.

NOTE: "Approved means" refers to protection methods approved by the authority having jurisdiction. For example, the AHJ could decide that sections of metal or heavy-duty plastic plumbing pipe can be used as sleeves. See the definition of "Approved" in Article 100.

Boxes or conduit bodies aren't required at the ends of raceways used to provide physical protection of cables. However, a fitting must be provided on the end(s) of the conduit or tubing to protect the cable from abrasion [300.15(C)]. When short sections of metal raceways are used to protect NM cable from physical damage, they aren't required to be grounded [250.86, Exception No. 2]. Note that the *NEC* doesn't define *short*. In doubtful situations, the authority having jurisdiction (AHJ) should be consulted to determine whether grounding is required.

Installations in Exposed Locations. Type NM cable installed exposed must closely follow the surface of the building finish or be on running boards.

• *Unfinished Basements.* Type NM cables in sizes 6/2 or 8/3 and larger can be stapled directly to the lower edges of joists. But 334.12(C) requires smaller cables to be run either through bored holes or on running boards (Figure 4.7).

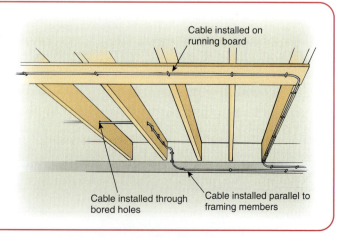

FIGURE 4.7 Nonmetallic-sheathed cables in unfinished basements must be installed through bored holes, along running boards, or parallel to framing members if they are smaller than 6/2 or 8/3.

Cable installed on running board

Cable installed through bored holes

Cable installed parallel to framing members

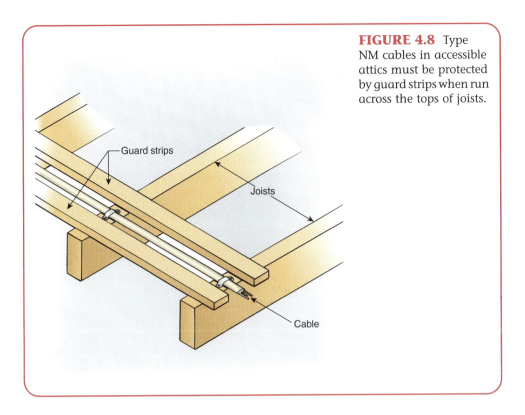

FIGURE 4.8 Type NM cables in accessible attics must be protected by guard strips when run across the tops of joists.

Guard strips

Joists

Cable

- *Accessible Attics.* Type NM cables run across the top of floor joists must be protected by guard strips when an attic is accessible by a permanent ladder or stairs (Figure 4.8). This rule protects the cable from damage by being walked on. When there is a scuttle hole (access hatch) but no permanent ladder or stairs, guard strip protection is only required within 6 ft of the edge of the hole. When cables are installed parallel to the sides of rafters, joists, or studs, guard strips aren't required [334.23].

Even in exposed locations, NM cables that run parallel to framing members or through bored holes must comply with the 300.4(A) rule requiring that they be kept back 1¼ in. from the face of the stud or other framing member unless supplemental protection is provided for the cables. The reason for this requirement is because unfinished basements and attics often are converted into finished recreation rooms, home offices, and guest bedrooms.

Type NM Cables Entering Boxes

The *Code* rules for NM cables entering boxes can be summarized as follows:

- When nonmetallic-sheathed cables enter boxes, the sheath must extend into the box at least ¼ in. so that unprotected conductors are not exposed to damage [314.17(C)].
- Cable connectors used with Type NM cables can be either separate fittings or connector devices that are an integral part of boxes. Multiple cables are permitted to enter a box through a single opening [314.17(C), Exception].
- Nonmetallic-sheathed cables must be secured to boxes. Separate metal connectors are used when the cable enters metal boxes. Most plastic boxes are manufactured with integral cable clamps. These clamps are small, springy flaps through which the NM cable is pushed into the box that prevent it from being pulled out again.

- Type NM cable is permitted to enter a nonmetallic box no larger than 2¼ in. × 4 in. (a single-gang device box) without being secured when it is fastened within 8 in. of the box. In other words, NM cable can enter a box without a connector if the cable is fastened closer to the box than usual [314.17(C), Exception].

- At least 6 in. of free conductor (measured from the end of the nonmetallic sheath) must be left at each box for splices of the connections of switches, receptacles, and luminaires (lighting fixtures). Each conductor must also be long enough to extend 3 in. outside the box. Although these are the *Code* minimums specified in 300.14, leaving longer wire ends at boxes and cabinets often makes termination much easier. Many installers prefer to leave 10–12 in. of free conductor, rather than 6 in.

Derating Bundled Cables

When more than two NM cables are bundled together and pass through wood framing that will be fire- or draft-stopped using thermal insulation or sealing foam, the ampacity of each conductor must be adjusted in accordance with Table 310.15(B)(2)(a), shown in Exhibit 4.1 [334.80].

In residential wiring, this ampacity derating is only required in places where two or more NM cables are stapled together lengthwise in contact with each other along a stud or framing member and where the cavity will later be filled with thermal insulation or sealing foam. Derating doesn't apply to NM cables run exposed or in wall spaces that aren't going to be filled with an insulating/sealing material.

Fire-stopping is rarely used in one- and two-family dwellings. Thermal insulation is typically used only in exterior walls and ceilings of upper stories, particularly in colder climates.

Explanation. This derating table seems to apply only to "conductors in a raceway or cable." However, *NEC* 310.15(B)(2)(a) explains that this table applies where multiconductor cables are stacked or bundled for longer than 24 in. without maintaining spacing and are not installed in raceways. Also, both the *Code* text and table title refer to "more than three current-carrying conductors in a raceway or cable."

EXHIBIT 4.1

NEC Table 310.15(B)(2)(a) Adjustment Factors for More Than Three Current-Carrying Conductors in a Raceway or Cable	
Number of Current-Carrying Conductors	**Percent of Values in Tables 310.16 through 310.19 as Adjusted for Ambient Temperature if Necessary**
4–6	80
7–9	70
10–20	50
21–30	45
31–40	40
41 and above	35

Since each Type NM cable used in residential wiring contains two or more current-carrying conductors, this means that two NM cables stacked or bundled together contain a minimum of *four* current-carrying conductors.

ARMORED CABLE (TYPE AC)

Type AC cable consists of two, three, or four insulated conductors, 14 AWG through 1 AWG, that are individually wrapped in waxed paper. These conductors are enclosed in a spiral flexible metal sheath (either steel or aluminum) that also serves as an equipment grounding conductor [250.118(8)]. The cable has a 16-gauge aluminum bonding strip running its full length to improve the metal sheath's performance as a grounding path. Type AC cable doesn't contain a separate equipment grounding conductor (Figure 4.9). Conductor sizes are 14 AWG through 1 AWG for copper, and 12 AWG through 1 AWG for aluminum or copper-clad aluminum.

Armored cable is manufactured in three varieties:

1. *Type ACTH* has conductors rated 75°C with thermoplastic insulation.
2. *Type ACTHH* has conductors rated 90°C with thermoplastic insulation.
3. *Type ACHH* has conductors rated 90°C with thermosetting insulation.

Installing Armored Cable

The *NEC* provides a number of basic rules for installing armored cables.

Securing. Type AC cable must be secured in place at intervals no greater than 4½ ft and within 12 in. from every box, cabinet, or termination [320.30]. Staples or cable ties are most commonly used. Simply draping the cable over air ducts, pipes, lower members of bar joists, and ceiling grid tees is not permitted.

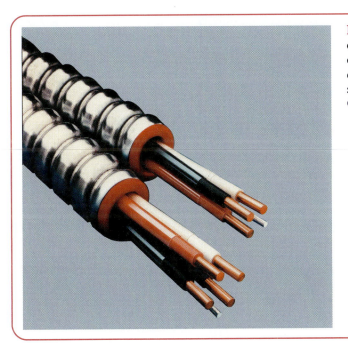

FIGURE 4.9 Type AC cable includes insulated conductors and a bare aluminum grounding strip. (Courtesy of AFC Cable Systems, Inc.)

Type AC cable can be used unsupported where fished (Figure 4.10), in lengths up to 2 ft at terminations where flexibility is needed, and in whips up to 6 ft long for connecting to luminaires (lighting fixtures) or equipment installed within accessible ceilings [320.30(D)].

When armored cable is run horizontally through holes or notches in framing members spaced not more than 4½ ft apart, it is considered to be adequately supported. Ties aren't required where the horizontal cable passes through these members. However, the cable must still be secured within 12 in. of each outlet.

Bending Radius. The radius of the curved inner edge of a Type AC cable bend shall not be less than five times the diameter of the cable. The diameters of 14/2, 14/3, 12/2, and 12/3 cables (the sizes most often used for branch circuits in dwellings) range from about ½ to ⅝ in. In practical terms, the minimum radius to the *inner edge of the cable* can't be less than 2½ in. for smaller AC cables and 3⅛ in. for larger AC cables [320.24].

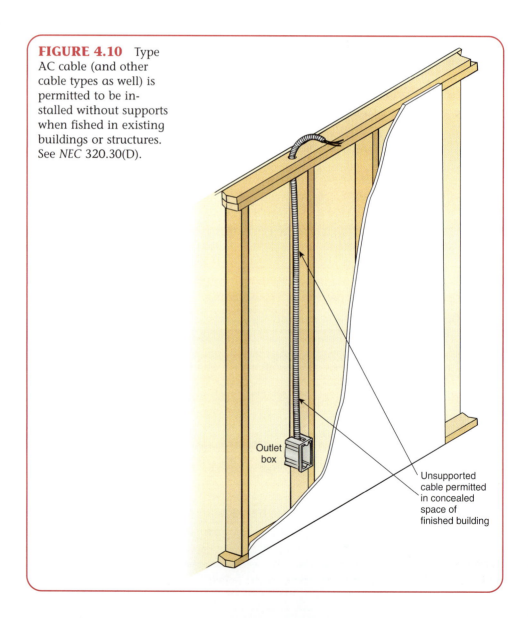

FIGURE 4.10 Type AC cable (and other cable types as well) is permitted to be installed without supports when fished in existing buildings or structures. See *NEC* 320.30(D).

Outlet box

Unsupported cable permitted in concealed space of finished building

Protection Against Physical Damage (Wood Framing). Armored cables run parallel to framing members must be installed so that the edge of the cable is not closer than 1¼ in. to any edge of the framing member where screws or nails are likely to penetrate. Generally, this requirement applies to *vertical* framing members such as studs [300.4(D)].

If this distance can't be maintained, the AC cable must be protected by a steel plate, sleeve, or other guard at least ¹⁄₁₆ in. thick. The reason for these rules is to prevent the cable from being damaged by drywall screws or other fasteners used to attach wall finishes.

When holes are drilled through vertical wood studs or other framing members for running AC cables, they must be drilled at least 1¼ in. from the nearest edge or a steel protector must be provided [300.4(A)] to prevent the cable from being damaged.

When drilling through floor joist members, the edge of a bored hole should be kept at least 2 in. from the bottom chord of the joist. Prefabricated trusses cannot be drilled.

Protection Against Physical Damage (Metal Framing). Factory-made openings in steel studs are located to keep cables more than 1¼ in. from the edge of the stud. When holes are field-cut, they must comply with the same rules as holes drilled through wood framing members. Insulating bushings or grommets aren't required when AC cable is run through openings in steel studs, since the steel armor provides adequate protection against damaging the conductors inside.

Installations in Exposed Locations. Type AC cable installed exposed must closely follow the surface of the building finish or be on running boards.

  • *Beneath Joists.* AC cables can to be stapled directly to the lower edges of joists, where they are supported at each joist and located to avoid physical damage [320.15].

  • *Accessible Attics.* Type AC cables run across the top of floor joists, or across the face of rafters or studding within 7 ft of the floor or floor joists, must be protected by guard strips when an attic is accessible by a permanent ladder or stairs, in order to protect the cable from damage when walked on or from being crushed by items stored in the attic. When there is a scuttle hole (access hatch) but no permanent ladder or stairs, guard strip protection is only required within 6 ft of the edge of the hole [320.23(A)]. If cables are run through holes bored in framing members, or installed parallel to the sides of rafters, joists, and studs, guard strips aren't needed.

Even in exposed locations, AC cables that run parallel to framing members or through bored holes must comply with the 300.4(A) rule requiring that they be 1¼ in. from the face of the stud or other framing member unless supplemental protection is provided for the cables. The reason for this requirement is because unfinished basements and attics often are converted into finished recreation rooms, home offices, and guest bedrooms.

Type AC Cable in Dropped or Suspended Ceilings. Type AC cable runs in dropped ceiling spaces must be properly secured and supported as required by 300.11(A).

Terminations. Armored cables must terminate in boxes or fittings listed for use with AC cable. An insulating bushing (often called a "red head" or "red devil" in the field) must be installed at each termination. The bushing is inserted between the conductors and the armor to protect the conductor insulation from damage by sharp edges of cut metal.

A good practice for holding insulating bushings in place is to bend back the cable's aluminum bonding strip over the bushing and back-wrap the strip into the convolutions of the armor (Figure 4.11). Connectors and clamps for use with AC cable are designed so that the insulating bushing is visible for inspection after installation.

Cutting Armored Cables

There are three methods for cutting the armor of Type AC cable:

1. Rotary armor cutter
2. Hacksaw
3. Wire cutters, such as diagonal cutters or lineman pliers

Cable manufacturers recommend the rotary cutter method because it avoids damaging the insulated conductors inside the cable.

Rotary Armor Cutter. Rotary cutters are designed specifically for safely cutting AC cable armor. They are available in various sizes to accommodate a range of cable sizes. The appropriate rotary cutting tool for the size of cable to be cut should be selected. The cutter has an adjustable anvil that, when adjusted properly, secures the cable in the tool when the handle is squeezed. A few turns of the handle make a cut through the cable armor without a risk of damage to the insulated conductors. The severed armor is then slid off the conductors.

Hacksaw. Type AC cable can be cut using a sharp hacksaw blade having at least 24 teeth per inch that is installed in a heavy-duty frame. The blade must be taut in the frame. The AC cable should be secured in a vise or supported on a block of wood. One of the armor convolutions should be cut at an approximate 60-degree angle, taking care not to cut any deeper than necessary to avoid damage to the insulated conductors.

Wire Cutters. Using a wire cutter to remove AC cable armor requires either breaking the armor or unwinding it. Breaking the armor involves bending the cable at the point where the cut is desired and twisting the armor slightly so the cutting pliers can be inserted between the conductors and the armor. The armor must be twisted back into the convolutions before it is inserted into the connector. At this stage, care

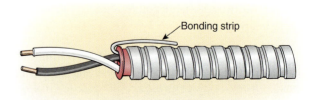

FIGURE 4.11 Aluminum bonding strip is bent back to help hold anti-short bushing in place until Type AC cable connector is installed.

Bonding strip

should be used to avoid damaging the cable, since the armor often has a sharp edge from the cutting process.

Derating Cables in Thermal Insulation

Type AC cable installed in thermal insulation must have conductors rated 90°C. However, the ampacity for cables installed in these applications must be that of 60°C conductors [320.80]. Thus, when AC cable is installed in spaces that will be filled with thermal insulation—such as a ceiling space below an attic—type ACHH or ACTHH cables must be used, but the ampacity is read from the 60°C column of Table 310.16.

BOX FILL AND SELECTION

The *Code* has conductor fill rules intended to ensure that boxes don't become overcrowded. Trying to jam too many conductors into a too-small box creates a number of potential problems:

- Difficulty installing wiring devices in boxes and/or installing covers
- Conductor stress when wires are bent too sharply
- Dangerous heat buildup
- Insulation damage, which may cause faults between conductors or from conductors to metal boxes
- Pulling or popping off of twist-on wire connectors, exposing live conductors

Section 314.16 requires that "boxes and conduit bodies shall be of sufficient size to provide free space for all enclosed conductors." Ensuring that a large enough box has been selected for a particular purpose depends on two concepts: *total box volume* and *total box fill.*

Total Box Volume

The volumes of common box sizes are listed in Table 314.16(A), along with the maximum number of conductors (all of the same size) permitted in each box. Some boxes and components are marked with their internal volumes. When a box is marked with a larger volume than the standard ones listed in Table 314.16(A), the larger volume can be used instead of the table value.

Total box volume is determined by adding up the volumes of the box itself plus any attachment such as a raised cover—often called a "plaster ring" or "mud ring" in the field (Figure 4.12)—a box extension, a domed (as opposed to flat) box cover, or a luminaire canopy [314.16(A)].

Total Box Fill

NEC Table 314.16(A), shown in Exhibit 4.2, addresses many common house-wiring situations, because frequently all the conductors in a given box are the same AWG size. Conductors that originate outside the box and are spliced or terminated within the box are counted as a single volume allowance. Conductors that pass through the

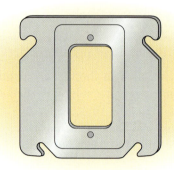

FIGURE 4.12 Raised box covers are also called "plaster rings" or "mud rings." When marked with their cubic-inch volume, the volume of the cover is added to the volume of the box to determine total cubic-inch volume.

EXHIBIT 4.2

NEC TABLE 314.16(A) Metal Boxes

Box Trade Size			Minimum Volume		Maximum Number of Conductors*						
mm	in.		cm³	in.³	18	16	14	12	10	8	6
100 × 32	(4 × 1¼)	round/octagonal	205	12.5	8	7	6	5	5	5	2
100 × 38	(4 × 1½)	round/octagonal	254	15.5	10	8	7	6	6	5	3
100 × 54	(4 × 2⅛)	round/octagonal	353	21.5	14	12	10	9	8	7	4
100 × 32	(4 × 1¼)	square	295	18.0	12	10	9	8	7	6	3
100 × 38	(4 × 1½)	square	344	21.0	14	12	10	9	8	7	4
100 × 54	(4 × 2⅛)	square	497	30.3	20	17	15	13	12	10	6
120 × 32	(4¹¹/₁₆ × 1¼)	square	418	25.5	17	14	12	11	10	8	5
120 × 38	(4¹¹/₁₆ × 1½)	square	484	29.5	19	16	14	13	11	9	5
120 × 54	(4¹¹/₁₆ × 2⅛)	square	689	42.0	28	24	21	18	16	14	8
75 × 50 × 38	(3 × 2 × 1½)	device	123	7.5	5	4	3	3	3	2	1
75 × 50 × 50	(3 × 2 × 2)	device	164	10.0	6	5	5	4	4	3	2
75 × 50 × 57	(3 × 2 × 2¼)	device	172	10.5	7	6	5	4	4	3	2
75 × 50 × 65	(3 × 2 × 2½)	device	205	12.5	8	7	6	5	5	4	2
75 × 50 × 70	(3 × 2 × 2¾)	device	230	14.0	9	8	7	6	5	4	2
75 × 50 × 90	(3 × 2 × 3½)	device	295	18.0	12	10	9	8	7	6	3
100 × 54 × 38	(4 × 2⅛ × 1½)	device	169	10.3	6	5	5	4	4	3	2
100 × 54 × 48	(4 × 2⅛ × 1⅞)	device	213	13.0	8	7	6	5	5	4	2
100 × 54 × 54	(4 × 2⅛ × 2⅛)	device	238	14.5	9	8	7	6	5	4	2
95 × 50 × 65	(3¾ × 2 × 2½)	masonry box/gang	230	14.0	9	8	7	6	5	4	2
95 × 50 × 90	(3¾ × 2 × 3½)	masonry box/gang	344	21.0	14	12	10	9	8	7	4
min. 44.5 depth	FS — single cover/gang (1¾)		221	13.5	9	7	6	6	5	4	2
min. 60.3 depth	FD — single cover/gang (2⅜)		295	18.0	12	10	9	8	7	6	3
min. 44.5 depth	FS — multiple cover/gang (1¾)		295	18.0	12	10	9	8	7	3	
min. 60.3 depth	FD — multiple cover/gang (2⅜)		395	24.0	16	13	12	10	9	8	4

*Where no volume allowances are required by 314.16(B)(2) through (B)(5).

box without a splice or connection also count as a single volume allowance. Small wires that originate within the box and don't leave it (such as pigtails for receptacles and jumper wires between different devices) are not counted when determining box fill.

NEC Table 314.16(B), shown in Exhibit 4.3, gives the volume allowance for each conductor in sizes 18 AWG through 6 AWG. It must be used when different conductor sizes are installed together in the same box. Note that the values given are not the volume actually occupied by each conductor, but the volume of *free space* required for safe installation of each conductor.

Items other than current-carrying conductors are also counted when determining box fill. These include wiring devices, equipment grounding conductors smaller than 14 AWG, support fittings such as studs and hickeys, and cable clamps inside the box. Small fittings such as locknuts and bushings don't need to be counted.

Up to four luminaire (lighting fixture) wires that enter a box from a domed luminaire (fixture) canopy also don't need to be counted if they are smaller than 14 AWG, since the domed canopy provides additional volume for wiring space.

Table 4.1, taken from the *National Electrical Code® Handbook,* summarizes all items that need to be counted when determining total box fill.

Determining Box Size

For a given application, the *total box fill* must be less than *the total box volume* of the box selected plus any accessories. The following three examples show how to determine the box size required for several different wiring situations.

Example 1: Box with Conductors Only, All the Same Size. Determine whether the 12 AWG conductors shown in Figure 4.13 are permitted to be installed in a standard 4 in. $\times$ 1½ in. square box (21.0 in.^{3}) containing no wiring devices and no other items such as fixture studs, cable clamps, switches, receptacles, or equipment grounding conductors.

1. Three conductors pass through the box without splices or connections.
2. Six conductors enter the box and are spliced using twist-on connectors.
3. The total conductor count for this box is *nine.*
4. *NEC* Table 314.16(A), shown in Exhibit 4.2, indicates that the maximum fill for a 4 in. $\times$ 1½ in. square box is nine 12 AWG conductors.

NEC TABLE 314.16(B) Volume Allowance Required per Conductor			EXHIBIT 4.3

Size of Conductor (AWG)	Free Space Within Box for Each Conductor	
	cm^3	in.^{3}
18	24.6	1.50
16	28.7	1.75
14	32.8	2.00
12	36.9	2.25
10	41.0	2.50
8	49.2	3.00
6	81.9	5.00

TABLE 4.1 Summary of Items Contributing to Box Fill

Items Contained Within Box	Volume Allowance	Based on [see Table 314.16(B)]
Conductors that originate outside box	One for each conductor	Actual conductor size
Conductors that pass through box without splice or connection (*less than 12 in. in total length)	One for each conductor	Actual conductor size
*Conductors 12 in. or greater that are looped and unbroken (see 300.15 for exact measurement)	Two for a single (entire) unbroken conductor	Actual conductor size
Conductors that originate within box and do not leave box	None (these conductors not counted)	n.a.
Fixture conductors [per 314.16(B)(1), Exception]	None (these conductors not counted)	n.a.
Internal cable clamps (one or more)	One only	Largest-sized conductor present
Support fittings (such as fixture studs, hickeys)	One for each type of support fitting	Largest-sized conductor present
Devices (such as receptacles, switches)	Two for each yoke or mounting strap	Largest-sized conductor connected to device or equipment
Equipment grounding conductor (one or more)	One only	Largest equipment grounding conductor present
Isolated equipment grounding conductor (one or more) [see 250.146(D)]	One only	Largest isolated and insulated equipment grounding conductor present

*2005 *Code* change.
Source: *National Electrical Code® Handbook,* 2005 edition, Commentary Table 314.1

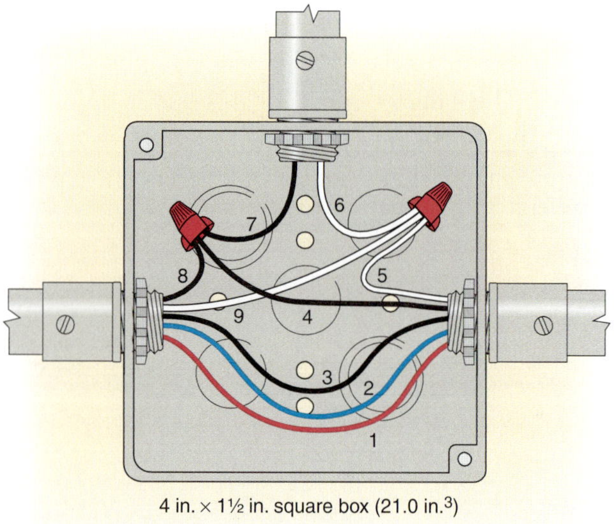

FIGURE 4.13 This 4 in. × 1½ in. square box contains no fittings or devices, such as fixture studs, cable clamps, switches, receptacles, or equipment grounding conductors.

4 in. × 1½ in. square box (21.0 in.^{3})

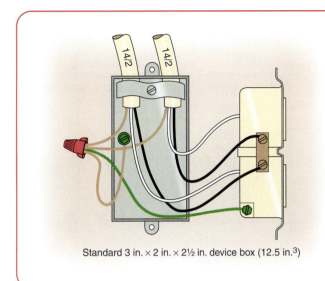

FIGURE 4.14 This device box contains components and conductors requiring deductions in accordance with *NEC* 314.16.

Standard 3 in. × 2 in. × 2½ in. device box (12.5 in.³)

5. The box shown is adequately sized.

Example 2: Box with Duplex Receptacle and Conductors All the Same Size. Determine whether a 3 × 2 × 2½ in. device box (12.5 in.³) is large enough to contain a duplex receptacle and four 14 AWG conductors, as shown in Figure 4.14.

1. Table 4.2 shows the volume allowances for the various items installed in this box.

2. The total box fill is 16 in.³

3. The box shown is too small.

TABLE 4.2 Total Box Fill for Example No. 2

Items Contained Within Box	Volume Allowance	Unit Volume Based on Table 314.16(B) (in.³)	Total Box Fill (in.³)
4 conductors	4 volume allowances for 14 AWG conductors	2.00	8.00
1 clamp	1 volume allowance (based on 14 AWG conductors)	2.00	2.00
1 device	2 volume allowances (based on 14 AWG conductors)	2.00	4.00
Equipment grounding conductors (all)	1 volume allowance (based on 14 AWG conductors)	2.00	2.00
Total			16.00

Source: *National Electrical Code® Handbook*, 2005 edition, Commentary Table 314.2

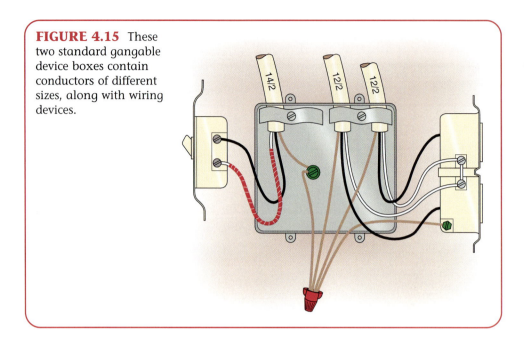

FIGURE 4.15 These two standard gangable device boxes contain conductors of different sizes, along with wiring devices.

4. The minimum size device box required for this installation is $3 \times 2 \times 3\frac{1}{2}$ (18.0 in.^{3}).

Example 3: Box with Switch, Duplex Receptacle, Switch, and Different-Sized Conductors. Determine whether two $3 \times 2 \times 2\frac{3}{4}$ in. device boxes ganged together (28 in.^{3}) are large enough to contain a duplex receptacle, a wall switch, four 12 AWG conductors, and two 14 AWG conductors, as shown in Figure 4.15.

1. Table 4.3 shows the volume allowances for the various items installed in this box.
2. The total box fill is 26 in.^{3}
3. The box shown is adequately sized.

GENERAL REQUIREMENTS FOR INSTALLING BOXES

Nonmetallic and Metal Boxes

Nonmetallic boxes are the type most commonly used in residential wiring. They are typically used with nonmetallic-sheathed cable (Type NM). However, nonmetallic boxes are also permitted to be used with metal raceways or metal-armored cables (such as Type AC) when they have internal bonding means or provisions for attaching equipment grounding conductors [314.3, Exceptions No. 1 and No. 2].

Metal boxes may be used with any wiring method, but are required to be grounded [314.4]. They are most commonly used with Type AC cable or metal raceways.

Gangable Boxes. Some 3 in. $\times$ 2 in. metal device boxes can be "ganged" by removing the side panel(s) and screwing the boxes together to create a single larger

TABLE 4.3 Total Box Fill for Example No. 3

Items Contained within Box	Volume Allowance	Unit Volume Based on Table 314.16(B) (in.^{3})	Total Box Fill (in.^{3})
6 conductors	2 volume allowances for 14 AWG conductors	2.00	4.00
	4 volume allowances for 12 AWG conductors	2.25	9.00
2 clamps	1 volume allowance (based on 12 AWG conductors)	2.25	2.25
2 devices	2 volume allowances (based on 14 AWG conductors)	2.00	4.00
	2 volume allowances (based on 12 AWG conductors)	2.25	4.50
Equipment grounding conductors (all)	1 volume allowance (based on 12 AWG conductors)	2.25	2.25
Total			26.00

Source: *National Electrical Code® Handbook*, 2005 edition, Commentary Table 314.3

box that accommodates two, three, or more wiring devices. After the switches or receptacles are installed, the installation is finished with a two-gang, three-gang, or larger cover plate.

Mounting Methods for Boxes

Brackets. Boxes are commonly mounted in walls and ceilings by attaching them to studs or joists with external brackets built onto the box that can be screwed or nailed to the structural member. Nails and screws are also permitted to pass through the interior of a box as described in *NEC* 314.23(B)(1), but this method reduces the usable wiring space inside the box. Using boxes with external mounting brackets is usually preferable.

Braces. When a box must be located between studs or joists, it can be installed using metal, polymeric, or wooden braces (Figure 4.16). Wood strips must be at least 1 in. × 2 in. in cross-section [314.23(B)(2)].

Finished Surfaces. Even in new construction, it's sometimes necessary to go back and add a box in a completed partition. There are also situations where structural members aren't spaced conveniently for standard box mounting techniques. In these cases, so-called remodeling or "old work" boxes can be used. See Chapter 17, Old Work, for more information about installing electrical boxes in finished surfaces.

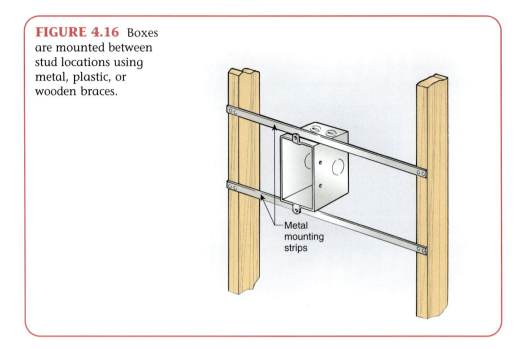

FIGURE 4.16 Boxes are mounted between stud locations using metal, plastic, or wooden braces.

Metal mounting strips

Boxes Mounted in Noncombustible Material

In walls or ceilings with a surface of concrete, gypsum, plaster, or other noncombustible material, boxes must be installed so that the front edge of the box will not be set back from the finished surface more than ¼ in. [314.20].

This requirement is intended to make sure that the standard mounting screws supplied with a luminaire (lighting fixture), wiring device, or other equipment are long enough to secure the equipment to its box. In cases where boxes are recessed too deeply into walls, the screw connection may not be mechanically secure. In the case of metallic boxes, the screw connection may not ensure adequate bonding continuity between the equipment and the box.

Boxes Mounted in Combustible Material

In walls and ceilings constructed of wood or other combustible surface material, boxes must be flush with the finished surface or must project beyond it [314.20]. The reason boxes aren't permitted to be recessed into combustible materials is so that, in case of a fault or poor connection, electrical sparking or flying bits of molten metal will be contained within the box and not come into contact with materials that can burn.

This requirement regarding combustible materials also applies to electrical boxes installed in ceilings or walls made of gypsum wallboard, often called by the trade names Drywall® or Sheetrock® in the field. Although gypsum is fireproof, the paper layer on the surface is combustible. Therefore, boxes must be installed so that they are flush with the surface of drywall or project from it.

Box Extensions. Sometimes installation situations occur where a wall covering is thicker than the electrician expected when roughing in the boxes. For example, custom wood paneling might be installed that's thicker than standard ⅝ in. wallboard. When this situation occurs and an electrical box ends up recessed into combustible material, a box extension can be installed to correct the problem.

NOTE: When box extensions are installed, they increase the total volume of the box for wiring [314.16(A)].

How Far Can Boxes Project from Walls?

The *NEC* doesn't specify how far electrical boxes are permitted to project beyond the wall surface. But as a practical matter, boxes shouldn't stick out more than ⅛ in., or the box and/or cover plate can create a hazard for snagging clothing and other materials.

In the case of device boxes, minor projections can often be covered with an extra-deep "goof plate." However, for the sake of appearance, all cover plates within the same room or area of a dwelling should normally be the same design.

In the case of wall or ceiling boxes for luminaires (lighting fixtures), ceiling fans, or other equipment, projecting boxes may be covered up by the equipment canopy.

Feathering. In situations where, for reasons beyond the electrician's control, boxes stick out beyond the wall surface more than ⅛ in., the usual remedy is to gradually build up or "feather" the surrounding wall area using drywall compound (called *mud* in the field). Feathering creates a new surface that a device cover plate or attached equipment can bear against securely, without leaving a gap.

Gaps Not Permitted

Frequently, holes cut in drywall and other materials are larger than necessary for the boxes installed. Wall surfaces must be patched so there are no gaps or open spaces greater than ⅛ in. at the edge of a box or fitting [314.21] to maintain the fire rating of the wall as required by building codes.

Installing Wiring Devices in Boxes

Receptacles, snap switches, dimmers, and similar components are all considered "wiring devices." Most wiring devices have screw terminals for attaching branch-circuit conductors. Figure 4.17 shows the *right* way to terminate conductors on wiring device terminal screws. Figure 4.18 illustrates several *wrong* ways to terminate branch-circuit conductors at wiring devices.

Screwless Terminals. Some receptacles and switches rated 15 amperes have push-in terminals for 14 AWG conductors. With this type of device, solid 14 AWG copper conductors are inserted into holes in the back of the device and gripped by spring-loaded terminals. Product listing standards don't allow push-in terminals to be used on wiring devices rated 20 amperes that attach to 12 AWG conductors.

TUBULAR RACEWAYS

Cable wiring methods are the ones most commonly used in residential construction. Raceway wiring methods are also used for one- and two-family dwellings in a few localities where required by local electrical codes, such as in Chicago. But in most parts of the country, raceways aren't much used in dwellings. Typically, they are

FIGURE 4.17 *Right* way to terminate conductors on wiring devices.

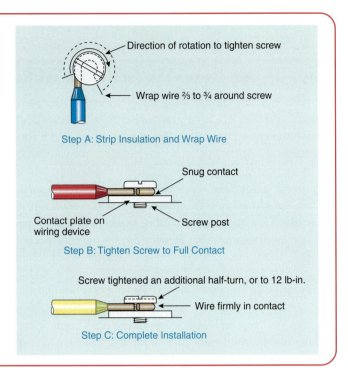

FIGURE 4.18 *Wrong* ways to terminate conductors on wiring devices.

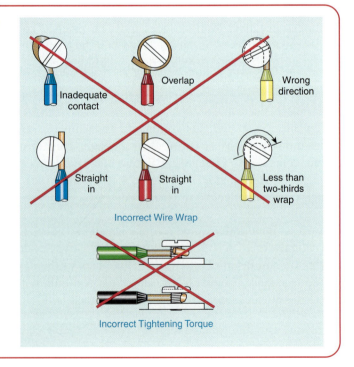

limited to particular locations where cables and conductors needs extra protection: outdoors, underground, and wherever wiring needs to be protected from physical damage.

> ### *General Guidelines for Installing Raceways*
>
> - Raceways should be cut carefully and the ends reamed to create a smooth surface that won't damage conductor insulation when wires are pulled into them. Where raceways containing conductors 4 AWG or larger enter a box or enclosure, a smooth fitting must be installed to protect wires from abrasion [300.4(F)] (Figure 4.19).
>
> - Raceways must be securely supported, at intervals that vary from one type to another. Horizontal runs of raceways can be supported by openings in framing members.
>
> - Bends must be carefully made so that raceways aren't damaged and the internal cross-sectional area isn't reduced. There can be a maximum of four right-angle bends (a total of 360 degrees) between pull points such as boxes, conduit bodies, and panelboard cabinets. This total includes offsets at boxes and enclosures. An offset with two 10-degree bends counts as 20 degrees as shown in Figure 4.20 [348.26 (Type FMC), 350.26 (Type LFMC), 352.26 (Type LFNC), 356.26 (Type RNC), 358.26 (Type EMT), and 362.24 (Type ENT)].
>
> - Generally speaking, raceway systems must be installed complete between outlets and other pull points before the conductors are pulled into them [300.18(A)].

Tubular raceways used in residential construction include electrical metallic tubing (Type EMT), flexible metal conduit (Type FMC), liquidtight flexible metal conduit (Type LFMC), liquidtight flexible nonmetallic conduit (Type LFNC), rigid nonmetallic conduit (Type RNC), and electrical nonmetallic tubing (ENT).

Electrical Metallic Tubing (Type EMT)

Electrical metallic tubing (Type EMT) is covered in Article 358 of the *NEC*. EMT is a thinwall raceway manufactured in both steel and aluminum, though the steel version is more common. It is joined with threadless set-screw or compression fittings (Figure 4.21).

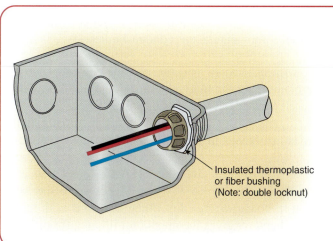

FIGURE 4.19 Insulating bushings are used to protect large conductors from chafing against conduit fittings.

Insulated thermoplastic or fiber bushing (Note: double locknut)

FIGURE 4.20 Raceways cannot have more than four right-angle bends (360 degrees maximum) between boxes or other pull and termination points. Since an offset with two 10-degree bends counts as 20 degrees, this installation has bends totaling 220 degrees.

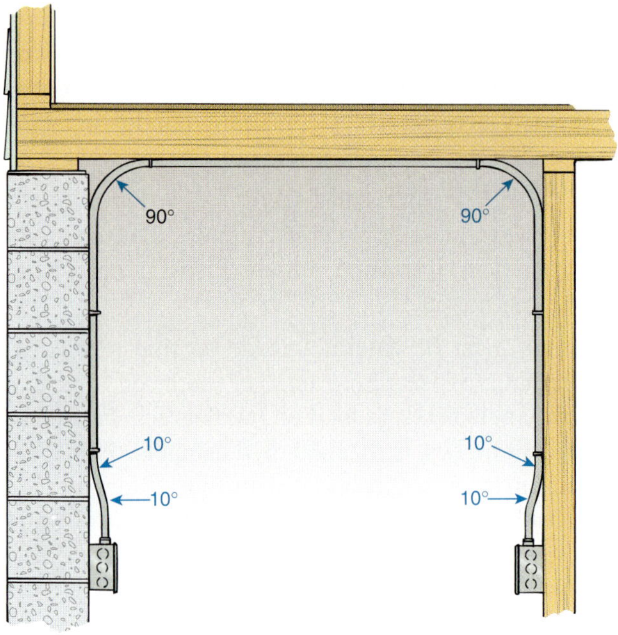

90°   90°

10°   10°

10°   10°

FIGURE 4.21 An electrical metallic tubing coupling is shown on the left. Electrical metallic tubing (EMT) is terminated at boxes using the connector on the right.

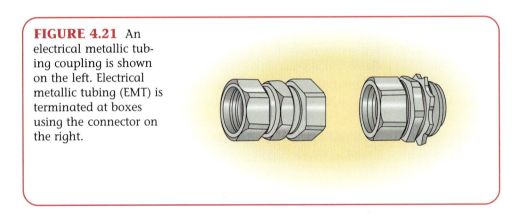

Uses. EMT can be used both indoors and outdoors, either exposed or concealed, in wet locations, and can be cast in concrete. It cannot be used to support luminaires (lighting fixtures), boxes, or other equipment except conduit bodies.

Installation. EMT can be cut with a hacksaw and the end reamed after cutting. Bends should be made using a conduit bender to avoid deforming the tubing. EMT must be securely fastened in place every 10 ft and within 3 ft of a box, cabinet, or other termination [358.30(A)] (Figure 4.22). Horizontal runs of electrical metallic tubing can be supported by holes or notches in framing members.

Raceway Sizes and Numbers of Conductors. Table C.1 of *NEC* Annex C is used to determine the maximum conductor fill for EMT when all wires are the same size (the most common situation in house wiring). For other situations with mixed conductor sizes, Table 4 of *NEC* Chapter 9 gives the maximum fill area for different sizes of EMT.

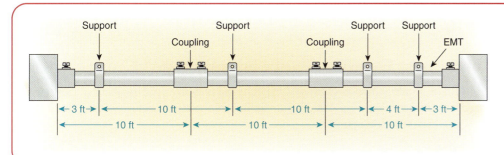

FIGURE 4.22 These minimum requirements apply to fastening electrical metallic tubing (EMT) in place, unless an exception applies.

Grounding. Electrical metallic tubing is permitted to serve as an equipment grounding conductor for circuits installed within it, when it forms a complete wiring system with metal boxes [250.118(4)]. However, in many applications, a separate equipment grounding conductor is run for convenience.

Flexible Metal Conduit (Type FMC)

Flexible metal conduit (Type FMC) is covered in Article 348. FMC is commonly used where flexibility is needed, either for ease of installation or to provide vibration isolation at connections to motorized equipment. It resembles Type AC cable without the wires, which are pulled in later.

Uses. FMC is only used indoors in dry locations, either exposed or concealed. Trade size ½ is the smallest FMC for general use. Trade size ⅜ can be used, in lengths up to 6 ft, to connect single appliances or luminaires (lighting fixtures) [348.20(A)(2)].

Installation. FMC can be cut with a hacksaw. It is attached to boxes, cabinets, and other equipment using special connectors. FMC must be securely fastened in place at the same intervals as Type NM and Type AC cables: every 4½ ft and within 12 in. of a box, cabinet, or other termination. Horizontal runs of flexible metal conduit can be supported by holes or notches in framing members [348.30(B)].

   FMC can be fished without securing or supporting it, and unsupported lengths up to 3 ft are also permitted where flexibility is necessary. Unsupported lengths up to 6 ft are permitted as whips to luminaires (lighting fixtures) [348.30(A), Exceptions No. 1, 2, 3, and 4].

Protection from Physical Damage. FMC run parallel to studs and other framing wood members must comply with the same rules as those for cable wiring methods. It must be installed so that the edge of the cable is no closer than 1¼ in. to any edge of the framing member where screws or nails are likely to penetrate [300.4(D)]. If this distance can't be maintained, the conduit must be protected by a steel plate, sleeve, or other guard at least ¹⁄₁₆ in. thick.

Raceway Sizes and Numbers of Conductors. Table C.3 of *NEC* Annex C is used to determine the maximum conductor fill for FMC when all wires are the same size (the most common situation in house wiring). For other situations with mixed conductor sizes, Table 4 of *NEC* Chapter 9 gives the maximum fill area for different sizes of FMC.

Grounding. Flexible metal conduit is permitted to serve as an equipment grounding conductor in lengths up to 6 ft where the conductors are protected at 20 amperes or less. A separate ground wire must be installed when required for all other applications [250.118(5) and 348.60].

Liquidtight Flexible Metal Conduit (Type LFMC)

Liquidtight flexible metal conduit (Type LFMC) is covered in Article 350 of the *NEC*. This type of raceway is similar to flexible metal conduit with a liquidtight, sunlight-resistant outer jacket.

Uses. In dwellings, LFMC is most often used for outdoor connections to equipment such as air-conditioning units, pump motors, and luminaires (lighting fixtures) in swimming pools and fountains. Trade size ½ is the smallest LFMC for general use. Trade size ⅜ can be used, in lengths up to 6 ft, to connect single appliances or luminaires (lighting fixtures) [350.20(A), Exception].

LFMC can also be direct-buried where listed and marked for the purpose and is sometimes used to enclose feeder conductors for an accessory structure such as a garage or pool equipment building. When direct-buried, it must have at least 24 in. of cover unless it meets the various conditions of *NEC* Table 300.5, shown in Exhibit 4.4. Requirements for underground raceway installations are covered in greater detail in Chapter 12, Outdoor Areas.

Installation. LFMC can be cut with a hacksaw. Bends can be made by hand without using special equipment (conduit benders), and minimum radii must comply with Table 2, Chapter 9 using the columns titled "Other Bends." LFMC is attached to boxes, cabinets, and other equipment using special connectors. It must be securely fastened in place at the same intervals as Type NM and Type AC cables: every 4½ ft and within 12 in. of a box, cabinet, or other termination [350.30(A)]. Horizontal runs of liquidtight flexible metal conduit can be supported by holes or notches in framing members.

LFMC can be fished without securing or supporting it, and unsupported lengths up to 3 ft are also permitted where flexibility is necessary. Unsupported lengths up to 6 ft are permitted as whips to luminaires (lighting fixtures) [350.30(A), Exceptions No. 1, 2, 3, and 4].

Protection from Physical Damage. LFMC run parallel to studs and other framing wood members must comply with the same rules as those for cable wiring methods. It must be installed so that the edge of the cable is no closer than 1¼ in. to any edge of the framing member where screws or nails are likely to penetrate [300.4(D)]. If this distance can't be maintained, the conduit must be protected by a steel plate, sleeve, or other guard at least ¹⁄₁₆ in. thick.

Raceway Sizes and Numbers of Conductors. Table C.7 of *NEC* Annex C is used to determine the maximum conductor fill for LFMC when all wires are the same size (the most common situation in house wiring). For other situations with mixed conductor sizes, Table 4 of *NEC* Chapter 9 gives maximum fill area for different sizes of LFMC.

Grounding. Liquidtight flexible metal conduit is permitted to serve as an equipment grounding conductor in lengths up to 6 ft where the conductors are protected at 20 amperes or less. A separate ground wire must be installed when required for all other applications [250.118(6) and 350.60].

EXHIBIT 4.4

NEC TABLE 300.5 Minimum Cover Requirements, 0 to 600 Volts, Nominal, Burial in Millimeters (Inches)

Location of Wiring Method or Circuit	Type of Wiring Method or Circuit									
	Column 1 Direct Burial Cables or Conductors		Column 2 Rigid Metal Conduit or Intermediate Metal Conduit		Column 3 Nonmetallic Raceways Listed for Direct Burial Without Concrete Encasement or Other Approved Raceways		Column 4 Residential Branch Circuits Rated 120 Volts or Less with GFCI Protection and Maximum Overcurrent Protection of 20 Amperes		Column 5 Circuits for Control of Irrigation and Landscape Lighting Limited to Not More Than 30 Volts and Installed with Type UF or in Other Identified Cable or Raceway	
	mm	in.	mm	in.	mm	in.	mm	in.	mm	in.
All locations not specified below	600	24	150	6	450	18	300	12	150	6
In trench below 50-mm (2-in.) thick concrete or equivalent	450	18	150	6	300	12	150	6	150	6
Under a building	0 (in raceway only)	0	0	0	0	0	0 (in raceway only)	0	0 (in raceway only)	0
Under minimum of 102-mm (4-in.) thick concrete exterior slab with no vehicular traffic and the slab extending not less than 152 mm (6 in.) beyond the underground installation	450	18	100	4	100	4	150 (direct burial) 100 (in raceway)	6 / 4	150	6
Under streets, highways, roads, alleys, driveways, and parking lots	600	24	600	24	600	24	600	24	600	24
One- and two-family dwelling driveways and outdoor parking areas, and used only for dwelling-related purposes	450	18	450	18	450	18	300	12	450	18
In or under airport runways, including adjacent areas where trespassing prohibited	450	18	450	18	450	18	450	18	450	18

Notes:
1. Cover is defined as the shortest distance in millimeters (inches) measured between a point on the top surface of any direct-buried conductor, cable, conduit, or other raceway and the top surface of finished grade, concrete, or similar cover.
2. Raceways approved for burial only where concrete encased shall require concrete envelope not less than 50 mm (2 in.) thick.
3. Lesser depths shall be permitted where cables and conductors rise for terminations or splices or where access is otherwise required.
4. Where one of the wiring method types listed in Columns 1–3 is used for one of the circuit types in Columns 4 and 5, the shallowest depth of burial shall be permitted.
5. Where solid rock prevents compliance with the cover depths specified in this table, the wiring shall be installed in metal or nonmetallic raceway permitted for direct burial. The raceways shall be covered by a minimum of 50 mm (2 in.) of concrete extending down to rock.

Liquidtight Flexible Nonmetallic Conduit (Type LFNC)

Liquidtight flexible nonmetallic conduit (Type LFNC) is covered in Article 356. This raceway type is similar to LFMC, but it is of completely nonmetallic construction. Liquidtight flexible nonmetallic conduit is manufactured in three different types:

1. *Type LFNC-A* has strength reinforcement between the core and cover.
2. *Type LFNC-B* has reinforcement built into the smooth conduit wall.
3. *Type LFNC-C* is of corrugated construction, without integral reinforcement.

Uses. Of the three, LFNC-B is most commonly used in residential construction, typically for outdoor connections to motorized equipment. *Flexible nonmetallic conduit* (FNMC) is another name for this raceway type [356.2(3), FPN]. Trade size ½ is the smallest used for most applications.

LFNC can be direct-buried where listed and marked for the purpose and is sometimes used to enclose feeder conductors for an accessory building such as a garage or pool-equipment building. When direct-buried, it must have at least 24 in. of cover unless it meets the various conditions of *NEC* Table 300.5, shown in Exhibit 4.4.

Installation. LFNC can be cut with a hacksaw. Bends can be made by hand without using special equipment (hot boxes), and minimum radii must comply with *NEC* Table 2, using the column "Other Bends," shown in Exhibit 4.5.

LFNC must be securely fastened in place every 3 ft and within 12 in. of a box, cabinet, or other termination. Horizontal runs of liquidtight flexible nonmetallic conduit are permitted to be supported by holes or notches in framing members [356.30(3)].

LFNC can be fished without securing or supporting, and unsupported lengths up to 3 ft are also permitted where flexibility is necessary. Unsupported lengths up to 6 ft are permitted as whips to luminaires (lighting fixtures) [356.30(1), (2), and (4)].

EXHIBIT 4.5

NEC Table 2 Radius of Conduit and Tubing Bends

Conduit or Tubing Size		One Shot and Full Shoe Benders		Other Bends	
Metric Designator	Trade Size	mm	in.	mm	in.
16	½	101.6	4	101.6	4
21	¾	114.3	4½	127	5
27	1	146.05	5¾	152.4	6
35	1¼	184.15	7¼	203.2	8
41	1½	209.55	8¼	254	10
53	2	241.3	9½	304.8	12
63	2½	266.7	10½	381	15
78	3	330.2	13	457.2	18
91	3½	381	15	533.4	21
103	4	406.4	16	609.6	24
129	5	609.6	24	762	30
155	6	762	30	914.4	36

Protection from Physical Damage. LFNC run parallel to studs and other framing wood members must comply with the same rules as those for cable wiring methods. It must be installed so that the edge of the cable is no closer than 1¼ in. to any edge of the framing member where screws or nails are likely to penetrate [300.4(D)]. If this distance can't be maintained, the conduit must be protected by a steel plate, sleeve, or other guard at least ¹⁄₁₆ in. thick.

Raceway Sizes and Numbers of Conductors. Table C.5 of *NEC* Annex C is used to determine the maximum conductor fill for LFNC-B when all wires are the same size (the most common situation in house wiring). For other situations with mixed conductor sizes, Table 4 of *NEC* Chapter 9 gives the maximum fill area for different sizes of LFNC-B.

Grounding. An equipment grounding conductor must be installed in liquidtight flexible nonmetallic conduit when required for the application.

Prewired LFNC-B. Liquidtight flexible nonmetallic conduit, Type B, is also available as a prewired assembly with factory-installed conductors (Figure 4.23). Prewired LFNC-B is used more often for commercial applications than residential construction.

Rigid Nonmetallic Conduit (Type RNC)

Rigid nonmetallic conduit (Type RNC), which is covered in Article 352, is a heavywall raceway manufactured in both fiberglass and polyvinyl chloride (PVC). The PVC type of RNC is used most frequently in residential construction, usually for limited purposes such as protecting direct-buried cables where they emerge from the ground [300.5(D)(1)]. *Schedule 40* PVC rigid nonmetallic conduit has the same dimensions as rigid metal conduit. The heavier *Schedule 80* has a smaller internal diameter, which reduces its conductor fill.

Electrical Nonmetallic Tubing (Type ENT)

Electrical nonmetallic tubing (Type ENT), covered in Article 362, is a pliable raceway that can be bent by hand without using heat. Both solvent-welded (glued) and snap-on fittings are manufactured for use with ENT. Though widely used in commercial electrical construction (it is also available in a prewired version), the use of ENT isn't very common in residences, where cable wiring methods are used more often.

Figure 4.24 illustrates supporting rules for electrical nonmetallic tubing installed in wood studs.

FIGURE 4.23 Liquidtight flexible nonmetallic conduit, Type B, is also available as a prewired assembly. (Courtesy of Carlon®, Lamson & Sessions)

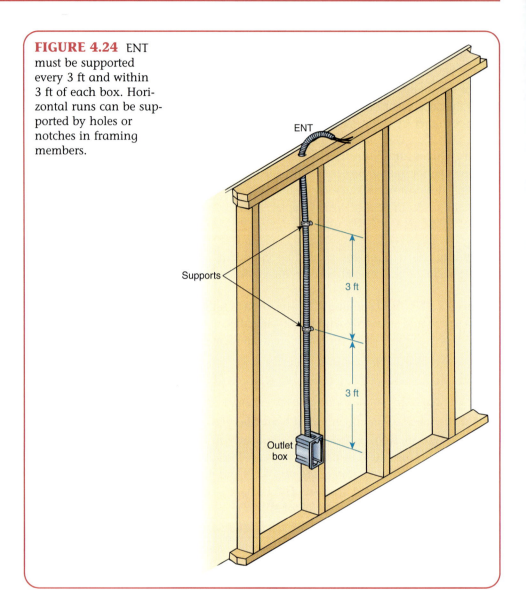

FIGURE 4.24 ENT must be supported every 3 ft and within 3 ft of each box. Horizontal runs can be supported by holes or notches in framing members.

ENT

Supports

3 ft

3 ft

Outlet box

SURFACE RACEWAYS

Surface raceways come in both metallic and nonmetallic versions. They are frequently known in the field by the trade name Wiremold®. Surface raceways are used more often in remodeling work than in new construction. However, there are situations where it may be desirable to use surface-mounted raceways of decorative design in finished areas of dwelling units. See Chapter 17, Old Work, for more information about installing surface raceways.

INSTALLING CONDUCTORS IN RACEWAYS

Once a raceway system has been completely installed, it's time to install the conductors [300.18(A)]. For short, straight runs between boxes using two 14 AWG or 12 AWG conductors, it may be possible simply to push the wires in at one box or conduit body and through the raceway to the next box. For longer runs or those with bends, fish tape is used.

Pulling Conductors Using Fish Tape

Fish tape is made of flexible, ⅛ in. wide springy steel, with a loop or "eye" at one end. Many installers wrap the place where the loop joins the main metal strip with electrical tape as shown in Figure 4.25 to help keep the loop from opening up under tension. The procedure for pulling conductors using fish tape follows.

Step 1. Push the fish tape through the raceway from one end, beginning at a box or other pull point until the loop appears at the next box.

CAUTION: Never push fish tape into an energized panelboard.

Step 2. Pull all the conductors in a particular segment of raceway (most often two, three, or four in house wiring) in a single operation.

Step 3. Cut wires to the length required, including free ends at boxes plus several extra inches to form the pull loop. It's often a good idea to cut wires about 2 ft longer than the anticipated need. Doing so will result in 10–12 in. of wire outside the box at each end for making connections.

Step 4. Strip off several inches of insulation from one end of the conductors, and push the bare wires through the loop, and twist them together (Figure 4.25). Some installers also use electrical tape to help hold the twisted wires together.

Step 5. Pull the fish tape in a strong, smooth motion, pulling the conductors through the raceway with it, and when the wires appear at the box where the tape was originally inserted, untwist the loop.

CAUTION: Don't pull too far, or you may accidentally pull the far end of the conductors into the raceway and have to repeat the whole operation.

Switch Loops

Device boxes for wall switches, dimmers, and similar controls should have only colored wires pulled to them. Usually black is used, though any color except white or gray is permitted. White and gray are reserved for grounded (neutral) conductors. Section 200.7(C) allows white or gray conductors in cables such as Type NM and Type AC to be reidentified for use as switch legs, but these colors are *not permitted* for individual conductors in raceways. Conductors with white or gray insulation cannot be used as switch legs [200.7(C)].

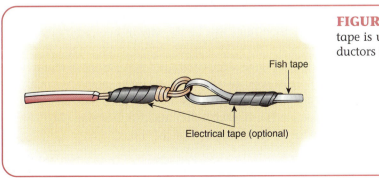

FIGURE 4.25 Fish tape is used to pull conductors into raceways.

Fish tape

Electrical tape (optional)

NOTES:

1. Special rules apply to identifying conductors of Type NM and Type AC cable assemblies that are used in switch loops. This important subject is covered in the "Switch Wiring" section of Chapter 5.

2. Pilot-light switches require a grounded (neutral) conductor in the switch box to provide 120-volt power for the light. This subject is also covered in the "Lighted Switches" section in Chapter 5.

USING OVERSIZED CONDUCTORS TO MINIMIZE VOLTAGE DROP

Although wiring runs in residences typically aren't as long as in large commercial buildings, voltage drop can still be a problem on longer circuits. The effects of reduced voltage include incandescent lights that burn more dimly, motors that run hotter, heaters that give reduced output, starting problems or flicker in fluorescent lamps, poor TV picture quality, and various operational problems for electronic equipment such as computers and stereos.

Voltage drop can be minimized by using oversized conductors on long homeruns (typically those longer than 50 ft) and even whole branch circuits (those over 100 ft in total length).

Article 210 permits 12 AWG conductors to be used on 15-ampere circuits and 10 AWG conductors on 20-ampere circuits. Section 210.3 states that "where conductors of higher ampacity are used for any reason, the ampere rating or setting of the specified overcurrent device shall determine the circuit rating."

Homeruns or Branch Circuits? In commercial electrical construction, job specifications sometimes call for a minimum conductor size of 10 AWG for all 20-ampere branch circuits. Wire binding screws on so-called "spec grade" wiring devices are listed to accept 10 AWG conductors, so this practice doesn't cause any problems with *Code* compliance.

However, 10 AWG conductors are larger and stiffer than 12 AWG conductors. These characteristics make it more difficult and time-consuming to do a "neat and workmanlike" job of connecting duplex receptacles and snap switches. So, whenever possible in residential wiring, it may be preferable to consider using oversized conductors for homeruns only, rather than entire branch circuits.

SPECIAL RULES FOR TWO-FAMILY DWELLINGS

There are no special wiring method rules that apply only in two-family dwellings. If there is a fire-rated wall, floor, or ceiling between two units in the same structure, any branch circuit or feeder penetrating that barrier is required to be firestopped [300.21]. However, two-family dwellings aren't normally separated by fire walls. Residential branch circuits are only permitted to supply loads located within, or associated with, each individual dwelling unit [210.25]. This subject is covered further in later chapters of this book.

CHAPTER REVIEW 4

MULTIPLE CHOICE

1. Nonmetallic-sheathed cables are permitted to be installed above dropped or suspended ceilings only
   A. When the ceiling has a 15-minute finish rating
   B. In dwellings
   C. When secured by independent support wires
   D. When it is Type NMC

2. Electrical metallic tubing (EMT), flexible metal conduit (FMC), liquidtight flexible metal conduit (LFMC), and rigid nonmetallic conduit (RNC) all
   A. Require the installation of a separate equipment grounding conductor
   B. Are permitted to be supported by horizontal openings in framing members
   C. Are tubular raceways
   D. Both B and C

3. The *Code* has conductor fill rules intended to ensure that boxes don't become overcrowded. Trying to jam too many conductors into a too-small box creates which of the following problems?
   A. Difficulty installing wiring devices in boxes, and/or installing covers
   B. Dangerous heat buildup
   C. Insulation damage, which may cause faults between conductors or from conductors to metal boxes
   D. All of the above

4. When most types of cables and raceways used for dwellings are run parallel to framing members or through bored holes, how far must they be kept from the face of the stud when protection is not provided for the cables or raceways?
   A. ¼ in.
   B. ¾ in.
   C. 1¼ in.
   D. 1¾ in.

5. Type AC cable armor can be removed by using
   A. Rotary armor cutter
   B. Hacksaw
   C. Electric metal saw
   D. Both A and B

6. The minimum radius of the curved inner edge of a Type NM or Type AC cable bend is not permitted to be
   A. Less than five times the diameter of the cable
   B. Less than ten times the diameter of the cable
   C. Less than 3 in.
   D. None of the above

7. When nonmetallic-sheathed cables pass through openings in steel studs and framing members, they must be protected from damage by sharp metal edges by installing
   A. Listed bushings
   B. Listed grommets
   C. Listed anti-short bushings
   D. Both A and B

8. The total volume of a $3 \times 2 \times 3\frac{1}{2}$ in. device box is
   A. 16 in.^{3}
   B. 18 in.^{3}
   C. 21 in.^{3}
   D. 27 in.^{3}

9. The following raceway type(s) is (are) rarely used in residential construction:
   A. Electrical metallic tubing (EMT)
   B. Flexible metal conduit (FMC)
   C. Liquidtight flexible metal conduit (LFMC)
   D. Rigid metal conduit (RMC)

10. Type NM cable must be secured as follows:
    A. At intervals no greater than 4½ ft
    B. Within 12 in. from every box, cabinet, or termination
    C. It can be installed unsupported in whips up to 6 ft long for connecting to luminaires (lighting fixtures) within accessible ceilings
    D. Both A and B

FILL IN THE BLANKS

1. Type AC cable is permitted to be used unsupported where fished, and in whips up to _____ long for connecting to luminaires (lighting fixtures) or equipment installed within accessible ceilings.

2. Raceways are permitted to bend a maximum of _____ degrees between pull points such as boxes, conduit bodies, and panelboard cabinets. This total includes offsets at boxes and enclosures.

3. Type ACTHH armored cable has conductors rated _____ °C with thermoplastic insulation.

4. When exposed Type NM cable passes through a floor, it must be enclosed in rigid metal conduit, intermediate metal conduit, electrical metallic tubing, Schedule 80 PVC conduit, listed metal or nonmetallic surface raceway, or other metal pipe extending at least _____ above the floor.

5. The tables in *NEC* Annex _____ can be used to determine the maximum conductor fill for raceways when all wires are the same size (the most common situation in house wiring).

6. Type NM cable is permitted to be used unsupported where fished, and in whips up to _____ long for connecting to luminaires (lighting fixtures) or equipment installed within accessible ceilings.

7. FNMC is an alternate designation for the raceway type also known as _____.

8. Volumes of common box sizes are listed in *NEC* Table _____, along with the maximum number of conductors (all of the same size) permitted in each box.

9. Conductors with _____ insulation cannot be used as switch legs, unless reidentified with tape or paint.

10. Type AC or Type NM cables run across the top of floor joists must be protected by _____ when the attic is accessible by a permanent ladder or stairs, or within 6 ft of a scuttle hole.

TRUE OR FALSE

1. Type AC cable consists of two or three insulated conductors, 12 AWG through 2000 kcmil, that are individually wrapped in waxed paper.

   _____ True   _____ False

2. Flexible metal conduit (FMC) is permitted to serve as an equipment grounding conductor in lengths up to 12 ft where the conductors are protected at 20 amperes or less.

   _____ True   _____ False

3. Increasing the size of conductors for long branch circuits and homeruns can help reduce voltage drop.

   _____ True   _____ False

4. At boxes and other terminations, nonmetallic-sheathed cables are required to be secured by cable clamps listed for the purpose.

   _____ True   _____ False

5. Electrical boxes are required to project a minimum of 1/16 in. beyond any combustible wall surface.

   _____ True   _____ False

6. At boxes and other terminations, armored cables are required to be secured by cable clamps listed for the purpose.

   _____ True   _____ False

7. *Total box volume* is determined by adding up the volumes of the box itself plus any attachment such as a plaster ring, extension ring, domed (as opposed to flat) cover, or luminaire canopy.

   _____ True   _____ False

8. Types NM, NMC, and NMS are all types of nonmetallic-sheathed cable defined in *NEC* Article 334.

   _____ True   _____ False

9. An insulating bushing (often called a "red head" in the field) must be installed at every cable termination to protect conductors from damage by sharp edges of cut metal, wood splinters, and similar items.

   _____ True   _____ False

10. Nonmetallic-sheathed cable contains conductors rated 90°C (194°F), but the ampacity of those conductors is based on the 60°C (140°F) column of *NEC* Table 310.16.

    _____ True   _____ False

CHALLENGE QUESTIONS

1. What is the reason for limiting most raceways to a total of 360° in bends between boxes or pull points?

2. Discuss the reasons that most residential wiring uses cable wiring methods rather than raceways.

3. The *Code* requires that most cable and raceway wiring methods be kept back 1¼ in. from the face of studs unless a protective steel plate is provided. What hazard is this installation technique guarding against?

4. What are some applications where raceways are commonly used in wiring one- and two-family dwellings?

5. Why does the *NEC* specify minimum bending radii for cable conductors?

6. When installing conductors in raceways, why is it important to pull all conductors through a particular raceway segment in the same operation?

7. What is the most significant difference, in terms of *Code* rules and compliance, between a building that contains two grade-level dwelling units separated by fire walls and a similar building with no fire walls between the dwelling units?

Lighting the Home

5

OBJECTIVES

After completing this chapter, the student will be able to understand the following:

- Required locations of lighting outlets and switches
- Basics of lamps and luminaires
- Thermal protection of luminaires (lighting fixtures)
- Boxes for luminaires and ceiling fans
- Switch control of lighting circuits

INTRODUCTION

This chapter covers principles of lighting in the home. Outdoor lighting is covered in Chapter 12, Outdoor Areas. The following major topics are included:

- Basics of lamps and luminaires
- Lighting outlets and switches
- Thermal protection of luminaires
- Bathroom lighting
- Closet lighting
- Lighting track
- Boxes for luminaires and paddle fans
- Types of switches and controls used in homes
- Installation of switches
- Grounding switches and faceplates
- Dimmer controls for homes
- Special rules for two-family dwellings

IMPORTANT NEC TERMS

Ampacity

Branch Circuit

Branch Circuit, Multiwire

Identified (as applied to equipment)

Labeled

Lighting Outlet

Lighting Track

Listed

Live Parts

Location, Damp

Location, Dry

Location, Wet

Luminaire

Receptacle Outlet

Switch, General-Use Snap

Utilization Equipment

BASICS OF LAMPS AND LUMINAIRES

Adequate, convenient lighting is one of the most important features of any home. People use their lighting more often than any other kind of electrical utilization equipment. As a general rule, more lighting outlets and wall switches are better than fewer. It costs only a few extra dollars to add another three-way switch during original construction, and probably less than a hundred dollars to add many kinds of luminaires (lighting fixtures). But it's expensive and inconvenient to go back after the fact and install another switch or luminaire once the walls and ceilings have been closed in.

Article 100 in the *National Electrical Code®* defines *luminaire* as follows.

> **Luminaire:** A complete lighting unit consisting of a lamp or lamps together with the parts designed to distribute the light, to position and protect the lamps and ballast (where applicable), and to connect the lamps to the power supply.

The maximum voltage allowed in dwellings is 120 volts [210.6(A)]. Figure 5.1 illustrates types of luminaires permitted to be supplied by 120-volt circuits, as specified in 210.6(B).

WARNING: NEC 600.32(I) prohibits neon lighting equipment (including artwork) with a secondary circuit voltage more than 1000 volts from being installed in or on dwellings.

Types of Luminaires

Products that fit into luminaires and produce light are typically sold as *bulbs* or *fluorescent tubes*, but both are actually lamps—technically, only the outer glass globe or tube is a bulb. The *NEC®* refers to them as lamps in Article 410 and elsewhere. Figure 5.2 illustrates lamp types used for general lighting.

Luminaires (lighting fixtures) installed inside of dwellings commonly use only a few lamp types: incandescent, fluorescent (including compact fluorescent), and tungsten-halogen.

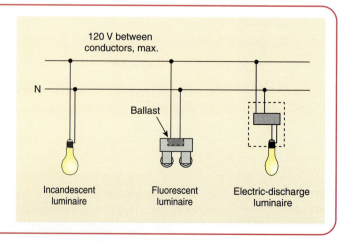

FIGURE 5.1 Luminaires rated not higher than 120 volts are permitted in and on dwellings, per *NEC* 210.6(A).

120 V between conductors, max.

N

Ballast

Incandescent luminaire

Fluorescent luminaire

Electric-discharge luminaire

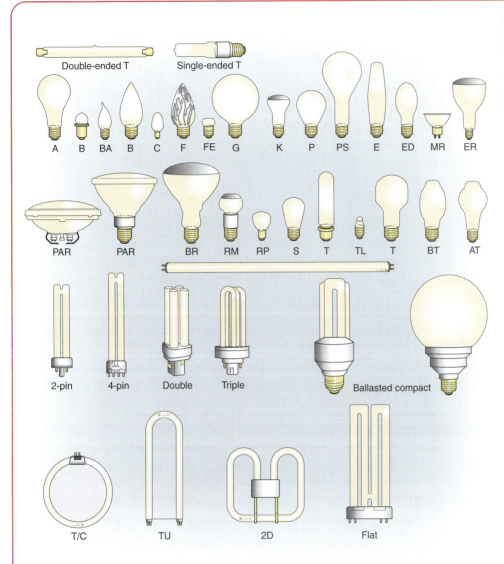

FIGURE 5.2 Lamps used for residential lighting come in many types and shapes. (Redrawn from *The IESNA Lighting Handbook,* 9th Edition, 2000, courtesy of the Illuminating Engineering Society of North America)

NOTE: High-intensity discharge (HID) luminaires (which include mercury vapor, metal halide, and high-pressure sodium) are often used outdoors for security lighting. Low-pressure sodium lamps, which aren't actually HID but have ballasts and resemble HID lamps in other ways, are also sometimes used for residential outdoor lighting. See Chapter 12, Outdoor Areas.

Incandescent Lamps

Incandescent lighting is the most common type used in homes. Incandescent lamps come in a wide variety of types, are inexpensive, and operate properly under nearly all conditions, including very low temperatures outdoors.

While frosted incandescent lamps give a soft, pleasing light that consumers are accustomed to, incandescent lighting is the least energy-efficient type in terms of light output per watt consumed. Larger lamps are generally more efficient than smaller ones. Three 60-watt lamps (180 watts total) produce 10 percent more light than five 40-watt lamps (200 watts total). In other words, decorative chandeliers and wall sconces that use multiple small-wattage lamps generally produce less light than luminaires designed to use a single large lamp of equivalent wattage.

Incandescent lamps have a relatively short life of 750 to 1000 hours when used at their rated voltage. So-called "long life" or "extended service" incandescent lamps are actually rated at higher voltages. A 125-volt lamp operated on a 120-volt circuit lasts up to twice as long as a conventional lamp, and a 130-volt lamp lasts up to three times longer. There's a catch, however: These higher-voltage lamps produce less than their rated light output when operated at lower voltages, and they're more expensive than ordinary 120-volt lamps.

Fluorescent Lamps

Fluorescent lamps last much longer than incandescent ones and produce far more light output per watt of power consumed. Unlike incandescent lamps, they have no filament. Instead, current flows in an arc through mercury vapor between contacts called *cathodes* at each end of the tubular lamp. The inside of the tube is coated with a powder called *phosphor* that glows when excited by ultraviolet radiation, thus producing visible light.

Fluorescent lamps require an auxiliary component called a *ballast* to operate. The ballast performs two functions:

1. It produces a jolt of high voltage to vaporize the mercury inside the lamp and start the arc from one end to the other.

2. Once the lamp is started, it limits current to the lower value needed for proper operation.

There are many different types of fluorescent lamps and ballasts. Figure 5.3 shows a single-lamp fixture with a preheat lamp and reactance ballast that requires an additional component called a *starter*. These days, luminaires using rapid start or instant start lamps or luminaires that don't require starters are much more widely used. Older types of fluorescent ballasts are known as *core-and-coil*, but newer electronic ballasts are becoming common.

Fluorescent lamps are available in a number of different "light colors," including white, cool white, warm white, deluxe cool white, and deluxe warm white. Because

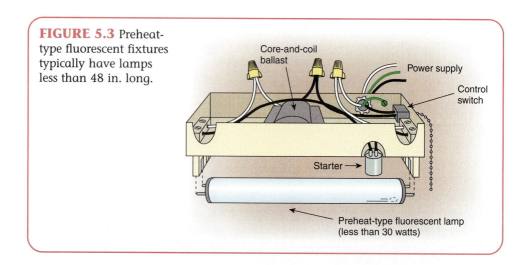

FIGURE 5.3 Preheat-type fluorescent fixtures typically have lamps less than 48 in. long.

Core-and-coil ballast

Power supply

Control switch

Starter →

Preheat-type fluorescent lamp (less than 30 watts)

consumers have traditionally preferred the warmer colors of incandescent lighting, and because long, tubular lamps haven't been compatible with many residential luminaire designs, fluorescent lighting has generally been used only in certain areas within dwellings, such as kitchens, basements and utility rooms, garages, and recreation/family rooms with lay-in suspended ceilings. However, a newer type of compact fluorescent lamp is starting to be more widely used in residences.

Compact Fluorescent Lamps. Compact fluorescent lamps come in a variety of shapes and styles, as shown in Figure 5.2. Some include integral ballasts and screw shells and are marketed as energy-saving replacements for incandescent lamps (a 7-watt compact fluorescent lamp produces nearly as much light as a 40-watt incandescent lamp). Other types of compact fluorescent lamps are intended to be used in luminaires specially designed for them.

Some types of compact fluorescent lamps, although intended to replace incandescent models, are actually too large to fit in the same luminaires. For new home construction, it generally makes more sense to select luminaires designed for use only with compact fluorescent lamps.

CAUTION: Remember that no fluorescent lamps of any kind can be used with conventional dimmers listed to control incandescent lamps.

Tungsten-Halogen Lamps

Tungsten-halogen lamps are a type of filament lamp (also called *quartz-halogen* and *quartz-iodide*) that uses a lamp-within-a-lamp design. They last up to 3500 hours and are more energy-efficient than conventional incandescent lamps, which reduces long-term energy costs. Tungsten-halogen lamps aren't physically interchangeable with other types of incandescent lamps and require special luminaires.

CAUTION: Because their outer glass bulbs develop very high temperatures in the range of 1000°F, tungsten-halogen lamps should not be used in luminaires designed or installed so that people or flammable materials such as curtains and furnishings might come into contact with them. Tungsten-halogen lamps used in torchère-style floor lamps that can easily be knocked over have been implicated in many burn injuries and fires, according to the U.S. Consumer Product Safety Commission (CPSC).

LIGHTING OUTLETS AND SWITCHES

Required Lighting Outlets

As illustrated in Figure 5.4, Section 210.70(A) requires the following locations in dwellings to have a lighting outlet:

- Every habitable room (living, dining, kitchen, bedroom, den/library, family/recreation, home office)
- Bathrooms
- Kitchens
- Hallways and stairways
- Attached garages, and detached garages with electric power

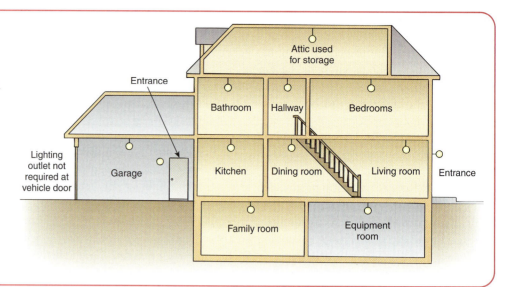

FIGURE 5.4 Lighting outlets are required in dwellings for habitable rooms, stairways, storage or equipment rooms, attached garages, and outdoor entrances.

- Attics, basements, utility rooms, and underfloor spaces used for storage or that contain equipment that requires servicing (such as equipment for heating, air-conditioning, and laundry; or water heaters)
- Entrances to a dwelling or attached garage with grade-level access (an outdoor light controlled by a switch inside the house is required)

NOTES:
1. *This requirement also applies to entrances that are reached by stairs from grade level.*
2. *A vehicle door in a garage is not considered an entrance [210.70(A)(2)(b)].*

Switch Control of Lighting Outlets

NEC requirements for switch control of residential lighting can be summarized as follows:

- *Each habitable room* is required to have at least one wall switch-controlled lighting outlet [210.70(A)(1)].
- *Interior stairways* with six or more risers must have a wall switch at every floor and at each landing that includes an entryway [210.70(A)(2)(c)].

CAUTION: Some local building codes require a lighting outlet for any stairway.

- *Outdoor entrances* to dwelling units and garages with electric power must have wall switch-controlled lighting outlets [210.70(A)(2)(b)].
- *Storage or equipment areas* are permitted to have a lighting outlet "containing a switch or controlled by a wall switch," but one point of control must be at the usual point of entry to these spaces [210.70(A)(3)]. Installing a wall switch is usually the most practical solution to this *Code* requirement. However, a lighting outlet located directly above an entry hatch from the floor below could be an incandescent lampholder or fluorescent fixture with an integral pull-chain switch.

- *Exterior stairways* aren't required to be lighted. However, an outdoor entrance at the top of an exterior stairway must have a wall switch-controlled lighting outlet, as explained previously.

Switch Locations. Because it's a safety standard, the *Code* doesn't always specify required locations for lighting switches. Although habitable rooms and outdoor entrances are required to have wall switch-controlled lighting outlets, the *NEC* doesn't specify where these switches must be located. Exceeding the minimum *Code* requirements often results in greater convenience and satisfaction for home-owners [90.1(B)]. The guidelines to use when locating light switches in dwelling units follow.

- *Point of Entry.* For each habitable room or other space with lighting, a wall switch inside that space is best located at the point of entry. Most often, this location is just inside a doorway on the side near the doorknob.

- *Multiway Switching.* Three- and four-way switches should be installed in all rooms and hallways with more than one entrance. Residents should be able to move throughout a dwelling without ever being in darkness or leaving lights turned on behind them. Not only do multiway switches improve safety and convenience, but providing additional switches may also help conserve energy by making it easier to turn off unneeded lighting.

- *Interior Stairways.* Interior stairways with six or more risers are required by the *Code* to have a wall switch at every floor and at each landing that includes an entryway [210.70(A)(2)(c)].

- *Outside Entrances.* A wall switch should be installed inside each exterior door to control the required outside lighting outlet. (Note: For improved convenience, some builders install three-way switches that allow outdoor lighting to also be controlled from the master bedroom or other central location.)

- *Attached Garages.* If a garage is attached to the house, it should be treated as another room. Installing three- or four-way switches to control the lighting inside the garage is a convenient—and safe—solution. There should be a wall switch inside the garage vehicle door, another inside the exterior personnel door (if any), and one at the door leading into the house.

Switch-Controlled Receptacles

NEC rules for switch-controlled receptacles can be summarized as follows:

- In habitable rooms other than kitchens and bathrooms, the lighting outlet required in each habitable room is permitted to be a receptacle outlet controlled by a wall switch [210.70(A)(1), Exception No. 1].

- If a switch-controlled receptacle is used to provide lighting in a dining room, breakfast room, or similar area, as shown in Figure 5.5, it must be in addition to the required receptacle outlets on 20-ampere small-appliance branch circuits [210.52(B)(1), Exception No. 1]. The small-appliance receptacle outlets aren't permitted to be switched.

- Switch-controlled receptacle outlets cannot be used in hallways, stairways, garages, or at entrances. These locations all require switch-controlled lighting outlets [210.52(A)(2)].

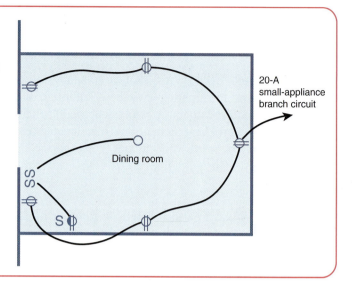

FIGURE 5.5 In rooms with receptacle outlets on 20-ampere small-appliance branch circuits, a switch-controlled receptacle must be supplied from a separate general-purpose branch circuit.

Typically, a duplex receptacle is wired so that the top half is controlled by a switch and the bottom half is live at all times (these are sometimes called "split-wired" receptacles in the field). A table lamp, floor lamp, or hanging lamp can then be plugged into a receptacle that is turned ON and OFF, while the other "always-on" receptacle is available for use by other appliances (Figure 5.6). The steps for wiring a duplex receptacle so that the top half is controlled by a switch and the bottom is always live are as follows:

Step 1. Break off the tab joining the gold-colored screws on the receptacle. Do not break the tab joining the white-silver-colored screws.

Step 2. Connect the receptacle branch-circuit conductors as shown in Figure 5.7.

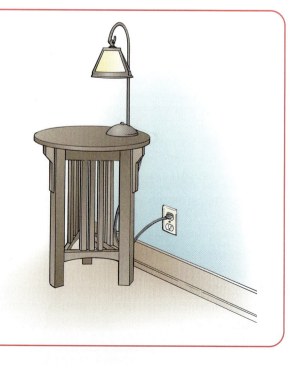

FIGURE 5.6 A wall switch-controlled receptacle is permitted to be the required lighting outlet in rooms other than kitchens or bathrooms.

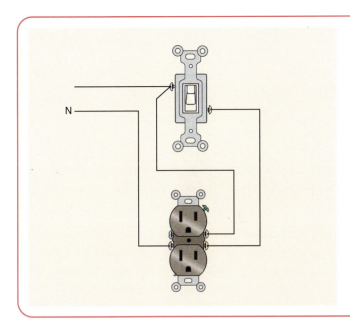

FIGURE 5.7 This receptacle has been split-wired so that the top half is controlled by a switch and the bottom half is always on.

Switch and Receptacle Heights

The *Code* doesn't have requirements for the height of wall switches and receptacles above the finished floor (AFF). The typical range is 48–54 in. for switches and 12–18 in. for receptacle outlets. Receptacles mounted above counters are typically mounted 6 in. above the counter surface or backsplash. All dimensions are measured to the centerline of the device box. The most important detail in new construction is to make the heights of each type of device uniform throughout the dwelling.

The Americans with Disabilities Act Guidelines (ADAG) require that wall switches in commercial construction be located not more than 48 in. AFF and receptacle outlets no lower than 18 in. AFF. Although the ADAG don't apply to residential construction, they should be followed when the electrical installer knows that a house or apartment is intended for occupancy by people with disabilities.

Mounting of Snap Switches

Section 314.20 requires that electrical boxes in noncombustible walls or ceilings such as plaster or gypsum wallboard be installed so that the front edge of the box is not more than ¼ in. back of the finished surface. Flush-type snap switches must be installed in these recessed boxes must so that the extension plaster ears are seated against the finished wall surface [404.10(B)]. The purpose of this *Code* rule is to ensure that switch handles protrude the proper distance through the installed switch plate.

THERMAL PROTECTION OF LUMINAIRES

The *NEC* has special safety rules to prevent overheating of luminaires. The hot metal shells of incandescent luminaires, as well as overheated ballasts of fluorescent and high-intensity discharge (HID) luminaires, pose a potential fire threat if they contact flammable materials. These flammable materials can include wood framing members and thermal insulation installed to conserve energy.

Although fiberglass insulation doesn't pose a fire hazard, cellulose insulation made from shredded paper is flammable. For this reason, 410.68 requires that luminaires installed in void spaces of walls and ceilings be constructed so that adjacent combustible materials aren't subjected to temperatures that exceed 90°C (194°F).

Incandescent and HID Recessed Luminaires

Incandescent and HID luminaires designed for void spaces that may be filled with insulation (either before or after installation of the luminaires) are required to have special construction features or thermal protection ratings. These features don't apply to incandescent and HID luminaires installed in poured concrete. They also don't apply to fluorescent luminaires, whose lamps don't reach dangerously high temperatures.

Thermal Protectors. Recessed incandescent and HID luminaires are required to have an integral thermal protector [410.65(C)] that opens automatically in case of overheating to cut off electricity to the lamp. Most thermal protectors (also called *thermal cutouts*) are bimetallic strips, similar to thermostat elements, that open when they heat up and close when they cool down. Thermal protectors do not disconnect the source of supply as a fuse, circuit breaker, or motor overload protector does. Instead, a thermal protector may cycle an overheating luminaire ON and OFF repeatedly, calling attention to the problem until it is repaired. Some luminaires, which are identified as "inherently protected" because of their design, construction, and thermal characteristics, aren't required to include a thermal protector [410.65(C), Exception No. 2].

Recessed Luminaires (Type IC)

Many recessed luminaires are identified as Type IC and are permitted by 410.66(A)(2) to be installed in contact with thermal insulation (Figure 5.8). If a recessed luminaire isn't rated Type IC, 410.66(B) prohibits insulation from being installed within 3 in. of the fixture, wiring compartment, or ballast (Figure 5.9).

Fluorescent Luminaires

Most fluorescent luminaires installed indoors are required to have integral thermal protection [410.73(E)(1)]. Listed Class P ballasts automatically disconnect the power

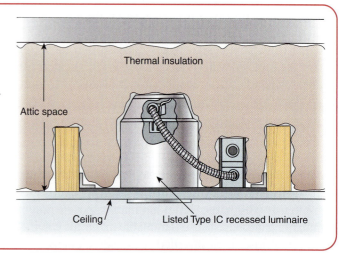

FIGURE 5.8 Listed Type IC recessed luminaires are suitable for use in insulated ceilings installed in direct contact with thermal insulation. (Redrawn from Thomas Lighting, a Genlyte Company)

Thermal insulation

Attic space

Ceiling

Listed Type IC recessed luminaire

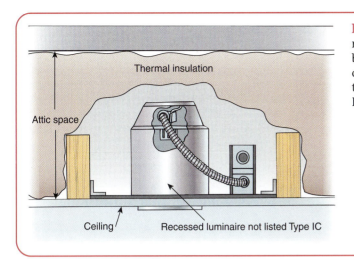

FIGURE 5.9 Thermal insulation cannot be installed within 3 in. of a recessed luminaire that is not listed Type IC.

supply when their case temperature exceeds 90°C (194°F) [410.68]. All new rapid-start and instant-start fluorescent luminaires that use 48-in. lamps come supplied with Class P ballasts.

Fluorescent luminaires with simple reactance ballasts and tubular lamps (the type shown in Figure 5.3) aren't required to have thermal protection. These luminaires are preheat-type units with starters and lamps not more than 36 in. long. Such luminaires are often used in residential applications such as under-cabinet lighting, lighting inside closets, or vanity lighting on the sides of medicine cabinets.

Branch-Circuit Conductors for Luminaires

Luminaires must be constructed or installed so that conductors in outlet boxes aren't subjected to temperatures higher than their temperature rating [410.11]. Many luminaires designed for flush or recessed mounting have a junction box remotely located on a mounting bracket that is connected to the fixture shell itself by factory-installed wiring, as shown in Figure 5.10. The remote junction box construction complies with the *Code* requirement and prevents heat damage to the insulation of branch-circuit conductors.

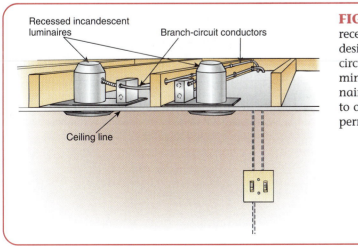

FIGURE 5.10 When recessed luminaires are designed for branch-circuit conductors terminating at each luminaire, no feed through to other luminaires is permitted.

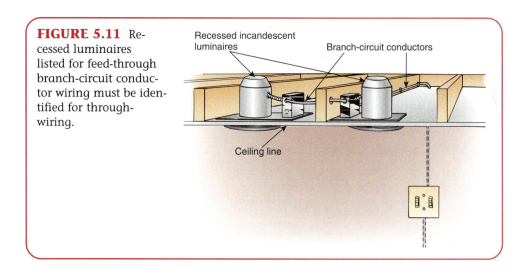

FIGURE 5.11 Recessed luminaires listed for feed-through branch-circuit conductor wiring must be identified for through-wiring.

Recessed incandescent luminaires

Branch-circuit conductors

Ceiling line

Most listed luminaires are designed to limit conductor temperature to 75°C (167°F). Nonmetallic-sheathed cable and armored cable, both widely used for residential wiring, have conductors rated 90°C (194°F), which provides a comfortable margin of safety.

Through-Wiring. Branch-circuit conductors themselves also generate heat, and outlet boxes that form an integral part of luminaires are generally sized only for the supply conductors to that single luminaire [410.11]. Figure 5.10 illustrates recessed incandescent luminaires listed for one set of supply conductors terminating in each outlet box. Figure 5.11 illustrates luminaires listed for feed-through branch-circuit wiring. The luminaires must be marked ''Identified for Through-Wiring.''

When fluorescent luminaires are mounted end-to-end, the conductors of a 2-wire or multiwire branch circuit are permitted to pass through the luminaires without any special listing [410.32]. Conductors run within 3 in. of a ballast must have insulation rated 90°C (194°F) [410.33].

Extra conductors cannot be run through the bodies of single fluorescent luminaires (that is, those not connected together) unless they are listed for the purpose and marked ''Suitable for Use as a Raceway.''

BATHROOM LIGHTING

To reduce the risk of electric shock in a normally wet location, 410.4(D) defines a ''bathtub and shower zone'' that extends 3 ft horizontally and 8 ft vertically from the top of the bathtub rim or shower stall threshold. As illustrated in Figure 5.12, the following are prohibited within this zone:

- Hanging or pendant luminaires
- Lighting track
- Paddle (ceiling) fans
- Cord-connected luminaires (not shown)

Surface-mounted or recessed luminaires and recessed exhaust fans are permitted to be located within the so-called ''bathtub zone.''

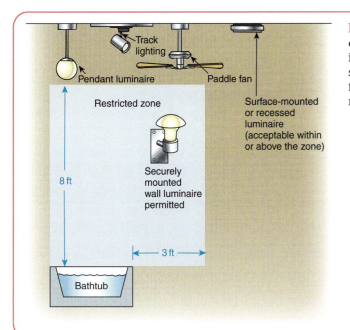

Track lighting

Pendant luminaire

Paddle fan

Restricted zone

Surface-mounted or recessed luminaire (acceptable within or above the zone)

8 ft

Securely mounted wall luminaire permitted

3 ft

Bathtub

FIGURE 5.12 Pendant luminaires, lighting track, and suspended (paddle) fans aren't permitted near a bathtub.

Damp and Wet Locations

Luminaires or other equipment located directly above a bathtub or within an enclosed shower stall must be identified for use in damp locations in order to comply with 110.11. In doubtful situations, the authority having jurisdiction (AHJ) should be consulted before installing luminaires, fans, or other items in a bathtub or shower space.

Light Switches

Switches aren't permitted to be installed within wet locations of bathtub or shower spaces unless they are part of a listed tub or shower assembly [404.4]. Such listed assemblies sometimes include light switches similar to those supplied with spa and hot tub units. These pneumatically operated switches prevent contact between users and potentially energized metal parts.

The *NEC* doesn't have a rule stating how far away from tubs and showers the switches must be installed. When possible, it's good practice to follow the "bathtub zone" rule by locating wall switches at least 3 ft horizontally from the rim of the bathtub or threshold of the shower stall. However, some bathrooms are too small to allow this safe installation practice, which is why the *Code* doesn't have an absolute rule on the subject (Figure 5.13).

Note that guidelines for installing receptacle outlets in bathrooms are covered in Chapter 8, The Bathroom.

CLOSET LIGHTING

The *Code* doesn't require lights to be installed inside closets, but has strict safety requirements when they are installed [410.8]. To minimize the danger of lamp

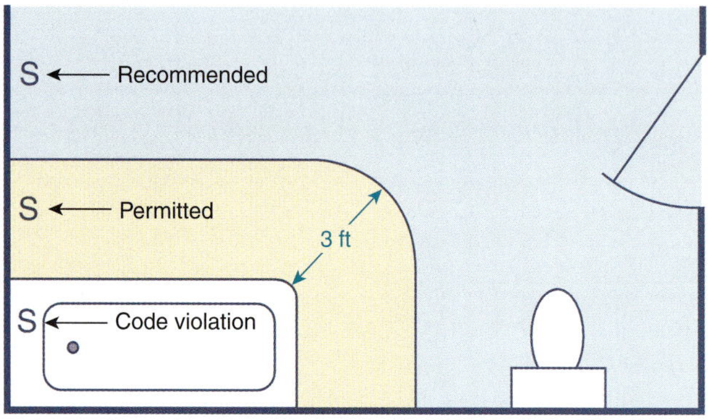

FIGURE 5.13 Wall switch locations in bathrooms are less rigidly regulated than the locations of luminaires.

S ← Recommended

S ← Permitted

3 ft

S ← Code violation

breakage and fire, only fluorescent luminaires and completely enclosed incandescent luminaires are permitted. The following are never permitted inside closets:

- Incandescent luminaires with open or partially enclosed lamps
- Pendant luminaires and lampholders of any kind

Required Clearances Inside Closets

Surface-mounted incandescent luminaires must have a minimum 12-in. separation from the "nearest point of a storage space." *NEC* Figure 410.8, shown in Exhibit 5.1, illustrates what is meant by "storage space" inside a typical closet.

Assuming a shelf depth of 12 in. and several inches for the depth of a surface-mounted incandescent luminaire (with a fully enclosed lamp) means, in practical terms, that no closet less than about 27 in. deep can have a surface-mounted incandescent luminaire installed inside it. However, fluorescent and recessed incandescent

EXHIBIT 5.1

NEC FIGURE 410.8 Closet Storage Space

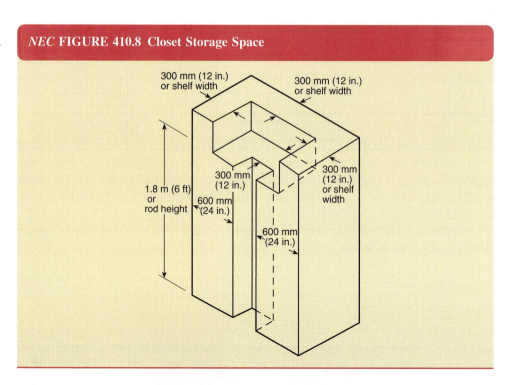

300 mm (12 in.) or shelf width

300 mm (12 in.) or shelf width

300 mm (12 in.) or shelf width

300 mm (12 in.)

1.8 m (6 ft) or rod height

600 mm (24 in.)

600 mm (24 in.)

luminaires require only a 6-in. separation from storage space. Another practical option for providing illumination in shallow closets is to install a ceiling-mounted luminaire of any type just outside the closet.

Walk-In Closets

Walk-in closets are more like small rooms. Typically, there is sufficient space for a surface-mounted ceiling luminaire with adequate clearances from storage space. A wall switch can be provided either inside or outside the door of the walk-in closet. A door-jamb switch can also be used to turn the light on automatically whenever the door is opened.

LIGHTING TRACK

Article 410, Part XV, defines *lighting track* as follows.

> **Lighting Track:** A manufactured assembly designed to support and energize luminaires (lighting fixtures) that are capable of being readily repositioned on the track. Its length can be altered by the addition or subtraction of sections of track.

Lighting track (more often called *track lighting* in the field) is used for decorative and accent lighting, to illuminate artwork and collectibles, and for similar applications. Lampholders (more often called *heads*) can easily be added to, removed from, and repositioned along the track.

Lighting track is manufactured with conductors that are, in effect, miniature copper busbars running its length. When a lighting head is positioned on the track, it is typically twisted into place (or a switch or lever is flipped) to bring the terminals on the base of the lighting head into electrical contact with the busbars.

Locations for Lighting Track

Lighting track is normally installed indoors, exposed, in dry locations. The *Code* prohibits lighting track in the following locations [410.101(C)]:

- Where subject to physical damage
- In damp or wet locations
- Where concealed, or extended through walls or partitions
- Less than 5 ft above the finished floor, unless it operates at less than 30 volts
- In a zone extending 3 ft horizontally and 8 ft vertically from the top of a bathtub rim or shower threshold

Installing Lighting Track

A summary of the *NEC* rules for installing lighting track follows.

- *Mounting and Supporting.* Lighting track must be securely mounted, with a single section having two supports and additional sections installed in a continuous row having one additional support each [410.104]. Lighting track is available for both surface and pendant mounting.

FIGURE 5.14 This lighting track has been installed with a center-feed power supply.

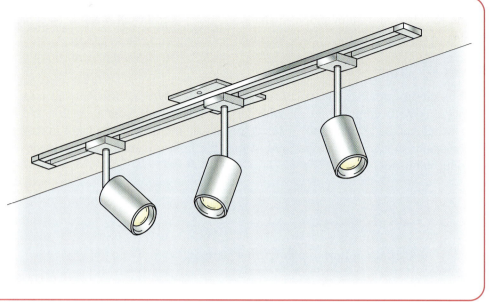

• *Electrical Connections.* Lighting track can be either end-feed or center-feed. Typically, it's supplied with components that allow either type of connection to be made (Figure 5.14). Lighting track must be grounded at the outlet box, and the track sections are required to be coupled securely enough to maintain grounding continuity [410.105(B)]. Ends of track sections are required to be protected by insulating caps, which are supplied by the lighting track manufacturer [410.105(A)].

• *No Field Modifications.* The definition of lighting track states that "its length can be altered by the addition or subtraction of sections of track" [410.100]. This definition means that lighting track can't be cut to length in the field. Doing so also violates the product's listing. The reason for this prohibition is that track sections, safety end caps, and power-feed modules are designed to mate together securely.

CAUTION: Field modifications could damage track sections and their continuous conductors, creating potential safety problems such as poor fit, poor grounding, insecure power connections, and the possibility of accidentally exposing live conductors or energizing non–current-carrying parts of metal track.

Load Calculations for Lighting Track

When lighting track is installed in occupancies other than dwellings, a load of 150 volt-amperes must be included for every 2 ft of length in branch-circuit, feeder, and service calculations. However, lighting track installed in dwelling units is considered part of the general lighting load of 3 volt-amperes per square foot. No additional load needs to be included in residential calculations [220.12].

BOXES FOR LUMINAIRES AND PADDLE FANS

Boxes at Luminaire (Lighting Fixture) Outlets

NEC 314.27 states that boxes used at luminaire outlets "shall be designed . . . so that a luminaire (lighting fixture) may be attached." Many outlet boxes intended

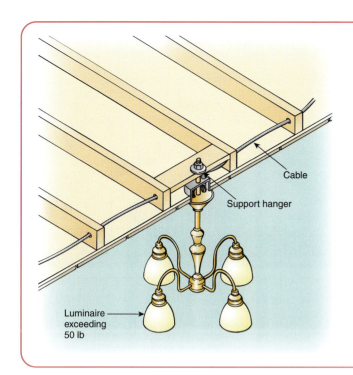

FIGURE 5.15 A wooden brace can be used to independently support a heavy luminaire as required by *NEC* 314.27(B). It must meet the size specifications of *NEC* 314.23(B)(2).

Cable

Support hanger

Luminaire exceeding 50 lb

for use with lighting have accessories known as *hickeys* or *luminaire (fixture) studs* that allow convenient attachment of a luminaire and help support its weight. Sometimes these hickeys or fixture studs come supplied as part of the box, or sometimes they're furnished with the luminaire itself, and they can also be purchased as accessories for installation in the field [410.16(D)].

Lightweight Wall-Mounted Luminaires. Wall-mounted luminaires weighing up to 6 lb are permitted to be supported directly to outlet boxes or plaster rings with at least two No. 6 (or larger) screws [314.27(A), Exception]. The use of standard device boxes is an easy and inexpensive way to mount lightweight wall-mounted luminaires such as sconces.

However, lightweight ceiling-mounted luminaires *are not* permitted to be supported by device boxes in this way. Boxes for ceiling-mounted lighting outlets must comply with 314.27, which requires that they be "designed or installed so that a luminaire (lighting fixture) may be attached." This generally means that a hickey, stud, or similar fitting has been provided.

Ceiling-Mounted Luminaires. Ceiling outlet boxes are permitted to support luminaires weighing up to 50 lb. Heavier luminaires must be supported by outlet boxes listed for the weight to be supported or must be supported independently of the box (Figure 5.15) [314.27(B)]. Installation and support of boxes are covered in more detail in Chapter 4, Wiring Methods.

Boxes for Ceiling-Suspended (Paddle) Fans

Listed outlet boxes used to support paddle fans must be identified for the use [314.27(D), 422.18]. Standard luminaire-type outlet boxes can't be used to support paddle fans, which have several characteristics that make them incompatible with ordinary outlet boxes:

FIGURE 5.16 Ceiling-suspended (paddle) fans must be supported with a box identified for such use. (Courtesy of RACO Interior Products, Inc.)

- *Weight.* A typical paddle fan is heavier than most lighting fixtures.
- *Torque and Vibration.* The twisting motion when the motorized fan starts and vibration while it's running tend to loosen conventional boxes.
- *Larger Screws.* Standard outlet boxes for luminaires typically have holes tapped to accept No. 8-32 screws, while larger ceiling fans are installed using No. 10-32 screws.

Several acceptable methods of supporting paddle fans meet applicable *Code* requirements in Articles 314 and 422:

1. Lighter-weight paddle fans—up to 35 lb—must be supported by outlet boxes identified for the purpose (Figure 5.16).
2. Heavier paddle fans—up to 70 lb—must be supported by listed outlet boxes or outlet box systems identified for the purpose and marked with the maximum allowable weight.
3. Ceiling fans of any weight can be installed using standard outlet boxes if they are supported independently of the box [422.18]. Figure 5.17 shows one way to provide this support.
4. A different type of ceiling fan box is shown in Figure 9.5 of Chapter 9, The Bedroom.

Recommended Installation Practice for Ceiling-Mounted Boxes

Where ceiling-suspended (paddle) fans are installed at the time of original house construction, an installer must use the correct type of box/support system for the weight of the fan to be supported. However, long after the original construction has been completed, homeowners themselves frequently replace existing luminaires with paddle fans in bedrooms, dens, family rooms, and other areas.

For this reason, it is good practice to install boxes of the type identified for supporting paddle fans up to 35 lb at ceiling-mounted lighting outlets in habitable

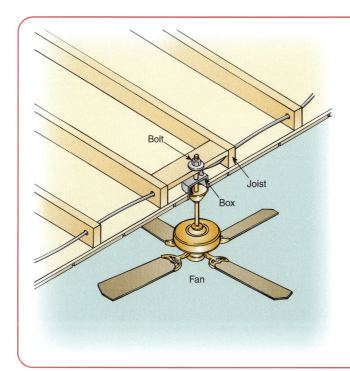

FIGURE 5.17 This ceiling fan is mounted on a standard outlet box with an independent means of support.

Bolt

Joist

Box

Fan

rooms, as shown in Figure 5.16. The extra cost to provide these special outlet boxes during original construction is very low compared to the expense and trouble of retrofitting adequate support for paddle fans after construction is completed.

In addition, most homeowners aren't aware of the *NEC* requirements for safe installation of paddle fans and are unlikely to have their installations inspected. Thus, providing ceiling fan boxes during the original construction process can help avoid possible safety problems later on.

TYPES OF SWITCHES AND CONTROLS USED IN HOMES

Section 404.14 lists detailed performance requirements for ac-only and ac/dc general-use snap switches. These devices are also called *wall switches* in 210.70 (and *toggle switches* in the field). Important points for each type are summarized in the sections that follow.

Alternating-Current General-Use Snap Switches

Most switches used in residential wiring are alternating-current general-use snap switches (Figure 5.18). In addition to voltage and current ratings, they are marked "ac only." Most switches are dual rated for 120 and 277 volts, which allows them to be used on either 120-volt lighting circuits in homes or 277-volt lighting circuits in commercial construction. All snap switches that offer "soft" or "quiet" operation are ac general-use snap switches [404.14(A)].

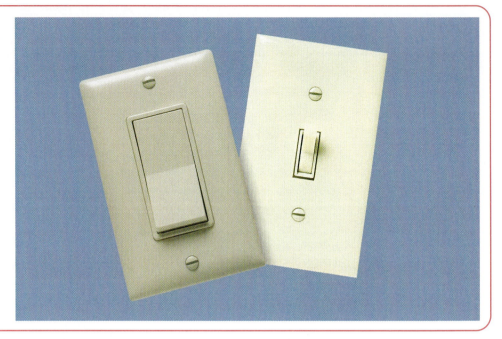

FIGURE 5.18 Most lighting outlets in dwellings are required to be wall-switch controlled. (Courtesy of Pass & Seymour/Legrand®)

Alternating-Current or Direct-Current General-Use Snap Switches

These switches are typically of heavier construction than ac-only snap switches. They are available rated in horsepower for controlling motor loads. The ac/dc switches use springs to break their contacts quickly, which minimizes arcing when they are used to interrupt dc circuits. This feature makes them harder and noisier to operate than ac-only switches. For this reason, ac/dc switches aren't often used in residential wiring [404.14(B)].

CO/ALR Snap Switches

Alternating-current snap switches rated 15 and 20 amperes, and designed for connection to either copper or aluminum conductors, are listed and marked "CO/ALR" [404.14(C)]. These switches are typically used for replacement purposes in older houses with aluminum branch-circuit wiring. Aluminum wiring in sizes 14 AWG and 12 AWG is not installed in newly constructed homes. Larger sizes are sometimes used to supply appliances such as ranges, water heaters, and electric clothes dryers.

Lighted Switches

Snap switches with indicator lights are used in residential wiring for two purposes:

1. *Pilot light switches* indicate the state of a load or outlet that can't be seen from the switch location (such as for attic lighting). The pilot light is ON when the lighting or other controlled equipment is ON. Pilot light switches have three terminals, and a grounded (neutral) conductor is required in the device box to power the light.

2. *Glow switches* serve as a night light that helps users to locate the switch in the dark. This type of switch normally has a glowing handle that is off when the

lighting is on (and vice-versa). The small neon lamp is in series with the phase conductor and doesn't require a grounded (neutral) conductor to operate.

Timer Switches, Sensors, and Detectors

Controls other than wall switches are sometimes used to control lighting outlets at dwelling units. All of the following are considered *wiring devices,* the same as snap switches, dimmers, and receptacles.

Timer Switches. Timer switches are occasionally used in residential wiring. One fairly common application is the use of a one- or two-minute timer to control an infrared heat lamp in a bathroom. Timer switches must be installed according to all the same *Code* rules required for snap switches.

Occupancy Sensors. Section 210.70(A)(1), Exception No. 2 permits lighting outlets in habitable rooms of dwellings to be controlled by occupancy sensors. However, they must either be used in addition to a wall switch or be installed at a customary wall switch location and equipped with a manual override. Occupancy sensors aren't often used to control residential lighting.

Detectors. Photoelectric detectors and motion detectors are often used to automatically control outdoor lighting at dwellings.

Combination Switch/Receptacles

Wiring devices that combine a single-pole switch on the same yoke (strap) with a single receptacle (Figure 5.19) are sometimes installed in areas such as storage spaces or mechanical equipment rooms. A grounded (neutral) conductor is required in the device box to power the receptacle. These combination devices can be installed so that the switch controls the receptacle, or by breaking off the tab joining the gold-colored screws, the device can be wired so that the receptacle is independent of the switch and is always energized.

CAUTION: Combination switch/receptacles cannot be installed at locations in dwellings that are required to have ground-fault circuit-interrupter (GFCI) protection for receptacle outlets. However, combination switch/receptacles are permitted in bedrooms, because NEC 210.12(B) requires that entire branch circuits supplying 15- and 20-ampere outlets in bedrooms be protected by an arc-fault circuit-interrupter (AFCI)—as opposed to each individual outlet having AFCI protection.

INSTALLATION OF SWITCHES

Residential switch wiring can be divided into three parts according to the type of connection used for the switch:

1. *Single-pole switches* control a luminaire or other equipment from one location.
2. *Three-way switches* control a luminaire or other equipment from two locations.
3. *Four-way switches* control a luminaire or other equipment from three or more locations.

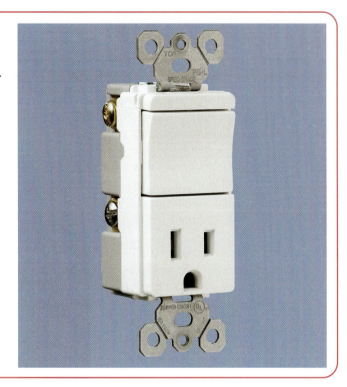

FIGURE 5.19 This wiring device combines a single-pole switch with a single receptacle. (Courtesy of Pass & Seymour/Legrand®)

Single-Pole Switches

A single-pole switch has two terminals and is marked with ON and OFF positions. It is used to control a load from one location. Single-pole switches are available in the following types:

- ac only
- ac/dc
- CO/ALR-rated

Three-Way Switches

Three-way switches are used to control a load from two locations. The term *three-way switch* is somewhat misleading, because three-way switches actually have three wiring terminals that can be connected in two different positions (Figure 5.20). One terminal is called the *common* and is attached to the hot (ungrounded) conductor. The other two terminals, which are lighter in color, are called *traveler terminals*. Two conductors, called *travelers,* run from these terminals to the traveler terminals of a second three-way switch. The switched conductor runs from the darker common terminal of the second three-way switch to the outlet or other controlled load.

Three-way switches, which don't have marked ON and OFF positions, are available in the following types:

- ac only
- ac/dc
- CO/ALR

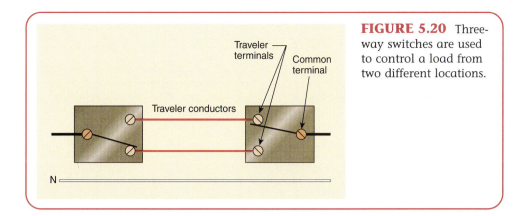

Traveler terminals

Common terminal

Traveler conductors

N

FIGURE 5.20 Three-way switches are used to control a load from two different locations.

Number of Three-Way Switches. Two three-way switches are always used together to control a luminaire, other load, or switched receptacle from two different locations (see Table 5.1). For example, three-way switches are often used at the top and bottom of a stairway, to provide convenient lighting control.

Four-Way Switches

One or more four-way switches can be used with a pair of three-way switches to control an outlet or load from three or more locations. The switches have four terminals that can be connected in two different positions (Figure 5.21). Two sets of traveler conductors run between three- and four-way switches.

Four-way switches, which don't have marked ON and OFF positions, are available in the following types:

- ac only
- ac/dc
- CO/ALR

Required Number of Three-Way Switches Needed with Four-Way Switches. As Table 5.1 shows, a pair of three-way switches is always used with one or more four-way switches to control a luminaire, other load, or switched receptacle from three or more different locations. In other words, three-way switches are installed at the first two control locations, and then a four-way switch is installed at each additional control location.

TABLE 5.1 Number of Two-, Three-, and Four-Way Switches per Number of Control Locations

Control Locations	Three-Way Switches Needed	Four-Way Switches Needed
2	2	0
3	2	1
4	2	2
5	2	3

FIGURE 5.21 Four-way switches are used to control a load from three or more locations.

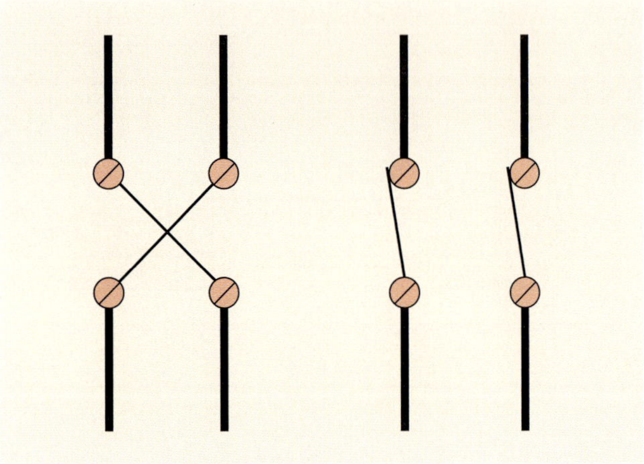

Switch Wiring

A couple of general considerations apply to *all* switch wiring. First, switching always takes place in the ungrounded (hot) conductor. Grounded conductors (neutrals) are never permitted to be switched [404.2(A), 404.2(B)]. However, when using 2-wire Type NM or Type AC cables, conductors with white insulation are used as part of switch loops or as travelers. This color coding is explained in the sections that follow.

Single-Pole Switches. There is no *Code* rule specifying whether the source of supply for single-pole switches should be at the switch box or at the controlled equipment, such as a luminaire. Either of the wiring methods that follows is acceptable.

 • *Power Feed at Load.* The branch circuit runs to the load, such as a lighting outlet. A switch loop drops down to the single-pole switch. When 2-wire Type NM cable is used for the switch loop, the white conductor must be permanently re-identified as described in 200.7(C)(2). Black electrician's tape is often used for this purpose (Figure 5.22).

CAUTION: When two-wire cable is used for a switch loop, the white wire is permitted to be used as the supply to the switch but NOT as the return wire from the switch to the outlet or load. The switched ungrounded (hot) conductor at the load must always have insulation that is a color other than white or gray, to identify it clearly for reasons of safety [200.7(C)(2)].

 • *Power Feed at Switch.* The branch circuit runs first to the switch box and then up to the lighting outlet. Both black conductors are connected to switch terminals, and the white conductors are spliced together. No reidentification of conductors is required with this wiring scheme (Figure 5.23).

Section 404.15(B) requires that switches with a marked OFF position must disconnect all ungrounded conductors. In other words, single-pole switches aren't permitted to be installed "backward," so that the controlled outlet or load is ON when the switch handle indicates OFF.

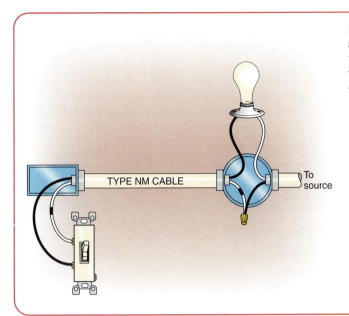

FIGURE 5.22 A single-pole switch controls a lighting outlet, with power feed at the outlet.

TYPE NM CABLE

To source

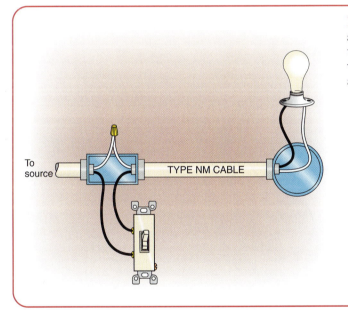

FIGURE 5.23 A single-pole switch controls a lighting outlet, with power feed at the switch.

To source

TYPE NM CABLE

Three-Way Switches. Three-way switches, which are used to control an outlet or load from two locations, have three terminals. Two are called *traveler terminals*, and the third, which is distinguished by a darker color, is called the *common*. Two traveler conductors run between three-way switches, along with the ungrounded (neutral) conductor. So in practical terms, there are always three conductors between a pair of three-way switches.

There are a number of different ways to wire three-way switches to a lighting outlet or other controlled load. Figures 5.24, 5.25, and 5.26 illustrate the following typical wiring arrangements for three-way switching.

• ***Three-Way Switching, Power Supply at Switch Box.*** In Figure 5.24, a 2-wire cable runs to the first switch box. The black wire is connected to the common

FIGURE 5.24 Three-way switching can be installed with the power supply at the first switch.

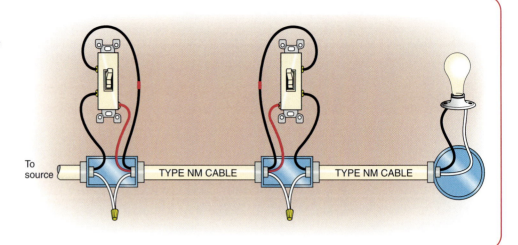

FIGURE 5.25 Three-way switching can also operate with the power supply at the load.

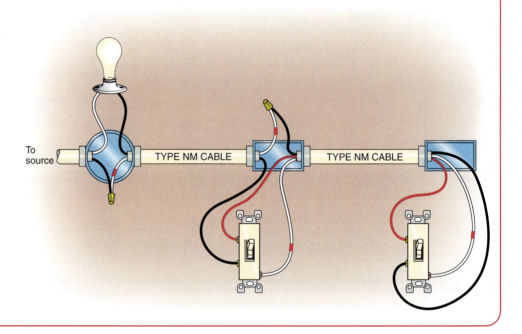

terminal, and the white wire is spliced to the white wire of the next cable, a three-wire cable that runs between the two switches. The black and red wires serve as travelers, and the white wire is spliced to the white wire of the next cable, a 2-wire cable that runs from the common terminal of the second switch to the outlet. The black and white wires are called the "switch loop."

OPTIONAL COLOR CODING: Some installers would mark the black wire used as a traveler between the two three-way switches with red tape or paint, so that both travelers are red conductors. This marking makes the installation easier to troubleshoot and repair in the future. However, there is no NEC rule that requires such reidentification.

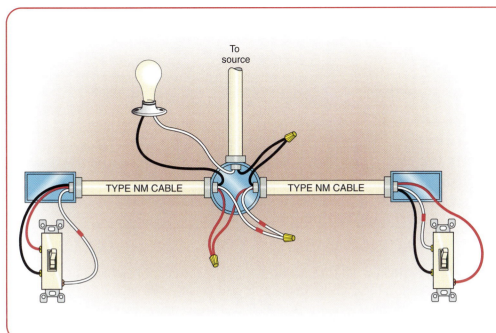

FIGURE 5.26 Three-way switching can be wired as shown with the power supply at the load between the two switches.

- ***Three-Way Switching, Power Supply at Load.*** In Figure 5.25, a 2-wire cable runs to the outlet. The white wire connects to the luminaire, and the black wire is spliced to the white wire of the next cable, a 2-wire cable that runs to the first three-way switch. At the first three-way switch, the black wire connects to the common terminal, and the white wire is spliced to the black wire of the next 3-wire cable. This 3-wire cable runs to the second three-way switch, where the wires are connected as shown in Figure 5.25.

REQUIRED COLOR CODING: White wires used as the supply to the switch in single-pole, three-way, or four-way loops are required to be permanently reidentified [200.7(C)(2)]. Electrician's tape or paint is commonly used for this purpose. Also, this installation has been wired so that the switched "hot" conductor at the load is a black wire. Reidentified white conductors are not permitted for this purpose [200.7(C)(2)].

- ***Three-Way Switching, Power Supply at Luminaire Between Two Switches.*** In Figure 5.26, a 2-wire cable runs to the outlet. Three-wire cables run from the outlet to each three-way switch. The white wire connects to the luminaire, and the black wire is spliced to the black wire of a 3-wire cable running down to one of the three-way switches. The other two conductors in that 3-wire cable serve as travelers between the switches. The remaining connections are as shown in Figure 5.26.

REQUIRED COLOR CODING: White wires used as the supply to the switch in single-pole, three-way, or four-way loops are required to be permanently reidentified [200.7(C)(2)]. Electrician's tape or paint is commonly used for this purpose. The installation shown in Figure 5.26 has been wired so that the switched "hot" conductor at the lighting outlet is a black wire, since reidentified white wires are not permitted for this purpose [200.7(C)(2)]. Red would also be an acceptable color (any insulation color except white, gray, or green is acceptable), but in Figures 5.24, 5.25, and 5.26 all travelers are red and all hot wires at outlets are black. Keeping house wiring as consistent as possible makes it easier to troubleshoot and repair in the future.

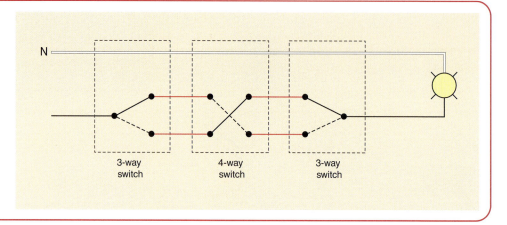

FIGURE 5.27 Four-way switching, with the power supply at the last switch, allows control from three locations.

Four-Way Switches. Four-way switches are used to control an outlet or load from three or more locations. One or more four-way switches are always used with a pair of three-way switches. Four-way switching is essentially the same as three-way, except that the two traveler wires between the three-way switches are also connected to the four-way switch(es).

In Figure 5.27, the power supply is at the last switch. This configuration avoids the need to reidentify the white wire with colored tape or paint. However, all the previously discussed cautions regarding reidentification of white wires, and the warnings against the use of white wires as the return conductor from a switch to the load, also apply to four-way switching.

GROUNDING SWITCHES AND FACEPLATES

Section 404.9(B) requires that "snap switches, including dimmer and similar control switches, shall be effectively grounded and shall provide a means to ground metal faceplates." Switches are manufactured with equipment grounding terminals, normally a green hexagonal screw, for connection to the equipment grounding conductor of the wiring method used (Figure 5.28). Typically, the grounding conductor is the bare copper equipment grounding conductor included in Type NM-B cable.

When a metal faceplate is installed on a switch box, it is grounded by connection to the metal yoke of the grounded snap switch by means of two No. 6-32 screws. Switches and other wiring devices are sold with small cardboard washers holding these screws in place. The washers should be removed before installing a metal faceplate, to ensure that it is grounded as required by the *Code*.

CAUTION: Snap switches without green grounding screws are widely available, but the Exception to 404.9(B) states they are only permitted for replacement purposes. Switches without grounding screws cannot be used in new construction.

DIMMER CONTROLS FOR HOMES

Dimmers are often used to control incandescent lighting in homes. Fluorescent dimmers that control luminaires equipped with special dimming ballast are also

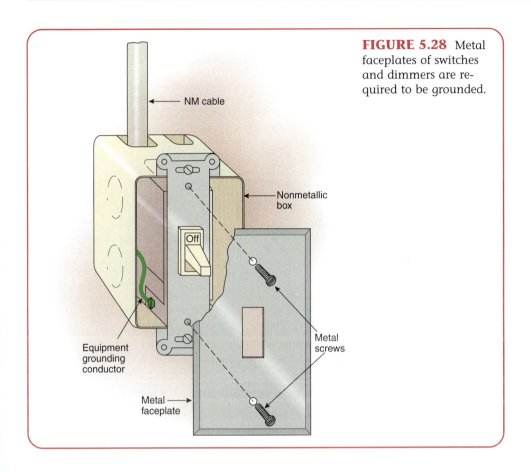

FIGURE 5.28 Metal faceplates of switches and dimmers are required to be grounded.

NM cable

Nonmetallic box

Off

Metal screws

Equipment grounding conductor

Metal faceplate

available but are rarely installed in dwellings. Dimmers can be used anywhere the *Code* requires a wall switch and in general must comply with all applicable rules for switches.

Dimmers don't have marked ON and OFF positions. Some types of dimmers, such as mechanical rotary or slide designs, connect and disconnect the phase or hot conductor when the slide or rotary switch is pushed inward. But many solid-state dimmers don't actually disconnect a conductor. Instead, they reduce the current to a level where the controlled luminaire doesn't operate, although a tiny "trickle" current may still be flowing in the switch loop.

Incandescent Dimmers

Dimmers work by controlling voltage. Reducing the voltage to an incandescent lamp results in lower light output. Incandescent dimmers come in two types: rheostat and electronic. The older rheostat type of dimmer typically has a rotary knob that turns to vary the lighting level and pushes in to turn the controlled luminaires (lighting fixtures) ON or OFF. Newer electronic dimmers may have either rotary or slide controls (Figure 5.29).

Dimmer Ratings.   Dimmers must have wattage ratings sufficient for the incandescent luminaires they control. Typical load ratings are 600, 1000, 1500, and 2000 watts. A 600-watt dimmer is typically used to control one or several luminaires. A large chandelier with many lamps might require a larger 1000-watt model. A 1500-watt

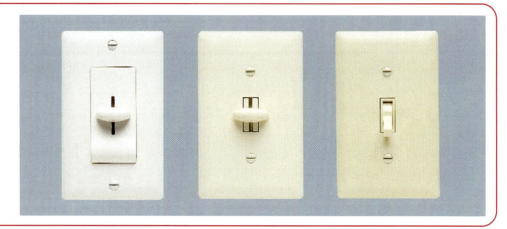

FIGURE 5.29 Dimmers come in a variety of designs. (Courtesy of Pass & Seymour/Legrand®)

dimmer can control an entire 15-ampere branch circuit of lighting, while a 2000-watt dimmer can control an entire 20-ampere branch circuit.

No Receptacles or Fans. Dimmers are permitted to control only permanently installed incandescent luminaires, unless listed for the control of other loads (such as fluorescent lighting) [404.14(E)]. This rule means they cannot be used to control a receptacle outlet in the same way as a snap switch [210.70(A)(1), Exception No. 1]. Because dimmers reduce voltage, home appliances such as televisions, stereos, and computers could be damaged if plugged into a "switched receptacle" controlled by a dimmer.

Incandescent dimmers also can't be used as speed controls for ceiling-suspended (paddle) fans. Fan controls are covered in Chapter 9, The Bedroom.

No Fluorescent Lamps. Incandescent dimmers cannot be used to control fluorescent lamps. This prohibition includes compact fluorescent lamps (CFLs) with integral ballasts and screw shells that are designed to replace standard incandescent lamps in the same luminaires. Listed CFLs of this type are typically marked "Not for Use with Dimmers" or "Not for Use in Luminaires Controlled by a Dimmer."

Some fluorescent fixtures with special ballasts can be used with special fluorescent dimmers, but this type of luminaire is seldom used in one- and two-family dwellings.

Box Sizes for Dimmers. Incandescent dimmers rated up to 2000 watts normally fit into single-gang boxes. However, they are physically larger than conventional single-pole, three-way, and four-way switches. For this reason, it may be necessary to select a deeper device box than usual to provide adequate wiring space required by 314.16.

Electronic Dimmers. Dimmers and other wiring devices should never be installed while a circuit is energized. However, this standard safety precaution takes on extra importance when installing electronic dimmers, because they may be ruined and not work properly if installed on an energized circuit.

NOTE: Working "hot" is never a safe practice and should be avoided whenever possible to minimize the danger of shock.

Circuit Breakers Used as Switches

Most residential lighting is controlled by wall switches or by devices such as dimmers and timers that are installed in the same locations and function in place of wall switches. However, if circuit breakers are used as the only switches to control lighting circuits (e.g., for outdoor lighting, luminaires in a workshop or accessory building, etc.), 240.83(D) establishes two special requirements:

- Circuit breakers used in fluorescent lighting circuits must be listed and marked SWD (switching duty) or HID.
- Circuit breakers used in high-intensity discharge lighting circuits must be listed and marked HID.

SPECIAL RULES FOR TWO-FAMILY DWELLINGS

- Residential branch circuits are only permitted to supply loads located within, or associated with, each individual dwelling unit. Thus, lighting for shared areas such as an interior stairway or hall, laundry room, or attic must be supplied from a house loads panel [210.25].
- Lighting outlets for interior stairways at two-family dwellings are sometimes controlled by a time clock, as permitted by the Exception to 210.70(A)(2). This requirement avoids the need for a wall switch at each stairway landing containing an entryway when the levels are separated by six or more risers.

5   CHAPTER REVIEW

MULTIPLE CHOICE

1. The highest voltage permitted by the *NEC* for lighting in dwellings is
   A. 120 volts
   B. 208 volts
   C. 240 volts
   D. 277 volts

2. The following stairways are required to have a wall switch at every floor or landing that includes an entry:
   A. All interior stairways
   B. Exterior and interior stairways with six or more risers
   C. Interior stairways with six or more risers
   D. Both A and B

3. The following are not permitted to be located within a zone extending 3 ft horizontally and 8 ft vertically from the top of the bathtub rim or shower stall threshold:
   A. Hanging or pendant luminaires
   B. Track lighting
   C. Ceiling-suspended (paddle) fans
   D. All of the above

4. Most listed luminaires are designed to limit the temperature to which supply conductors are exposed to a maximum of
   A. 60°C (140°F)
   B. 75°C (167°F)
   C. 90°C (194°F)
   D. None of the above

5. An incandescent lamp rated 130 volts will operate longest on a circuit operating at
   A. 115 volts
   B. 120 volts
   C. 208 volts
   D. 240 volts

6. Electric-discharge lighting sources include the following lamp types:
   A. Fluorescent
   B. Metal halide
   C. High-pressure sodium
   D. All of the above

7. Switch-controlled receptacle outlets are permitted to be used as the required lighting outlets in
   A. Kitchens
   B. Living rooms
   C. Hallways
   D. Garages

8. Many flush- and recessed-mounted incandescent luminaires use remote junction box construction to
   A. Comply with *readily accessible* requirements in the *NEC*
   B. Make installation easier
   C. Prevent overheating of the supply conductors
   D. Extend lamp life

9. When lighting track is installed in dwellings, the following load must be used for purposes of branch-circuit, feeder, and service calculations:
   A. 150 volt-amperes per 2 ft of length
   B. 180 volt-amperes per 2 ft of length
   C. 1500 volt-amperes
   D. None of the above

10. Switches with pilot lights
    A. Are used to indicate when a controlled luminaire or other load is ON or OFF
    B. Need a grounded conductor in the switch box to power the light
    C. Are required for mechanical equipment spaces in dwellings
    D. Both A and B

FILL IN THE BLANKS

1. Ceiling-suspended (paddle) fans weighing more than _____ lb must be independently supported. Lighter-weight paddle fans must be supported by outlet boxes identified for the purpose.

2. Track lighting that operates at 30 volts or higher is not permitted to be located less than _____ above the finished floor.

3. Class P ballasts are associated with the following type of electric-discharge luminaires: _____.

4. Three-way switches control a lighting outlet from _____ different locations.

5. So-called "long life" or "extended service" incandescent lamps are rated at _____ operating voltages than conventional incandescent lamps.

6. _____ lamps use lamp-within-a-lamp construction to minimize temperatures on the exterior bulb or glass envelope.

7. General-use snap switches rated 15 and 20 amperes and designed for connection to either copper or aluminum conductors are listed and marked _____.

8. Equipment, materials, or services included in a list published by an organization that is acceptable to the authority having jurisdiction and concerned with evaluation of products or services is (are) considered to be _____.

9. The inside of a fluorescent lamp is coated with a powder called _____ that glows when excited by ultraviolet light, thus producing visible light.

10. Recessed-mounted incandescent and HID luminaires installed in _____ are not required to have integral thermal protection.

TRUE OR FALSE

1. Switch-controlled receptacle outlets are permitted to function as lighting outlets in all habitable rooms of a dwelling.
   _____ True   _____ False

2. Fluorescent luminaires mounted end-to-end must be marked "Identified for Through-Wiring" if the conductors of other branch circuits pass through them.
   _____ True   _____ False

3. Luminaires equipped with integral thermal protectors are marked Type IC.
   _____ True   _____ False

4. Luminaires in clothes closets must contain a switch, or be controlled by a switch, with the point of control located at the usual point of entry to the closet.
   _____ True   _____ False

5. Receptacles located within a zone extending 3 ft horizontally and 8 ft vertically from the top of the bathtub rim or shower stall threshold are required to have ground-fault circuit-interrupter protection for personnel.
   _____ True   _____ False

6. Single-pole wall switches have marked ON and OFF positions.
   _____ True   _____ False

7. Lighting track is permitted to be cut to length in the field when listed for this application.
   _____ True   _____ False

8. A *luminaire* is a complete lighting unit consisting of a lamp or lamps together with the parts designed to distribute the light, to position and protect the lamps and ballast (where applicable), and to connect the lamps to the power supply.
   _____ True   _____ False

9. The 2005 *National Electrical Code* prohibits the use of pull-chain switches to control luminaires (lighting fixtures).
   _____ True   _____ False

10. Section 600.32(I) prohibits neon lighting in dwellings.
    _____ True   _____ False

CHALLENGE QUESTIONS

1. Why do you think the Americans With Disabilities Act Guidelines (ADAG) require that wall switches be mounted no higher than 48 in. AFF and receptacle outlets no lower than 18 in. AFF?

2. What do the following features or ratings of luminaires have in common: Class P ballasts, integral thermal protectors for recessed incandescent and HID luminaires, and recessed luminaires identified as Type IC?

3. Name a feature or piece of auxiliary equipment that all of the following luminaire types have in common: fluorescent, mercury vapor, metal halide, high-pressure sodium, and low-pressure sodium.

4. What are some reasons why fluorescent lamps, including compact fluorescent lamps, are becoming more common in homes?

5. There are two general ways of installing power and switch wiring to lighting outlets. (a) Run branch circuit conductors to the outlets with "drops" to the switch boxes. (b) Run branch circuit conductors to switch locations and then run switched conductors to the lighting outlet. Discuss the general advantages and disadvantages of each approach.

6. The *Code* requires that every habitable room in a dwelling have at least one lighting outlet controlled by a wall switch but doesn't specify where these switches must be located. What's the reason for this?

7. What do you think is the reason that the *NEC* doesn't specify color coding for traveler conductors between three-way and four-color switches?

The Living Room

6

INTRODUCTION

This chapter covers the rules of the *National Electrical Code*® that apply to wiring in the living room areas of a typical home. The following major topics are included:

- Basic concepts
- Receptacle outlets in the living room
- Lighting and switching in the living room
- Recommended heights for wall-mounted outlets
- Circuiting for the living room
- Special rules for two-family dwellings

BASIC CONCEPTS

Most *National Electrical Code* rules apply to many different types of buildings. However, some requirements apply to one- and two-family dwellings only. Many of these rules are concentrated in Article 210, although there are rules for residential wiring in other articles as well. In this chapter, we begin looking at the *Code* rules and how they apply to residential wiring in the living room areas of the home.

People frequently refer to *receptacles* and *outlets* as if they were the same thing, but the two terms have different meanings (Figures 6.1 and 6.2). *NEC* Article 100 contains three key definitions that apply.

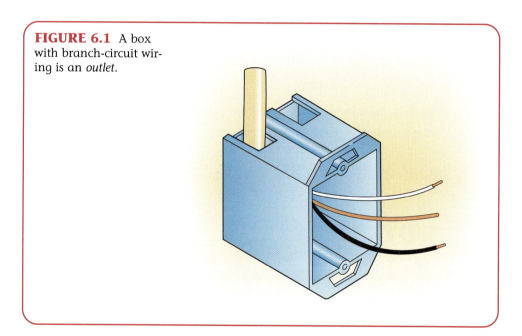

FIGURE 6.1 A box with branch-circuit wiring is an *outlet.*

FIGURE 6.2 An outlet where one or more receptacles are installed is a *receptacle outlet.*

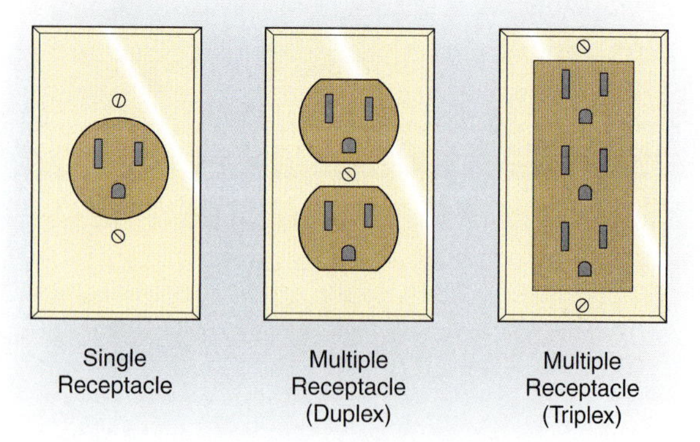

Single Receptacle

Multiple Receptacle (Duplex)

Multiple Receptacle (Triplex)

> **Outlet:** A point on the wiring system at which current is taken to supply utilization equipment.
>
> **Receptacle:** A contact device installed at the outlet for the connection of an attachment plug. A single receptacle is a single contact device with no other contact device on the same yoke. A multiple receptacle is two or more contact devices on the same yoke.
>
> **Receptacle Outlet:** An outlet where one or more receptacles are installed.

Receptacle and Branch-Circuit Ratings

General-purpose 120-volt branch circuits installed in dwellings are protected by either 15- or 20-ampere overcurrent devices. Receptacles rated 15 amperes are permitted to be installed on both 15- and 20-ampere branch circuits. However, receptacles rated 20 amperes are permitted to be installed *only* on 20-ampere branch circuits, according to Table 210.21(B)(3), shown in Exhibit 6.1.

The reason that 15-ampere overcurrent devices can be used is that, although the branch circuit may be rated 20 amperes and use 12 AWG conductors, individual loads plugged into each receptacle (such as table lamps, home entertainment equipment, and vacuum cleaners) generally don't exceed the rating of a 15-ampere receptacle.

Receptacles rated 125 volts, 20 amperes are often called *T-slots*. They accept attachment plugs rated either 15 amperes or 20 amperes (Figure 6.3). T-slot receptacles are commonly used in commercial construction, but very rarely in homes. Nearly all general-use receptacles in homes are rated 15 amperes. Other types of receptacles used for laundry, cooking, and HVAC (heating, ventilating, and air-conditioning) equipment are discussed later in chapters devoted to those subjects.

RECEPTACLE OUTLETS IN THE LIVING ROOM

The *Code* rules specifying the spacing of receptacle outlets in dwelling units apply to most areas of the home, including the living room. However, other factors can influence their location in living rooms. Floor-to-ceiling windows or sliding glass doors may need to be taken into account, along with fireplaces, bay windows, and baseboard heaters. Convenience is also an important consideration in planning for receptacle outlets in the living room.

NEC TABLE 210.21(B)(3) Receptacle Ratings for Various Size Circuits   **EXHIBIT 6.1**

Circuit Rating (Amperes)	Receptacle Rating (Amperes)
15	Not over 15
20	15 or 20
30	30
40	40 or 50
50	50

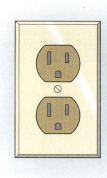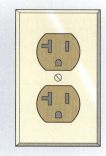

FIGURE 6.3 Nearly all general-purpose receptacles used in residential wiring are rated 15 amperes (left); 20-ampere T-slot receptacles (right) are seldom used in homes.

Receptacle Spacing

One of the most familiar of all *NEC* requirements is the so-called "12-foot rule" for locating receptacles in dwelling units. Actually, it's a 6-foot rule: 125-volt, 15- and 20-ampere receptacle outlets must be installed so that no point along the floor in any wall space is more than 6 ft, measured horizontally, from another receptacle outlet in that space [210.52(A)(1)].

The "12-foot-rule" applies to receptacle outlets located in living rooms, dining rooms, bedrooms, parlors, libraries, dens, family rooms, recreation rooms, kitchens, sunrooms, and similar areas of dwellings.

The reasoning behind this requirement is that many attachment cords on lamps, televisions, and other portable household appliances are 6 ft long. Thus, the requirement that no point along a wall be more than 6 ft from the nearest receptacle allows most appliances to be plugged in without using extension cords.

Minimizing Use of Extension Cords. Extension cords indoors are seen as inherently hazardous, since they can create a tripping hazard. Also, when receptacles aren't conveniently placed to serve lamps and appliances, homeowners sometimes run extension cords underneath rugs or carpets. In effect, the homeowners are using these flexible cords as a substitute for permanent wiring methods, a practice that the *NEC* discourages for safety reasons [400.8].

Extension cords aren't designed for this kind of hard service, so when they are located beneath rugs—where people can walk on them or furniture may inadvertently be placed on top of them—the insulation and/or conductors may be damaged, creating potential shock and fire hazards.

Narrow Walls. Any wall space 2 ft or wider must have its own receptacle outlet [210.52(A)(2)(1)]. The intent of this rule is to prevent appliance cords or extension cords from being laid across doorways, where they present a tripping hazard and could be damaged if a door closes on them.

Glass Doors and Floor-to-Ceiling Windows. The standard 12-foot rule for spacing receptacles also applies to wall space occupied by fixed (nonsliding) glass panels [210.52(A)(2)(2)]. Floor receptacle outlets can be used in this situation if they are located within 18 in. of the fixed panel [210.52(A)(2)(3)]. Sliding panels are treated as doors and aren't counted when measuring to determine required spacing of receptacle outlets along walls.

Figures 6.4 and 6.5 show two different receptacle layouts for a living room with a floor-to-ceiling picture window. The layout in Figure 6.4 uses a floor receptacle. In Figure 6.5, the receptacle outlets have been spaced differently (but still according to the *Code*) to avoid the need for a floor receptacle.

Floor Receptacles. Floor-mounted receptacles aren't used much in residential construction. When floor receptacles are installed in a dwelling unit, they must be used with boxes specifically listed for this application [314.27(C)].

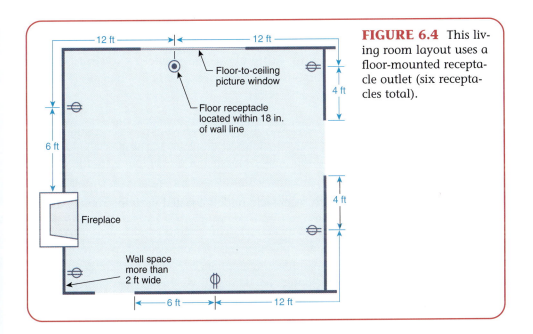

FIGURE 6.4 This living room layout uses a floor-mounted receptacle outlet (six receptacles total).

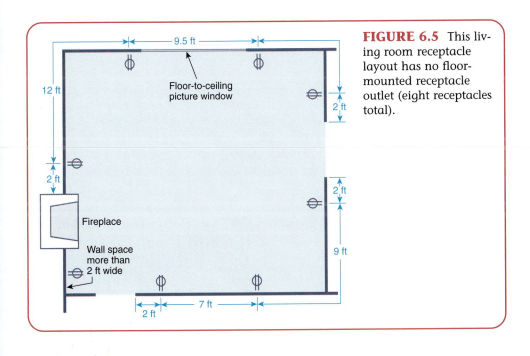

FIGURE 6.5 This living room receptacle layout has no floor-mounted receptacle outlet (eight receptacles total).

Floor-mounted receptacles come in both raised and floor models. So-called "tombstone" or "doghouse" receptacle housings present a tripping hazard and may interfere with floor cleaning, while the flush type is prone to getting dirt in its slots. The most common use for floor-mounted receptacles is in commercial construction for offices with underfloor duct power distribution systems.

Additional Receptacle Outlets. As with all *NEC* requirements, the receptacle spacing rules in 210.52(A) are *minimum* requirements to ensure electrical safety. Additional receptacles can always be provided for extra convenience or where required for a particular purpose. For example, receptacles are sometime centered below windows to allow plugging in holiday lighting such as electric candle lamps.

However, even when more than the minimum number of receptacle outlets is installed in a room, they must still be located so that no point in any wall space is more than 6 ft, measured horizontally, from another receptacle outlet. Some living rooms have receptacles located 6 to 12 in. above the mantel to serve decorative lamps. If located more than 5½ ft above the floor, these can't be counted as part of the required receptacle outlets [210.52(A)(1)].

Baseboard Heaters. To meet receptacle spacing rules, electric baseboard heaters are available with integral receptacle modules (Figure 6.6). These receptacles are not permitted to be connected to the heater circuit but are supplied by the normal branch circuit(s) feeding other living room receptacles [210.52, 424.9]. Circuit requirements for electric baseboard heaters are covered in Chapter 14, HVAC Equipment and Water Heaters.

Although there is no *Code* rule to this effect, listing requirements for permanently installed electric baseboard heaters prohibit locating them beneath wall receptacles [210.52, FPN]. The reason is to avoid having appliance or lamp cords draped across the heater, where the insulation could be damaged by high-temperature electric heating elements. Hot water-type baseboard heaters are permitted to be located below wall-mounted receptacles, because they operate at lower temperatures.

Importance of Listing Requirements. The FPN to 210.52 is a good illustration of *NEC* 110.3(B): "Listed or labeled equipment shall be installed and used in accordance with any instructions included in the listing or labeling." In effect, this

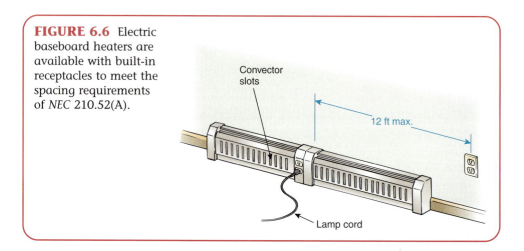

FIGURE 6.6 Electric baseboard heaters are available with built-in receptacles to meet the spacing requirements of *NEC* 210.52(A).

Convector slots

12 ft max.

Lamp cord

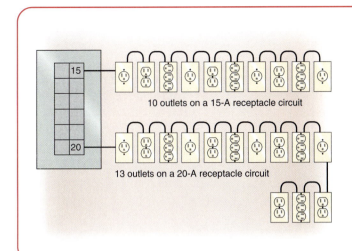

FIGURE 6.7 The *NEC* rules limiting the maximum number of outlets permitted on 15- and 20-ampere branch circuits don't apply to dwellings but offer guidance to installers.

rule makes equipment listing conditions an enforceable part of the *National Electrical Code.*

Maximum Number of Receptacles on a Branch Circuit. The *NEC* doesn't define a maximum number of receptacle outlets that can be installed on a single branch circuit in a dwelling. The reason is that receptacle circuits in homes tend to be lightly loaded compared to those in commercial buildings.

However, some jurisdictions have their own rules limiting the number of receptacle outlets on a single circuit in residential construction. Sometimes a jurisdiction's requirement is the same as the *NEC* rule for commercial occupancies—that is, 10 outlets on a 15-ampere branch circuit or 13 outlets on a 20-ampere branch circuit, based on a load of 180 volt-amperes per receptacle outlet (Figure 6.7). If in doubt, the authority having jurisdiction (AHJ) should be consulted.

NOTE: It is important to understand that the receptacle loading rules of 220.14(I) don't normally apply to dwellings. Receptacle outlets installed in dwelling units are considered part of the general lighting load of 3 volt-amperes per square foot, as shown in NEC Table 220.3. They aren't considered to have a load of their own, and there is no specified number that is allowed on a single branch circuit.

LIGHTING AND SWITCHING IN THE LIVING ROOM

Section 210.70(A)(1) requires that a minimum of one lighting outlet, controlled by a wall switch, be installed in every habitable room. In rooms other than kitchens and bathrooms, this outlet is permitted to be a receptacle outlet controlled by a wall switch rather than a lighting outlet. Some rooms have both: a dedicated lighting outlet and one or more switch-controlled receptacles.

Lighting Types and Locations. The *NEC* doesn't specify locations for the lighting outlets (either those with permanently installed luminaires or switched receptacles) required by 210.70(A)(1). Few living rooms in homes built today have a lighting

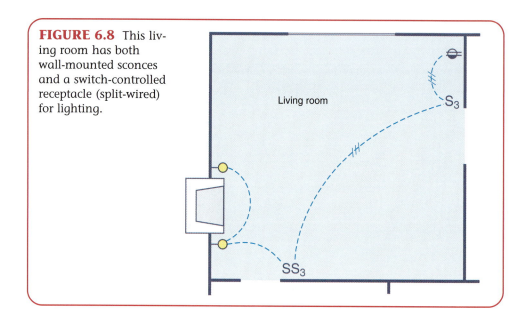

FIGURE 6.8 This living room has both wall-mounted sconces and a switch-controlled receptacle (split-wired) for lighting.

Living room

outlet with a surface-mounted or pendant luminaire (lighting fixture) located in the middle of the ceiling.

Instead, other solutions are used, depending on decorative considerations. These solutions include ceiling-mounted recessed can lights, ball spots, cove lighting around the sides of a room, wall sconces, track lighting, and receptacle outlets controlled by wall switches. Such solutions allow the homeowner to vary the lighting to suit the mood or time of day.

The living room shown in Figure 6.8 has wall sconces on either side of the fireplace and a switched receptacle into which a table lamp is plugged. Other common lighting solutions are illustrated in later chapters that cover other rooms in a typical residence.

Switching. Wall switches are normally located at the entrance(s) of a room. For safety's sake, the homeowner should not have to walk through a dark room to reach a switch. Typically, when there is more than one entrance to a room, switches are provided for some or all of the lighting outlets at each entrance. These additional switches are for convenience, since there is no *Code* rule requiring multiple light switches in a room.

The living room shown in Figure 6.8 has three-way switches to control the split-wired receptacle, in which a table lamp will probably be plugged. A single-pole switch located at the entrance closest to the fireplace controls the wall sconces. Thus, there are two light switches at this entrance to the room. They are mounted in double-gang boxes and trimmed with double-gang faceplates (Figure 6.9).

Lightweight Luminaires (Lighting Fixtures). The wall-mounted sconces located above each end of a fireplace mantel are small luminaires weighing 6 lb or less, which means they are permitted to be supported on regular device boxes using No. 6 or larger screws. Typically, such lightweight luminaires are supplied with 6-32 screws (No. 6 screws with 32 threads per inch) [314.27(A), Exception].

Heavier or ceiling-mounted luminaires are required to use boxes designed for lighting fixture attachments [314.27(A), 314.27(B)] and offering adequate support [410.16].

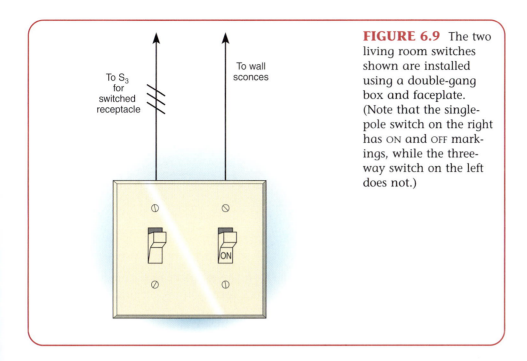

FIGURE 6.9 The two living room switches shown are installed using a double-gang box and faceplate. (Note that the single-pole switch on the right has ON and OFF markings, while the three-way switch on the left does not.)

RECOMMENDED HEIGHTS FOR WALL-MOUNTED OUTLETS

There are no *Code* rules that require a particular mounting height for wall-mounted receptacles, switches, or lighting outlets. In general, mounting heights are seen as a design or esthetic consideration, not a safety issue.

Many installers use the heights shown in Table 6.1, which comply with the Americans with Disabilities Act Guidelines (ADAG). Technically, the ADAG apply only to commercial construction, not residences—unless a dwelling unit is intended

TABLE 6.1 Recommended Heights for Wall-Mounted Outlets

Outlet	Height (in.)
Switches, dimmers, fan controls	48
Receptacle outlets, general	18
Receptacle outlets, kitchen, utility room, workbenches, etc.	42, or 6 above countertop or backsplash
Telephones, general	18
Intercom/music stations	48
Telephones, wall mounted	60
Thermostats	60
Security system keypads	48
Kitchen clock hangers	6 below ceiling; above doors, center the clock outlet between door trim and ceiling
Doorbells, buzzers, chimes	78
Lighting	78, typical; varies depending on size and design of luminaire (lighting fixture)

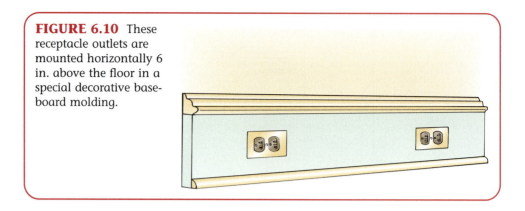

FIGURE 6.10 These receptacle outlets are mounted horizontally 6 in. above the floor in a special decorative baseboard molding.

and designed for use by the handicapped. However, the ADAG mounting heights are starting to become "universal" guidelines for mounting electrical equipment.

Consistency. Whatever heights are used for wall-mounted outlets, consistency is important. All receptacle outlets within the same room or area of a home should be located at the same mounting height. The same holds true for switches, dimmers, fan speed controls, and similar items within the same room or area. Consistent mounting height helps create a neater appearance that doesn't seem visually cluttered.

Sometimes, of course, special architectural or interior design treatments will require that receptacle outlets and other items be mounted at different heights than those shown in Table 6.1 or mounted in nontraditional ways (Figure 6.10).

CIRCUITING FOR THE LIVING ROOM

Deciding on Branch Circuits

Most general-purpose branch circuits installed in dwellings for lighting and receptacles use 14 AWG conductors protected by 15-ampere overcurrent devices (circuit breakers or fuses). The *Code* requires 20-ampere branch circuits for only three special purposes: small-appliance branch circuits, laundry branch circuits, and bathroom branch circuits [210.11(C)].

There is little cost difference between installing 15-ampere branch circuits and 20-ampere branch circuits, and 15-ampere duplex receptacles can be connected on 20-ampere branch circuits, as explained earlier in this chapter.

However, 14 AWG wires are smaller and slightly more flexible than 12 AWG wires, making them easier to handle in crowded outlet boxes. Also, as previously noted, most general-purpose branch circuits in residences are lightly loaded. For these reasons, most installers use 15-ampere branch circuits throughout houses, apartments, and condominiums, except where other branch-circuit ratings are specifically required.

Laying Out Branch Circuits

The *Code* specifically permits lighting and other loads to be connected together on 15- or 20-ampere branch circuits in homes, except in a few special cases [210.23(A)].

Small-appliance branch circuits, laundry branch circuits, and bathroom branch circuits must supply only receptacle outlets [210.23(A), Exception].

In residences, lighting and receptacles are normally wired together on the same branch circuits. This practice is different from commercial construction, where lighting and receptacles are typically segregated onto different circuits fed from different panels. Of course, lighting operates at 277 volts in many commercial buildings, while the convenience receptacles are limited to 120 volts [210.6(A)].

Lighting and receptacles in different rooms are permitted to be mixed together on the same branch circuit(s), and this is a normal practice. A single room in a typical house or condominium doesn't have enough load to require a separate 15-ampere branch circuit.

However, for ease of future maintenance and troubleshooting, it is a good idea to lay out branch circuits in dwellings as simply and logically as possible. It's much more difficult to figure out what's going on later when the drywall is up and all the wiring is hidden!

Small-Appliance, Laundry, and Bathroom Branch Circuits. Lighting and receptacle outlets in living rooms and other general areas of the home *cannot* be connected on the small-appliance branch circuits required to serve receptacles in the kitchen and dining room, the branch circuit(s) for bathroom receptacle outlets, or the branch circuit for laundry receptacle outlets [210.11(C)].

However, living room lighting and/or receptacles can be connected to lighting outlets in kitchens, dining rooms, bathrooms, laundry rooms, and other areas of the house. There is no *NEC* rule prohibiting this practice, and it's very common in residential wiring.

Circuit Diagram. The plan in Figure 6.11 shows the circuit diagram for the living room. Note that the outdoor ground-fault circuit-interrupter (GFCI) receptacle is connected to the living room circuit, while two living room receptacles are connected to a circuit that continues in another room of the house.

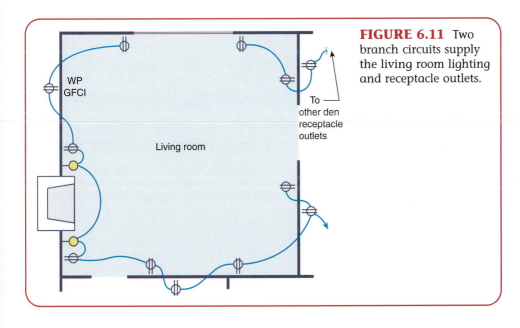

FIGURE 6.11 Two branch circuits supply the living room lighting and receptacle outlets.

General Guidelines for Laying Out Branch Circuits

- Only lighting, receptacles, and other loads from adjacent rooms should be connected together (Figure 6.12). A single outlet from some other part of the house shouldn't be tacked onto the same branch circuit just because it seems convenient to do so during original construction. It will seem less convenient later, when an electrician is trying to troubleshoot that outlet and figure out which circuit supplies it. For the same reason, mixing loads from different floors on the same branch circuit should be avoided.

- Many installers feel it is a good practice to have more than one branch circuit supplying the lighting and receptacles in a room or other area of the home. Then, if a circuit breaker trips or a circuit needs to be turned off for maintenance reasons, energized receptacles are still available in the same room to supply portable lamps and power tools.

- Individual branch circuits should be considered for utilization equipment such as electric heaters and ceiling fans, even when not required by the *Code*. If these items then need maintenance or repairs, the lighting and receptacles in the room will still be operating, making service easier.

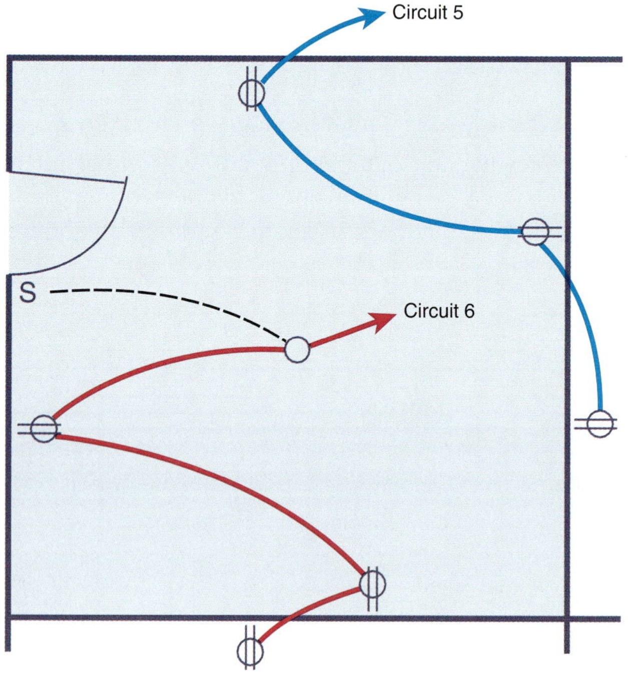

FIGURE 6.12 Lighting and receptacles from adjacent rooms can be circuited together.

SPECIAL RULES FOR TWO-FAMILY DWELLINGS

There are no special *Code* rules that apply to living rooms in two-family dwellings. Residential branch circuits are only permitted to supply loads located within, or associated with, each individual dwelling unit [210.25]. Also, occupants are required to have "ready access" to the overcurrent devices protecting conductors in their occupancy [240.24(B)]. This requirement means that a single duplex receptacle in Unit A cannot be connected to a branch circuit in an adjacent room of Unit B just because it saves wire or seems more convenient to do so.

6 CHAPTER REVIEW

MULTIPLE CHOICE

1. The minimum mounting height permitted by the *Code* for wall-mounted receptacles is
   A. 6 in.
   B. 12 in.
   C. 18 in.
   D. None of the above

2. The maximum number of duplex receptacles permitted on a 15-ampere branch circuit in a dwelling is
   A. 8
   B. 10
   C. 13
   D. None of the above

3. Listed or labeled equipment is required to be installed and used in accordance with
   A. Applicable *Code* rules
   B. Instructions included in the listing or labeling
   C. Special instructions of the AHJ
   D. Both A and B

4. Having more than one branch circuit supply lighting outlets and receptacle outlets in a room offers the following advantages:
   A. Tripping a circuit breaker doesn't result in losing all power to a room.
   B. If one branch circuit is turned off to service or replace an item such as a luminaire or receptacle, energized receptacles are still available for portable lamps and power tools.
   C. It provides better load diversity.
   D. Both A and B

5. A habitable room with a lighting outlet and three entrances is required by the *Code* to have how many wall switches or equivalent controls such as dimmers and occupancy sensors?
   A. One
   B. Two
   C. Three
   D. Two three-way switches

6. A lighting outlet with a permanently installed luminaire (lighting fixture) is required in the following room(s) of a home:
   A. Living room
   B. Kitchen
   C. Laundry room
   D. Both A and B

7. A single receptacle can have the following number(s) of contact devices on the same yoke:
   A. 1
   B. 2
   C. 3
   D. All of the above

8. Permanently installed electric baseboard heaters are prohibited from being located beneath wall-mounted receptacle outlets by
   A. *NEC* 210.13(A), Exception
   B. Baseboard heater listing instructions
   C. *NFPA 5000®*, *Building Construction and Safety Code®*
   D. All of the above

9. Receptacles located higher above the floor than the following are not permitted to be counted among the receptacle outlets required by 210.52(A)(1) and (2):
   A. 24 in.
   B. 36 in.
   C. 54 in.
   D. 66 in.

10. Floor receptacles used as required receptacle outlets along floor-to-ceiling windows or fixed panels of sliding glass doors are required to be installed within what distance of the wall line (fixed glass panel)?
    A. 6 in.
    B. 9 in.
    C. 12 in.
    D. 18 in.

FILL IN THE BLANKS

1. The 125-volt, 15- and 20-ampere receptacle outlets in living rooms must be installed so that no point along the floor in any wall space is more than _____ ft, measured horizontally, from another receptacle outlet in that space.

2. The *Code* requires 20-ampere branch circuits in dwellings for _____ circuits and _____ circuits.

3. Any wall space _____ or wider must have its own receptacle outlet.

4. Most 120-volt branch circuits in a typical dwelling unit are rated _____ amperes.

5. According to the definition in Article 100, a multiple (duplex) receptacle is considered to be the same as _____ receptacle(s).

6. Wall-mounted luminaires (lighting fixtures) weighing not more than _____ are permitted to be supported by standard outlet boxes using No. 6 or larger screws.

7. Small-appliance branch circuits serving receptacles in kitchens, dining rooms, pantries, and breakfast nooks are required to be protected by overcurrent devices rated _____.

8. When firewalls are penetrated by electrical cables or conduits, the openings are required to be _____.

9. A 20-ampere branch circuit requires _____ AWG conductors.

10. In habitable rooms, 125-volt, 15- and 20-ampere receptacle outlets must be installed so that no point along the floor in any wall space is more than _____ft, measured horizontally, from another receptacle outlet in that space.

TRUE OR FALSE

1. Built-in receptacles in electric baseboard heater units are supplied from one leg of the 240-volt heater circuit and the grounded (neutral) conductor.

_____ True   _____ False

2. Duplex receptacles mounted 5 ft above the floor can be counted among the receptacle outlets required by 210.52(A)(1) and (2).

_____ True   _____ False

3. A 125-volt, 15-ampere single receptacle can be installed on either a 15-ampere or 20-ampere branch circuit.

_____ True   _____ False

4. Section 210.25 permits certain common-area loads of two-family dwellings, such as outdoor receptacles, to be supplied from either unit.

_____ True   _____ False

5. In other than kitchens and bathrooms, the lighting outlet required by 210.70(A) is permitted to be a receptacle outlet controlled by a wall switch rather than a lighting outlet.

_____ True   _____ False

6. Small-appliance branch circuits, laundry branch circuits, and bathroom branch circuits are permitted to supply only receptacle outlets.

_____ True   _____ False

7. Living room receptacle outlets are permitted to be connected to small-appliance branch circuits for the kitchen and dining room, as long as the receptacles are rated 125 volts, 20 amperes.

_____ True   _____ False

8. The *National Electrical Code* permits lighting and other receptacle loads to be mixed together on the same branch circuits.

_____ True   _____ False

9. Living rooms in dwelling units are required to have lighting outlets with permanently installed luminaires (lighting fixtures).

_____ True   _____ False

10. Overcurrent protective devices (OCPD) for branch circuits may be either circuit breakers or fuses.

_____ True   _____ False

CHALLENGE QUESTIONS

1. Why doesn't the *Code* specify the mounting heights for wiring devices such as receptacles, wall switches, and dimmers?

2. What advantages are there in having more than one branch circuit supply lighting and receptacle outlets in a room?

3. What are some reasons why the *National Electrical Code* discourages the use of extension cords?

4. Why isn't ground-fault circuit-interrupter (GFCI) protection for personnel required for receptacle outlets in living rooms of dwellings?

5. In commercial construction, a maximum of 10 receptacle outlets are permitted on a 15-ampere branch circuit, and 13 receptacle outlets on a 20-ampere branch circuit. Why don't these same limitations apply to residences, and how is the receptacle load calculated for purposes of determining the service size?

6. The *National Electrical Code* permits lighting and other receptacle loads to be mixed together on the same branch circuits. This is commonly done in residential wiring but not in commercial wiring, where receptacles and lighting outlets are typically segregated onto different circuits. What are some reasons for this difference in wiring practices?

The Kitchen and Dining Room

7

OBJECTIVES

After completing this chapter, the student will be able to understand the following:

- Requirements for ground-fault circuit-interrupter (GFCI) protection on receptacle outlets
- Branch circuits used to supply cooking appliances
- Which kitchen appliances are permitted to be cord-and-plug connected
- Why multiwire branch circuits can't be used for kitchen counter receptacles
- Circuiting requirements for clock hanger outlets

INTRODUCTION

This chapter covers the fairly specialized *National Electrical Code*® rules that apply to how receptacle outlets and appliances are installed in kitchens, dining rooms, breakfast rooms, and pantries. The following major topics are included:

- Receptacle outlets in the kitchen and dining room
- Lighting and switching in the kitchen and dining room
- Ranges, wall-mounted ovens, and counter-mounted cooking units (cooktops)
- Other kitchen appliances
- Special systems
- Special rules for two-family dwellings

IMPORTANT NEC TERMS

Accessible, Readily (Readily Accessible)

Appliance

Attachment Plug (Plug Cap, Plug)

Automatic

Branch Circuit, Appliance

Branch Circuit, General-Purpose

Branch Circuit, Individual

Branch Circuit, Multiwire

Concealed

Cooking Unit, Counter-Mounted

Demand Factor

Device

Disconnecting Means

Equipment

Ground-Fault Circuit Interrupter

Lighting Outlet

Luminaire

Switch, General-Use Snap

RECEPTACLE OUTLETS IN THE KITCHEN AND DINING ROOM

What Is a Kitchen?

The term *kitchen* isn't defined in the *National Electrical Code*. Most people assume a kitchen to be a room or area with permanent facilities for washing, storing, and cooking food—in other words, a room with a sink, refrigerator, and cooking appliances that are permanently installed and/or occupy dedicated space. But other factors, such as the following, often need to be considered:

- A room that contains a sink plus a range, oven, or cooktop probably should be considered a kitchen, even if the refrigerator is physically located in an adjacent pantry.
- Some small apartments have a "kitchenette" consisting of a single wall area with a sink, a refrigerator, a range/oven, and counter space. These are considered kitchens and must comply with all applicable *NEC®* rules.
- An area of a basement that has a refrigerator or freezer plus a nearby laundry sink is not a kitchen.
- A wet bar is not a kitchen, even if it has a permanently installed refrigerator and a microwave oven.

The factor that determines if a room or area is a kitchen is whether it is the primary food-preparation area of a dwelling. In case of doubt, the authority having jurisdiction (AHJ) should be consulted.

Small-Appliance Branch Circuits

All receptacle outlets in the kitchen, dining room, pantry, and breakfast room of a dwelling must be supplied by two or more 20-ampere branch circuits [210.11(C)(1), 210.52(B)(1)]. The reason for this requirement is to provide sufficient capacity for high-load, cord-and-plug-connected appliances such as toasters, coffeemakers, slow cookers, and electric frying pans.

No small-appliance branch circuit is permitted to serve more than one kitchen [210.52(B)(3)]. So in a house that includes an accessory apartment or "granny flat," at least four small-appliance branch circuits are required.

- Each of the two kitchens requires a minimum of two 20-ampere small-appliance branch circuits to supply its receptacle outlets.
- Any of the four circuits are also permitted to supply receptacle outlets in an associated dining room, pantry, breakfast room, or similar area.

CAUTION: The two small-appliance branch circuits for each kitchen are not permitted to serve other loads such as lighting, exhaust hoods, permanently connected kitchen appliances, or receptacles in other rooms [210.52(B)(2)]. They can only serve receptacle outlets in the kitchen, dining room, pantry, and breakfast room. In addition to receptacles intended for small appliances, electric clock receptacles and receptacles that provide power to the electric equipment of gas ranges, ovens, and cooking units are also permitted on these circuits [210.52(B)(2), Exceptions No. 1 and No. 2].

Kitchen Countertops. Kitchen countertop receptacles must be supplied by at least two 20-ampere small-appliance branch circuits [210.52(B)(3)]. These can be the same two circuits that serve other receptacles, or they can be additional circuits.

NOTE: The NEC is not a design specification and contains minimum safety requirements, so there's never any objection to exceeding minimum Code rules.

Refrigerators. A receptacle outlet serving a refrigerator is permitted to be on one of the two 20-ampere small-appliance branch circuits that supplies kitchen countertop receptacles. However, 210.52(B)(1), Exception No. 2 also permits the refrigerator to be supplied by an individual branch circuit rated at least 15 amperes.

Clock Outlet. One or more receptacles installed for electric clocks (commonly called "clock hanger outlets") are permitted to be supplied by small-appliance branch circuits in the kitchen, dining room, breakfast room, pantry, and similar areas [210.52(B)(2), Exception No. 1].

Switched Receptacles. Small-appliance branch circuits are not permitted to serve lighting outlets. But in addition to the required receptacles on these circuits, switched receptacles can be located in the same rooms (kitchen, dining room, pantry, breakfast room) as long as they are supplied by a general-purpose branch circuit [210.52(B)(1), Exception No. 1].

Figure 7.1 illustrates two different layouts with GFCI protection that satisfy the *National Electrical Code* rules for receptacles and circuits in the kitchen, dining room, pantry, and breakfast room. Both layouts use 2-wire branch circuits to serve the kitchen counter receptacles. More detailed information on how GFCIs operate is provided in Chapter 8, The Bathroom.

Receptacle Outlets Serving Kitchen Countertops

The *NEC* provides numerous rules for spacing and locating countertop receptacle outlets (Figure 7.2). The wording of 210.52(C) states that these rules apply to receptacle outlets for counter spaces "in kitchens and dining rooms," although few dining rooms have counters. Similarly, it should be understood that these rules apply to counter spaces in pantries and breakfast rooms as well, even though the *Code* wording doesn't specify these locations.

The *NEC* rules for receptacle outlets serving kitchen countertops can be summarized as follows.

1. *Basic Rule.* No point along the wall line can be more than 24 in., measured horizontally, from a receptacle outlet [210.52(C)(1)]. One reason that kitchen counters have a "4-foot rule" for locating receptacles, rather than the "12-foot rule" used in other parts of dwellings, is that kitchen appliances such as toasters, coffeemakers, and blenders normally have short power cords.

2. *Narrow Walls.* Each wall along a counter space measuring 12 in. or wider must have a receptacle outlet [210.52(C)(1)].

3. *Separate Spaces.* Countertops separated by ranges, refrigerators, or sinks are considered separate countertop spaces. Each countertop space must comply with rules 1 and 2 [210.52(C)(4)].

4. *Islands.* Each island counter measuring 24 in. × 12 in. or larger must have a receptacle outlet [210.52(C)(2)]. Although only one receptacle per island is required

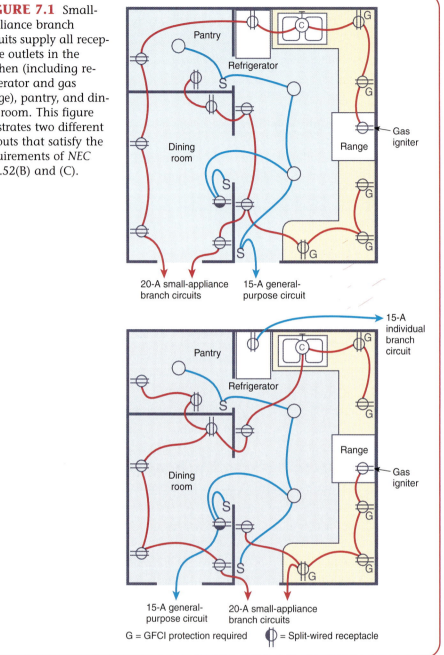

FIGURE 7.1 Small-appliance branch circuits supply all receptacle outlets in the kitchen (including refrigerator and gas range), pantry, and dining room. This figure illustrates two different layouts that satisfy the requirements of *NEC* 210.52(B) and (C).

by the *Code*, providing additional outlets on a large island may improve usability for homeowners. A guideline is the general kitchen counter rule of locating receptacle outlets approximately 4 ft apart.

5. *Peninsulas.* Each peninsula counter measuring 24 in. × 12 in. or larger must have a receptacle outlet [210.52(C)(3)]. Although only one receptacle is required on a peninsula, providing additional outlets on a long peninsula may be more convenient and improve usability for homeowners.

NOTE: Many installers follow the general kitchen counter rule of locating receptacle outlets approximately 4 ft apart.

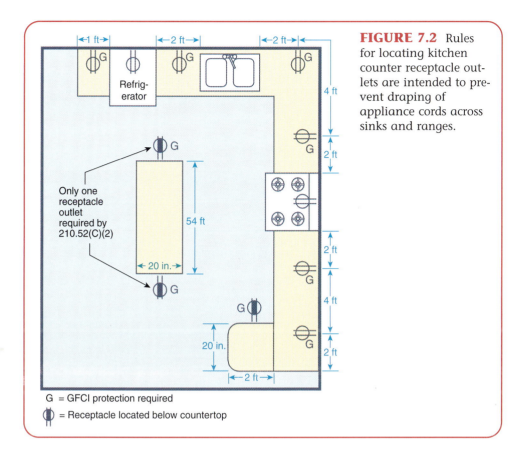

FIGURE 7.2 Rules for locating kitchen counter receptacle outlets are intended to prevent draping of appliance cords across sinks and ranges.

G = GFCI protection required

= Receptacle located below countertop

6. Accessibility. Kitchen counter receptacles must be readily accessible for connecting cord-and-plug-connected appliances. Receptacles that aren't accessible because they are blocked by major appliances, such as refrigerators or ranges, or receptacles that are located inside "appliance garages" (recessed spaces for smaller kitchen appliances such as microwaves and toaster ovens) don't meet the requirements of rules 1 through 5.

7. Face-Up Position Prohibited. Receptacles can't be installed in a face-up position in a countertop or similar surface [406.4(E)]. This rule is to prevent receptacles from collecting liquids, crumbs, and debris that could create a possible fire or shock hazard.

8. Not Behind Sinks or Ranges. Receptacle outlets aren't required on walls directly behind a sink or range. Exhibit 7.1 (*NEC* Figure 210.52), shows areas where countertop receptacles outlets aren't required.

The *NEC* also has rules for the mounting height of receptacle outlets located above countertops. These can be summarized as follows.

1. Basic Rule for Mounting Height. Receptacle outlets must be located above, but no more than 20 in. above, countertops [210.52(C)(5)]. A mounting height of 4 in. to 6 in. above the top edge of the backsplash, measured to the centerline of the device box, is very typical for kitchen counter receptacles.

2. Islands and Peninsulas Without Backsplashes. Receptacle outlets for islands and peninsulas without backsplashes may be mounted up to 12 in. below the level

EXHIBIT 7.1 | *NEC* Figure 210.52 Determination of Area Behind Sink or Range

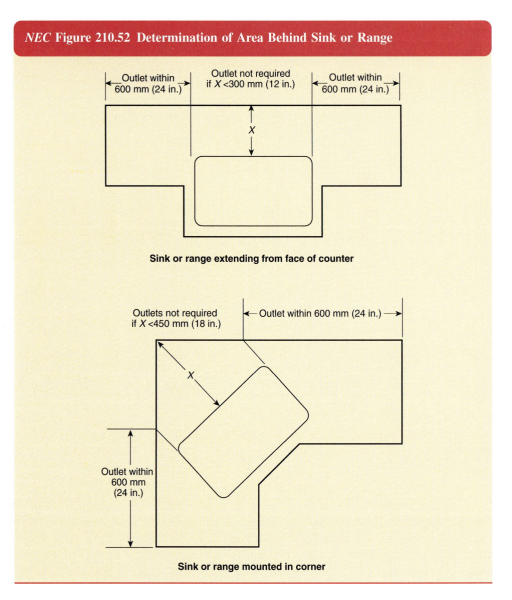

Sink or range extending from face of counter

Sink or range mounted in corner

of the countertop, as long as the counter doesn't extend more than 6 in. beyond its supporting base or cabinet [210.52(C)(5)]. However, when the edge of a counter extends more than 6 in. beyond its support, receptacles *are not permitted* to be located on that base or cabinet [210.52(C)(5), Exception]. (See "Why Receptacles Aren't Normally Mounted Below Countertops," which follows these rules.)

3. *Islands and Peninsulas with Cabinet Above.* When an island or peninsula countertop with no backsplash or other means for mounting a receptacle has a cabinet located above, within 20 in. of the countertop surface, the required receptacle outlet must be mounted in or on that cabinet [210.52(C)(5), Exception (b)].

NOTE: It is the responsibility of the architect or kitchen designer to provide a way to locate a receptacle outlet serving an island or peninsula countertop space that complies with applicable NEC rules. These rules are intended to prevent the use of extension cords in kitchens. In doubtful situations, the AHJ should be consulted.

4. *Handicapped Accessibility.* In dwellings intended for use by the physically impaired, receptacle outlets for all kitchen countertop spaces are permitted to be mounted up to 12 in. below the level of the countertop, as long as the counter doesn't extend more than 6 in. beyond its supporting base or cabinet [210.52(B)(1), Exception No. 1]. This mounting height is allowed to make receptacles easier to reach from a wheelchair.

Why Receptacles Aren't Normally Mounted Below Countertops. The rule requiring that receptacle outlets be mounted above kitchen countertops whenever possible, rather than below their edges is—like all *NEC* rules—based on safety. When a receptacle is mounted on a partition or cabinet below the edge of a countertop, the attachment cord of a cooking appliance such as a toaster, coffeemaker, or electric frying pan will have to hang over the edge to be plugged in. This arrangement creates a potential hazard, since children might yank on the cord and pull hot appliances onto themselves.

GFCI Protection. All 15- and 20-ampere receptacle outlets serving kitchen countertop surfaces are required to have GFCI protection for personnel [210.8(A)(6)]. (The subject of how to provide GFCI protection for receptacle outlets and the tradeoffs between different methods of doing so are covered in greater detail in Chapter 8, The Bathroom.) Installing a GFCI receptacle (Figure 7.3) at each kitchen countertop outlet is the most common method.

Electric Clocks. One or more receptacles installed for electric clocks are permitted to be supplied by small-appliance branch circuits in the kitchen, dining room, breakfast room, pantry, and similar areas [210.52(B)(2), Exception No. 2]. These recepta-

FIGURE 7.3 This 15-ampere duplex receptacle has integral GFCI protection. (Courtesy of Pass & Seymour/Legrand®)

FIGURE 7.4 Clock hanger outlets combine a recessed single receptacle with a means for supporting a wall-mounted clock. (Courtesy of Pass & Seymour/Legrand®)

cles are commonly called *clock hanger outlets* (Figure 7.4) and are installed 6 in. to 8 in. below the ceiling, measured to the center of the outlet box. They aren't required to have GFCI protection, since they don't serve countertop surfaces [210.8(A)(6)]. Clock hanger outlets can also be supplied by a general-purpose branch circuit, such as the one used for kitchen lighting, where it is more practical or convenient to do so.

Multiwire Branch Circuits Not Permitted. Multiwire branch circuits are frequently used in house wiring to economize on homerun wiring by sharing a grounded (neutral) conductor among either two or three phase conductors. Sometimes they are used to split-wire receptacles so that each half of a duplex receptacle is on a different circuit (Figure 7.5).

However, multiwire branch circuits should not be used to supply GFCI-protected receptacle outlets such as those on kitchen countertops. Ground-fault circuit interrupters work by comparing currents on the ungrounded and grounded (neutral) conductors of a 2-wire branch circuit, which should, under normal operating conditions, be equal. GFCIs don't work properly on circuits that share a grounded (neutral) conductor.

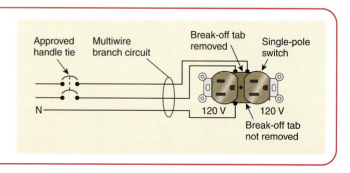

FIGURE 7.5 Multiwire branch circuits are sometimes used to supply split-wired receptacles.

Other Receptacle Outlets

Receptacles not serving countertop spaces in the kitchen, dining room, pantry, and breakfast room are located according to the same requirements as those in other areas of the house [210.52(A)(1)]. All 125-volt, 15- and 20-ampere receptacle outlets must be installed so that no point along the floor in any wall space is more than 6 ft from another receptacle outlet in the same space. Every wall space 2 ft or wider must have its own receptacle outlet. All other receptacle rules for general living spaces also apply and are discussed at greater length in Chapter 6, The Living Room.

GFCIs Not Required. Receptacle outlets that don't serve countertop spaces in the kitchen, dining room, pantry, and breakfast room aren't required to have GFCI protection for personnel. (Curiously, *all* 15- and 20-ampere, 125-volt receptacles installed in kitchens of buildings other than dwellings are required to be GFCI protected [210.8(B)(2)].)

Receptacle Ratings. Receptacles installed on 20-ampere small-appliance branch circuits should be rated 125 volts, 20 amperes, because it is likely that high-load appliances will be plugged into them. The subject of receptacle ratings is covered in greater detail in Chapter 8, The Bathroom.

LIGHTING AND SWITCHING IN THE KITCHEN AND DINING ROOM

Lighting

At least one lighting outlet with a fixed luminaire (lighting fixture) controlled by a wall switch must be installed in the kitchen [210.70(A)(1)]. Task lighting, such as additional luminaires over the sink and under the cabinets to illuminate countertop surfaces, is very common in kitchens. This permanently installed kitchen lighting must be supplied by general-purpose branch circuits, not by 20-ampere small-appliance circuits [210.52(B)(2)].

Cord-and-Plug-Connected Lighting. Although not installed during original construction, cord-and-plug-connected lighting is frequently used in kitchens. This type of lighting may include table lamps and office-style lamps used on countertops, swag-type decorative hanging lamps over a table, or small fluorescent fixtures installed under cabinets to provide brighter lighting. These types of lighting are permitted, like any other type of consumer appliance, to be plugged into kitchen receptacles supplied by 20-ampere branch circuits.

Switched Receptacles

In rooms other than kitchens, a receptacle outlet controlled by a wall switch may be used instead of a permanently installed lighting outlet. However, small-appliance branch circuits aren't permitted to serve lighting outlets [210.52(B)(2)].

For this reason, the dining room shown in Figure 7.1 has no lighting outlet supporting a permanently installed luminaire. Instead, it contains a switched receptacle outlet supplied by a 15-ampere general-purpose branch circuit [210.52(B)(1), Exception No. 1].

RANGES, WALL-MOUNTED OVENS, AND COUNTER-MOUNTED COOKING UNITS (COOKTOPS)

Wiring Methods

Ranges, wall-mounted ovens, and counter-mounted cooking units (cooktops) are permitted to be either hard-wired (permanently connected to the branch-circuit conductors) or cord-and-plug connected (Figure 7.6). When permanently connected, the branch-circuit breaker or fused switch is permitted to serve as the required disconnecting means where it is within sight of the appliance, or capable of being locked in the open position [422.16(B)(3), 422.31(B), 422.33(A)].

When cord-and-plug connections are used for cooking appliances, they must be accessible. Receptacles for household ranges are typically located at the rear base (i.e., at the wall near the floor), where they can be reached by removing a drawer at the bottom of the appliance [422.33(B)]. If the electric range doesn't have a removable drawer, the receptacle should be wall-mounted closer to the top of the appliance, where it can be reached by pulling the range away from the wall.

Branch-Circuit Ratings and Conductors

According to Table 220.55 of the *NEC*, shown in Exhibit 7.2, the maximum demand load is 8 kW for a single range, oven, or cooktop rated up to 12 kW. The minimum branch-circuit rating for an electric range rated 8 kW or more is 40 amperes [210.19(A)(3)]. For some larger electric ranges or ovens, 50-ampere branch circuits are required. In other words, most branch circuits serving electric cooking appliances are rated either 40 or 50 amperes at 240 volts and might include the following conductor ratings:

- Typical conductors for a 40-ampere circuit are three 8 AWG, Type TW copper or three 6 AWG, Type TW aluminum.

- Typical conductors for a 50-ampere branch circuit are three 6 AWG, Type TW copper or three 4 AWG, Type TW aluminum.

FIGURE 7.6 Receptacles rated 125/250 volts, 50 amperes, are often used to supply electric ranges. The receptacle on the left is for recessed mounting in a box. The receptacle on the right is surface mounted. (Courtesy of Pass & Seymour/Legrand®)

EXHIBIT 7.2

NEC **Table 220.55 Demand Factors and Loads for Household Electric Ranges, Wall-Mounted Ovens, Counter-Mounted Cooking Units, and Other Household Cooking Appliances over 1¾ kW Rating (Column C to be used in all cases except as otherwise permitted in Note 3.)**

| Number of Appliances | Demand Factor (Percent) (See Notes) | | Column C Maximum Demand (kW) (See Notes) (Not over 12 kW Rating) |
	Column A (Less than 3½ kW Rating)	Column B (3½ kW to 8¾ kW Rating)	
1	80	80	8
2	75	65	11
3	70	55	14
4	66	50	17
5	62	45	20
6	59	43	21
7	56	40	22
8	53	36	23
9	51	35	24
10	49	34	25
11	47	32	26
12	45	32	27
13	43	32	28
14	41	32	29
15	40	32	30
16	39	28	31
17	38	28	32
18	37	28	33
19	36	28	34
20	35	28	35
21	34	26	36
22	33	26	37
23	32	26	38
24	31	26	39
25	30	26	40
26–30	30	24	15 kW + 1 kW for each range
31–40	30	22	
41–50	30	20	25 kW + ¾ kW for each range
51–60	30	18	
61 and over	30	16	

1. Over 12 kW through 27 kW ranges all of same rating. For ranges individually rated more than 12 kW but not more than 27 kW, the maximum demand in Column C shall be increased 5 percent for each additional kilowatt of rating or major fraction thereof by which the rating of individual ranges exceeds 12 kW.
2. Over 8¾ kW through 27 kW ranges of unequal ratings. For ranges individually rated more than 8¾ kW and of different ratings, but none exceeding 27 kW, an average value of rating shall be calculated by adding together the ratings of all ranges to obtain the total connected load (using 12 kW for any range rated less than 12 kW) and dividing by the total number of ranges. Then the maximum demand in Column C shall be increased 5 percent for each kilowatt or major fraction thereof by which this average value exceeds 12 kW.
3. Over 1¾ kW through 8¾ kW. In lieu of the method provided in Column C, it shall be permissible to add the nameplate ratings of all household cooking appliances rated more than 1¾ kW but not more than 8¾ kW and multiply the sum by the demand factors specified in Column A or B for the given number of appliances. Where the rating of cooking appliances falls under both Column A and Column B, the demand factors for each column shall be applied to the appliances for that column, and the results added together.
4. Branch-Circuit Load. It shall be permissible to calculate the branch-circuit load for one range in accordance with Table 220.55. The branch-circuit load for one wall-mounted oven or one counter-mounted cooking unit shall be the nameplate rating of the appliance. The branch-circuit load for a counter-mounted cooking unit and not more than two wall-mounted ovens, all supplied from a single branch circuit and located in the same room, shall be calculated by adding the nameplate rating of the individual appliances and treating this total as equivalent to one range.
5. This table also applies to household cooking appliances rated over 1¾ kW and used in instructional programs.

Multiple Cooking Appliances on a Single Branch Circuit

The *Code* allows one cooktop and up to two wall-mounted ovens to be supplied from the same branch circuit when they are all located in the same room [Table 220.55, Note 4]. The maximum branch-circuit rating permitted is 50 amperes [210.19(A)(3), Exception No. 1].

Tap Conductors. Section 210.19(A)(3), Exception No.1 requires that the tap conductors to each individual cooking appliance must be rated a minimum of 20 amperes and cannot be longer than needed to service the appliance (Figure 7.7).

Reduced Neutral Conductor

The neutral conductor of a 120/240-volt, single-phase, 3-wire branch circuit supplying an electric range, wall-mounted oven, or cooktop can be smaller than the ungrounded (phase) conductors when the maximum demand is 8¾ kW, as computed according to Table 220.55, Column C, which is shown in Exhibit 7.2. The neutral ampacity must be at least 70 percent of the branch-circuit rating, and the conductor must be at least 10 AWG [210.19(A)(3), Exception No. 2].

Column C of Table 220.55 indicates that the maximum demand for one range rated 12 kW or less is 8 kW:

$$8 \text{ kW} = 8000 \text{ VA} \div 240 \text{ V} = 33.3 \text{ A}$$

Thus, the ungrounded range circuit conductors can be 8 AWG, Type TW copper, rated 40 amperes.

The neutral of this 3-wire circuit is permitted to be 10 AWG, which is smaller than 8 AWG but which has an ampacity (30 amperes) that is more than 70 percent

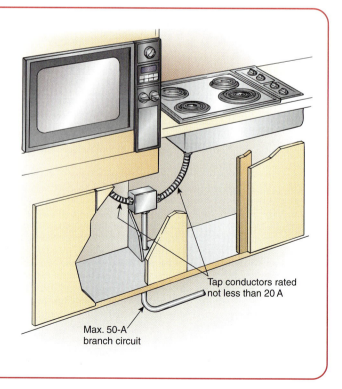

FIGURE 7.7 Tap conductors supplying electric ranges, wall-mounted electric ovens, and counter-mounted electric cooking units from a 50-ampere branch circuit must be rated not less than 20 amperes and cannot be longer than needed to service the appliance.

Tap conductors rated not less than 20 A

Max. 50-A branch circuit

of the 40-ampere rating of the phase conductors. The maximum neutral demand of a range circuit seldom exceeds 25 amperes. Current is drawn from the neutral only to power 120-volt lights, clocks, timers, and sometimes heating elements in their low-heat positions.

Microwave and Toaster Ovens

Microwave and toaster ovens are generally cord-and-plug connected.

Appliance Garage Receptacles. Receptacles located inside an appliance garage (such as for a microwave oven located above a range) or blocked by the appliance they serve aren't considered readily accessible. Therefore, they don't count as required countertop receptacles, even when located less than 20 in. above the counter surface [210.52(C)(5)].

Receptacles installed within an appliance garage cannot be supplied by the required 20-ampere small-appliance branch circuits [210.52(B)(2)]. They must be on separate branch circuits. Because they aren't considered kitchen countertop receptacles, receptacles installed in an appliance garage also aren't required to have GFCI protection [210.8(A)(6)].

Receptacle Ratings

Duplex receptacles are typically installed at most receptacle outlet locations in dwellings, even when only one receptacle is expected to be used. Both 15- and 20-ampere receptacles are permitted to be installed on branch circuits rated 15 and 20 amperes when those circuits supply two or more receptacles [210.21(B)(3)].

By contrast, a single receptacle on an individual branch circuit must have an ampere rating not less than that of the circuit [210.21(B)(1)]. This requirement means either a 15- or 20-ampere single receptacle can be installed on an individual branch circuit rated 15 amperes, but only a 20-ampere single receptacle can be installed on an individual branch circuit rated 20 amperes (Figure 7.8).

Gas Ranges and Ovens

Gas cooking appliances don't require branch circuits for power, and loads don't need to be included for them when determining service and panelboard sizes for

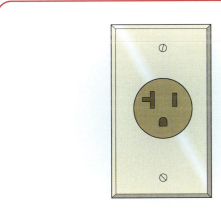

FIGURE 7.8 A single receptacle on a 20-ampere individual branch circuit must be rated 125 volts, 20 amperes. (NEMA configuration 5-20R is shown.)

dwellings. Gas ranges, ovens, and cooktops require small amounts of electric power for the operation of igniters, clocks, timers, and controls. These appliances normally have an attachment cord and plug that is connected to a receptacle on one of the required 20-ampere small-appliance branch circuits, as shown in Figure 7.1 [210.52(B)(2), Exception No. 2].

OTHER KITCHEN APPLIANCES

Kitchen Waste Disposers

Kitchen waste disposers are often cord-and-plug connected. They are equipped with grounding-type attachment plugs unless double-insulated, in which case they have 2-prong polarized plug caps [422.16(B)(1)]. The plug and receptacle are permitted to serve as the required disconnecting means for servicing the appliance [422.33(A)].

Waste disposer attachment cords are required to be 18 in. to 36 in. long. The reason for this product design requirement in the *NEC* is to avoid having a long cord tangled up with items such as cleaning supplies and pots and pans. Waste disposers are frequently plugged into a receptacle located below the kitchen sink (Figure 7.9).

The waste disposer receptacle is required to be accessible [422.16(B)(1)(4)] and isn't permitted to be supplied by one of the 20-ampere small-appliance branch circuits [210.52(B)(2)]. It isn't required to have GFCI protection for personnel, since it doesn't serve kitchen countertop surfaces [210.8(A)(6)].

Kitchen Waste Disposer Switch Control. Kitchen waste disposers are permitted to be controlled by a general-use snap switch, as long as the load doesn't exceed 80 percent of the switch's ampere rating [404.14(A)(3)]. Most residential models are rated ½ horsepower or less, about 3 amperes or less at 120 volts, which is well within the allowable 12 amperes for a load controlled by a 15-ampere snap switch. A general-use snap switch can be installed to control the receptacle to which a plug-

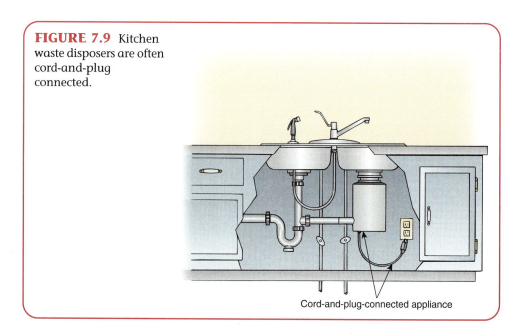

FIGURE 7.9 Kitchen waste disposers are often cord-and-plug connected.

Cord-and-plug-connected appliance

in-type kitchen waste disposer is connected. Other waste disposers are permanently connected, as shown in Figure 7.10.

Dishwashers and Trash Compactors

Built-in dishwashers and trash compactors are often cord-and-plug connected. They are equipped with grounding-type attachment plugs unless double-insulated, in which case they have 2-prong polarized plug caps [422.16(B)(2)]. The plug and receptacle are permitted to serve as the required disconnecting means for servicing the appliance [422.33(A)].

The attachment cords are required to be 3 ft to 4 ft long, and the receptacle is required to be located in the space occupied by, or adjacent to, the dishwasher or trash compactor. The reason for these two requirements is that dishwashers and trash compactors frequently are plugged into the same duplex receptacle as a kitchen waste disposer, with the attachment cord passing through a thin partition beneath the sink.

Receptacles for dishwashers and trash compactors are required to be accessible [422.16(B)(2)(5)] and aren't permitted to be supplied by the 20-ampere small-appliance branch circuits [210.52(B)(2)]. They aren't required to have GFCI protection, since they don't serve kitchen countertop surfaces [210.8(A)(6)].

Switch Control. Dishwashers and trash compactors normally have operating controls on the units themselves. So general-use snap switches and/or switch-controlled receptacles aren't required to turn these appliances on and off.

Microwave Ovens and Range Hoods

Most new kitchens have built-in microwave ovens. Sometimes these are combined with range hoods that contain exhaust fans; in other dwellings the microwave oven

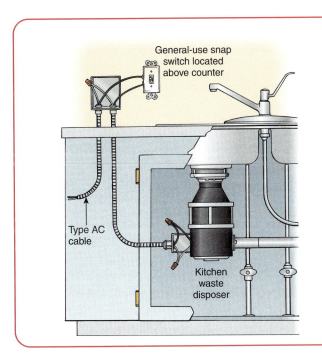

General-use snap switch located above counter

Type AC cable

Kitchen waste disposer

FIGURE 7.10 Some household kitchen waste disposers are hard-wired using Type AC or Type NM cable.

and range hood are separate. The *Code* rules for supplying power to microwave ovens and range hoods in residential kitchens are as follows.

Microwave Ovens. Some built-in microwave ovens are hard-wired on an individual branch circuit. Microwave ovens equipped with a cord and plug should be supplied with an individual branch circuit feeding a single receptacle, for the following reasons.

  • *General-Purpose Branch Circuits.* The load of utilization equipment fastened in place cannot exceed 50 percent of the rating of a branch circuit that also supplies lighting units, cord-and-plug-connected equipment not fastened in place, or both [210.23(A)(2)]. Since microwave ovens are typically rated 12 amperes or more, this means they can't be connected to general-purpose branch circuits.
  • *Small-Appliance Branch Circuits.* Likewise, microwave ovens aren't permitted to be supplied by the small-appliance branch circuits required to supply the receptacle outlets for countertop surfaces in dwelling unit kitchens [210.52(B)(2)].

Combination Microwave Ovens/Range Hoods. Wiring for these kitchen appliances follows the same *NEC* requirements as microwave ovens. As previously noted, only a 20-ampere single receptacle is permitted on an individual branch circuit rated 20 amperes. (See Figure 7.8.)

Range Hoods. Range hoods contain an exhaust fan and integral luminaire or lamp holder. Many are hard-wired; but they are also permitted to be cord-and-plug connected under the following conditions [422.16(B)(4)]:

  • The flexible cord is 18 in. to 36 in. long.
  • The single receptacle is accessible and located to avoid damage to the cord.
  • The single receptacle is supplied by an individual branch circuit.

SPECIAL SYSTEMS

Kitchens are increasingly a "nerve center" or principal gathering place in modern homes. Some kitchens include space for a home computer and fulfill much the same function as a small home office. For this reason, kitchens may include some or all of the following special systems:

  • *Telephone.* Traditionally, most kitchens have had a wall-mounted voice telephone outlet, but today that outlet might be supplemented by a second telephone outlet located at the computer countertop space to provide modem or DSL (digital subscriber line) access to the Internet.
  • *Security/Home Management.* Many homes with security or home management systems locate one of the keypads in the kitchen.
  • *Cable TV.* Many homeowners watch television in the kitchen, which means modern kitchen wiring may include a coaxial outlet for cable or satellite TV.
  • *Broadband.* A second coaxial outlet may be needed in the kitchen computer area for Internet access.

For more information about installing low-voltage communications outlets and control devices, see Chapter 15, Special Systems.

SPECIAL RULES FOR TWO-FAMILY DWELLINGS

There are no special *Code* rules that apply to wiring kitchens in two-family dwellings. Considerations that apply to one-family homes with two kitchens are covered in the section of this chapter entitled "Small-Appliance Branch Circuits."

MULTIPLE CHOICE

1. Receptacle outlets for refrigerators are allowed to be supplied by
   A. A small-appliance branch circuit
   B. An individual branch circuit rated 15 amperes
   C. An individual branch circuit rated 20 amperes
   D. All of the above

2. A small-appliance branch circuit is permitted to supply receptacle outlets in how many kitchens?
   A. One
   B. Two
   C. Not specified in the *Code*
   D. None of the above

3. Receptacles serving kitchen countertop areas are
   A. Required to be accessible
   B. Prohibited from being mounted in the face-up position
   C. Required to have GFCI protection for personnel
   D. All of the above

4. Small-appliance branch circuits are permitted to serve lighting outlets only in
   A. Kitchens
   B. Dining rooms
   C. Breakfast rooms
   D. None of the above

5. An appliance branch circuit is a branch circuit that
   A. Does not supply luminaires
   B. Supplies energy to one or more outlets to which appliances are to be connected
   C. Is protected by a dedicated overcurrent protection device
   D. Both A and B

6. Which statement about receptacle ratings is false?
   A. Both 15- and 20-ampere receptacles are permitted to be installed on branch circuits rated 15 and 20 amperes.
   B. Either a 15- or 20-ampere single receptacle can be installed on an individual branch circuit rated 15 amperes.
   C. Only a 20-ampere single receptacle is permitted to be installed on a branch circuit rated 20 amperes.

   D. Receptacle rating is not dependent upon branch-circuit rating.

7. In addition to supplying receptacle outlets serving kitchen countertop areas, small-appliance branch circuits are permitted to supply the following:
   A. Electric clock receptacles
   B. Receptacles that provide power to the electric equipment of gas ranges, ovens, and cooking units
   C. Receptacles for refrigerator/freezers
   D. All of the above

8. Switched receptacles are not permitted to serve as the required lighting outlet in the following room(s):
   A. Pantry
   B. Breakfast room
   C. Kitchen
   D. Dining room

9. Kitchen waste disposers and other motor loads are permitted to be controlled by a general-use snap switch under the following conditions:
   A. The switch is rated in horsepower.
   B. The load doesn't exceed 80 percent of the switch's rating.
   C. The switch is ac/dc as specified in Article 404.
   D. The disposer is cord-and-plug connected.

10. Which statement about built-in dishwashers and trash compactors with attachment cords and plugs is true?
    A. The plug and receptacle are permitted to serve as the required disconnecting means for servicing the appliance.
    B. The attachment cords are required to be 18 in. to 36 in. long.
    C. The receptacle must be located in the space occupied by the dishwasher or trash compactor.
    D. Both A and B.

FILL IN THE BLANKS

1. Kitchen receptacle outlets not serving countertop areas are required to be located according to the rules of _____.

2. _____ consisting of either two or three ungrounded conductors plus a grounded (neutral) con-

ductor cannot be used to supply GFCI-protected receptacle outlets.

3. Each kitchen wall space measuring _____ in. or wider must have a receptacle outlet installed.

4. Receptacle outlets located more than _____ in. above a kitchen countertop are not considered to be receptacles serving countertop surfaces.

5. Most branch circuits serving electric cooking appliances are rated either _____ amperes at 240 volts.

6. Built-in dishwashers and trash compactors aren't permitted to be supplied by _____ branch circuits.

7. Receptacle outlets serving _____ are required to have GFCI protection.

8. All 125-volt, 15- and 20-ampere receptacle outlets in dining rooms must be installed so that no point along the floor in a wall space is more than _____ from another receptacle outlet in the same space.

9. Receptacles aren't permitted to be installed in a _____ position in a countertop or similar surface, in order to prevent them from collecting liquid, crumbs, and other debris.

10. A countertop area that stands in the middle of a kitchen and isn't connected to another kitchen countertop is known as an _____.

TRUE OR FALSE

1. Kitchen counter peninsulas 48 in. × 12 in. or larger are required to have receptacle outlets located so that no point along the edge of the peninsula is more than 24 in., measured horizontally, from a receptacle outlet.

_____ True _____ False

2. All receptacle outlets located in kitchens are required to have GFCI protection for personnel.

_____ True _____ False

3. Switched receptacle outlets used in place of lighting outlets are permitted to be supplied by the small-appliance branch circuit in all rooms of dwellings except kitchens, as permitted by 210.70(A)(1).

_____ True _____ False

4. The neutral conductor of a 120/240-volt, single-phase, 3-wire branch circuit supplying an electric range, wall-mounted oven, or cooktop can be smaller than the ungrounded (phase) conductors when the maximum demand is 8¾ kW, as computed according to *NEC* Table 220.55, column C.

_____ True _____ False

5. An "appliance garage" is a dedicated space for storing non-fixed-in-place appliances when they are not in use.

_____ True _____ False

6. A receptacle that supplies power for the operation of igniters, clocks, timers, or controls on a gas cooking appliance is permitted to be connected to a 20-ampere small-appliance branch circuit.

_____ True _____ False

7. Range hoods and kitchen exhaust fans rated not more than ⅛ horsepower are permitted to be supplied by kitchen small-appliance branch circuits.

_____ True _____ False

8. Electric cooking appliances with a total demand exceeding 60 amperes are constructed so that they can be supplied by two branch circuits, each rated a maximum of 40 amperes.

_____ True _____ False

9. Ranges, wall-mounted ovens, and cooktops are permitted to be either hard-wired (permanently connected to the branch-circuit conductors) or cord-and-plug connected.

_____ True _____ False

10. Each wall along a counter space measuring 12 in. or wider must have a receptacle outlet.

_____ True _____ False

CHALLENGE QUESTIONS

1. Receptacle outlets serving kitchen countertop surfaces are required to have GFCI protection. Discuss why the *Code* doesn't mandate GFCI protection for other receptacle outlets on 20-ampere small-appliance branch circuits.

2. Most general-purpose branch circuits in dwellings are rated 15 amperes. Why does the *Code* require 20-ampere branch circuits for receptacle outlets in

kitchens, dining rooms, breakfast rooms, pantries, and similar areas of dwellings?

3. Clock hanger outlets are primarily installed in kitchens. What do you think is the primary reason that they have only a single receptacle, rather than duplex receptacles?

4. Small-receptacle branch circuits are rated 20 amperes. What are at least two reasons why 15-ampere receptacles are typically installed on these branch circuits?

5. Explain the difference between *connected loads* of household cooking appliances and *demand loads* for appliances rated over 1¾ kW, as defined in Table 220.19.

6. When multiple cooking appliances located in the same room are supplied by a single branch circuit, NEC 210.19(A)(3), Exception No. 1, requires that the tap conductors to each individual cooking appliance be rated 20 amperes or more and not be longer than needed to service the appliance. What is the reason for this latter requirement?

7. Explain why general-use snap switches are installed to control kitchen waste disposers but are not generally provided for dishwashers and trash compactors.

8. When cord-and-plug connections are used for kitchen cooking appliances, they are required to be accessible. What's the reason for this *Code* requirement?

The Bathroom

8

OBJECTIVES

After completing this chapter, the student will be able to understand the following:

- Receptacle outlets required in bathrooms
- Ground-fault circuit-interrupter (GFCI) protection for personnel
- Rules for locations of lighting, switches, and receptacles in bathrooms
- Requirements for installing hydromassage tubs
- Installation of exhaust fans and heat lamps in bathrooms

INTRODUCTION

This chapter covers the special *National Electrical Code*® rules that apply to an increasingly important area of modern homes, the bathroom, where the presence of moisture increases the risk of electric shock. The following major topics are included:

- Receptacle rules for bathrooms
- Ground-fault circuit interrupters
- Bathroom lighting
- Other rules for bathrooms
- Additional shock protection requirements
- Special rules for two-family dwellings

IMPORTANT NEC TERMS

Bathroom

Ground-Fault Circuit Interrupter

Hydromassage Bathtub

Lighting Outlet

Location, Damp

Location, Wet

Luminaire

Outlet

Receptacle

Receptacle Outlet

Switch, General-Use Snap

RECEPTACLE RULES FOR BATHROOMS

Combining electricity with water produces a powerful potential for electrocution. For this reason, the *NEC®* has a number of specific rules for receptacles used in areas of dwellings subject to water or moisture, including bathrooms. Article 100 defines a bathroom as follows.

> **Bathroom:** An area including a basin with one or more of the following: a toilet, a tub, or a shower.

This definition also covers powder rooms, which contain a basin and toilet. Thus, powder rooms must follow all *NEC* rules for wiring bathrooms.

Bathroom Branch Circuit

In dwelling units, at least one 20-ampere branch circuit must be provided to supply bathroom receptacle outlet(s) [210.11(C)(3)]. This circuit can supply receptacle outlets in more than one bathroom. But it isn't permitted to have any other outlets such as bathroom lighting, bathroom exhaust fan, or lights and receptacles in other rooms.

Circuit Supplying Single Bathroom. Where the 20-ampere circuit supplies only one bathroom, outlets for other equipment within that same bathroom, such as lighting or an exhaust fan, *are* permitted to be supplied by the bathroom branch circuit [210.1(C)(3), Exception].

However, some installers think it isn't a good idea to put other outlets on the same circuit with GFCI-protected bathroom receptacles. For example, a malfunctioning bathroom wall heater could cause a GFCI circuit breaker to trip. Similarly, bathroom lighting supplied through a GFCI-protected receptacle would go out if the GFCI tripped and thus increase the danger of falling for people in a tub or shower.

Required Receptacle Location(s)

At least one receptacle outlet is required within 3 ft of the outside edge of each basin (sink). The receptacle outlet must be located on a wall or partition adjacent to the basin or countertop [210.52(D)], as shown in Figure 8.1. Where two basins are located close together, a single receptacle outlet may be able to satisfy this requirement (Figure 8.2).

Receptacles on Vanities. A receptacle outlet installed on the wall or face of a basin cabinet (vanity) not more than 12 in. below the countertop also meets *Code* [210.52(D), Exception]. This rule allows alternative placement in bathrooms with mirrored walls, which may not have sufficient empty wall space to install a receptacle within 3 ft of the basin.

Other Bathroom Receptacles. Additional receptacles can also be installed in bathrooms. However, an outlet on a wall that isn't adjacent to the basin or built into a vanity doesn't meet the required location rule (for example, a receptacle outlet located on the wall *opposite* the sink in a very small powder room). A receptacle built into

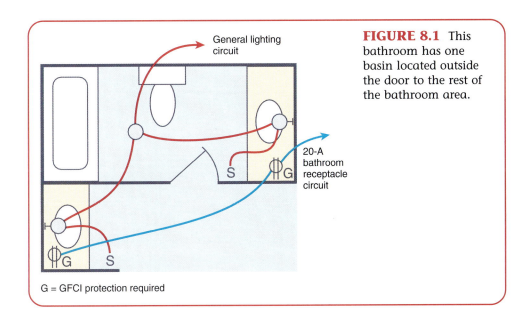

FIGURE 8.1 This bathroom has one basin located outside the door to the rest of the bathroom area.

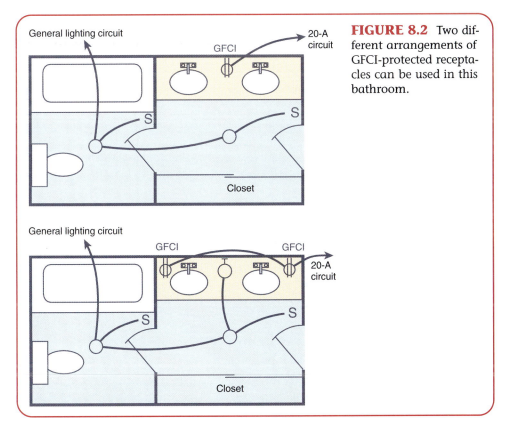

FIGURE 8.2 Two different arrangements of GFCI-protected receptacles can be used in this bathroom.

a medicine chest or installed inside a vanity also fails to meet the location requirement (Figure 8.3).

Prohibited in Bathtub and Shower Spaces. Receptacle outlets are prohibited within, or directly over, a bathtub or shower stall [406.8(C)]. This rule helps prevent the use of items such as shavers, radios, and hair dryers in wet locations of bathrooms.

FIGURE 8.3 A receptacle in the location shown doesn't satisfy *NEC* 210.52(D) because it isn't "on a wall or partition that is adjacent to the basin or basin countertop."

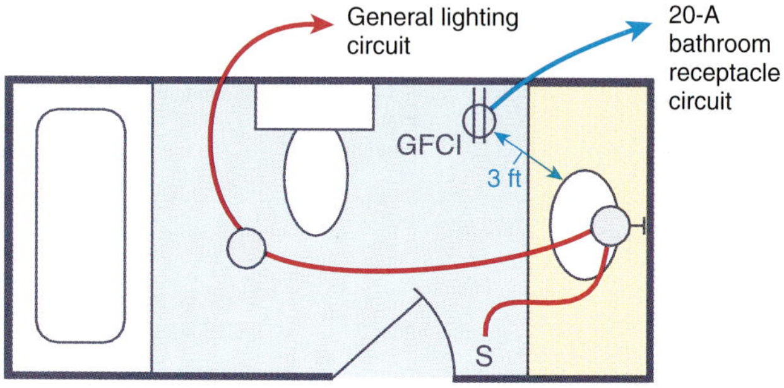

Even GFCI-protected receptacles aren't allowed to be installed in tub and shower spaces. The reason for this prohibition is that the unprotected line side conductors of GFCI-protected receptacles installed in bathtub and shower spaces could possibly become wet and create a shock hazard by energizing surrounding wet surfaces.

Section 406.8(C) doesn't define the dimensions of a "bathtub or shower stall." However, a safe approach is to follow 410.4(D), which prohibits the installation of several types of luminaires (and ceiling-suspended fans) within a zone measured 8 ft vertically and 3 ft horizontally from the bathtub rim or shower stall threshold (Figure 8.4).

Bathroom Receptacle Rating

Receptacles installed in bathrooms should always be rated 20 amperes. GFCI receptacles come in both 20- and 15-ampere models. If conventional receptacles are used in conjunction with a 20-ampere GFCI circuit breaker, then those conventional receptacles should also be rated 20 amperes for the reasons that follow.

- As explained in Chapter 6, The Living Room, *NEC* Table 210.21(B)(3) permits both 15- and 20-ampere receptacles to be supplied by 20-ampere circuits.
- *NEC* Table 210.21(B)(2), shown in Exhibit 8.1, specifies that when a branch circuit supplies two or more receptacles or outlets, a 15-ampere receptacle is

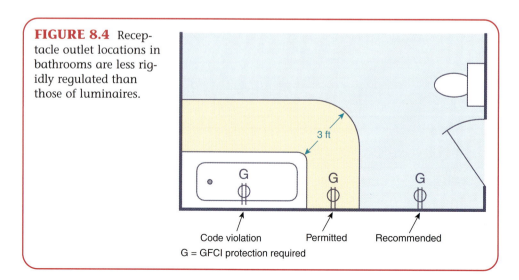

FIGURE 8.4 Receptacle outlet locations in bathrooms are less rigidly regulated than those of luminaires.

Circuit Rating (Amperes)	Receptacle Rating (Amperes)	Maximum Load (Amperes)
15 or 20	15	12
20	20	16
30	30	24

EXHIBIT 8.1

NEC TABLE 210.21(B)(2) Maximum Cord-and-Plug-Connected Load to Receptacle

not permitted to supply a total cord-and-plug-connected load exceeding 12 amperes (1800 volt-amperes).

• Since listed hair dryers rated up to 1875 watts are frequently used in bathrooms, all bathroom receptacles should be the 20-ampere type. *NEC* Table 210.21(B)(2) indicates that 20-ampere devices can supply cord-and-plug-connected loads up to 16 amperes (1920 volt-amperes).

NEC Table 210.21(B)(2), shown in Exhibit 8.1, lists maximum cord-and-plug-connected loads, such as hair dryers and curling irons, to a given receptacle.

GFCI Protection of Bathroom Receptacles

All 125-volt, 15- and 20-ampere receptacles installed in bathrooms must have ground-fault circuit-interrupter protection for personnel (Figure 8.5) [210.8(A)(1)]. No other outlets are permitted on the required 20-ampere bathroom branch circuit

FIGURE 8.5 This 15-ampere duplex receptacle has an integral GFCI that also protects downstream loads. (Courtesy of Leviton)

[210.11(C)(3)]. So when a GFCI circuit breaker or single GFCI receptacle is used to protect multiple downstream receptacles, they can *only* be receptacle outlets in bathrooms (including powder rooms).

GFCI-protected receptacle outlets in other locations such as kitchens, basements, garages, and outdoors aren't permitted to be on the same branch circuit with bathroom receptacles.

Circuit Supplying Single Bathroom. When a 20-ampere bathroom branch circuit supplies just one bathroom, outlets for other equipment within that bathroom are permitted to be on the same circuit [210.11(C), Exception]. For example, bathroom lighting, heater, and exhaust fan outlets are permitted to be on the same circuit with GFCI-protected receptacles if, and only if, the branch circuit supplies only that bathroom.

GROUND-FAULT CIRCUIT INTERRUPTERS

Article 100 defines a ground-fault circuit interrupter as follows.

> **Ground-Fault Circuit Interrupter (GFCI):** A device intended for the protection of personnel that functions to de-energize a circuit or portion thereof within an established period of time when a current to ground exceeds the values established for a Class A device.

This definition may seem a bit confusing, unless you know something about the history of GFCIs.

Background

GFCI circuit breakers were first introduced into the *Code* in 1968 to automatically disconnect underwater pool lighting fixtures if they experienced a fault to ground that could electrocute swimmers. These early swimming pool GFCI devices tripped at 20 milliamperes (mA). When GFCI circuit breakers and duplex receptacles for personnel protection were introduced a few years later, they were designed to trip at 5 milliamperes.

These new 5-milliampere GFCIs were designated "Class A" in the listing standard UL 943, and the older 20-milliampere units were designated "Class B." Class B GFCIs are no longer manufactured. All GFCI circuit breakers and receptacles sold today are the Class A type—including those used near swimming pools. (For more information, see Chapter 13, Swimming Pools, Hot Tubs, and Spas.)

How GFCIs Work

Figure 8.6 illustrates typical GFCI circuitry. The two line conductors pass through a sensor and are connected to a shunt-trip device. Under normal operating conditions, the currents in the line and grounded (neutral) conductors are equal. If one of the conductors comes into contact with a grounded object—either directly or through a person's body—the toroidal (doughnut-shaped) coil senses the unbalanced current and sends a signal to the shunt-trip device to open the circuit. GFCIs trip at a nominal

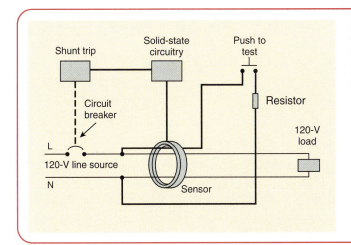

FIGURE 8.6 GFCIs work by sensing current unbalance between the line and grounded (neutral) conductors.

current imbalance of 5 milliamperes (the listing standards actually permit a range of 4 to 6 milliamperes).

Limitations of GFCIs. As important as GFCI protection is, it isn't foolproof. Although it opens the circuit quickly enough to prevent electrocution of a healthy adult, a GFCI doesn't prevent the sensation of receiving a shock. This limitation means that a shocked person could still be injured by slipping or falling while disoriented. Similarly, GFCIs only protect against phase-to-ground faults. They don't protect against faults between phase and grounded (neutral) conductors, or between two different phases.

No Green Wire Needed. The design of GFCI circuitry doesn't require the presence of an equipment grounding conductor. For this reason, 406.3(D)(3)(b) permits GFCIs to be used as replacements for conventional receptacles in older construction, where a grounding means doesn't exist in the wiring method.

Methods of Protecting Receptacles with GFCIs

There are three different ways to provide what the *NEC* calls "ground-fault circuit-interrupter protection for personnel" as it applies to receptacles:

1. GFCI circuit breaker
2. GFCI receptacle at each outlet
3. GFCI receptacles protecting downstream outlets

Each of these methods has its pros and cons, which are discussed in the sections that follow.

GFCI Circuit Breaker. This method can be used to protect entire branch circuits of receptacle outlets requiring GFCI protection, such as the required bathroom branch circuit or any circuit of miscellaneous outdoor/basement/crawl space/garage receptacles that are required by the *Code* to have GFCI protection.

PRO: Using a single circuit breaker to protect multiple receptacle outlets offers low initial cost, in terms of buying GFCI hardware. It's also fairly user-friendly, since the average homeowner knows to check the panelboard when a piece of electrical equipment isn't working.

CON: Using a single circuit breaker to protect multiple receptacles located in different parts of the house may require longer wiring runs to connect widely separated outlets together.

GFCI Receptacle at Each Outlet. This method, which is very common nowadays, involves installing a GFCI receptacle at each outlet requiring protection.

PRO: This approach is the most user-friendly, since homeowners can see immediately when a GFCI receptacle has tripped. It also requires no extra wiring. Receptacles can be circuited together in the most efficient, logical manner, providing GFCI protection only at those outlets where required by the Code.

CON: Installing a GFCI receptacle at each outlet results in a higher initial cost.

GFCI Receptacles Protecting Downstream Outlets. In this method, GFCI receptacles provide feed-through protection to ordinary receptacles in other bathrooms that are located "downstream" from the GFCI receptacle. This method is probably the least common method used in new construction.

PRO: Using one GFCI receptacle to protect multiple receptacle outlets offers low initial cost in terms of buying GFCI hardware.

CON: Using a single GFCI receptacle to protect multiple receptacles located in different parts of the house may require longer wiring runs to connect widely separated outlets together. Also, this approach is the least user-friendly of all three methods. Most homeowners can see and push the RESET button on a tripped GFCI receptacle. Similarly, most people understand that they should check their panelboard when a piece of electrical equipment has been de-energized. However, very few homeowners know enough about ground-fault circuit-interrupter technology to understand that they should check the GFCI receptacle outlet in the powder room when their master bath receptacle doesn't seem to be working.

Labeling. If a GFCI receptacle with a visible test button is *not* installed at each required outlet, a label that reads "GFCI-Protected Outlet" or equivalent wording should be applied to the face of the device. These labels are supplied with GFCI circuit breakers and receptacles.

BATHROOM LIGHTING

Areas of bathrooms outside of the tub or shower space aren't considered damp or wet locations, and luminaires (lighting fixtures) installed in these spaces aren't required to be listed for special environmental conditions. All sorts of decorative luminaires are commonly used in bathrooms, including wall sconces and fixtures with exposed, decorative lamps. The special rules for installing lighting in bathrooms may be summarized as follows.

- *Bathtub and Shower Spaces.* As noted in Chapter 5, Lighting the Home, hanging luminaires (lighting fixtures), pendant lighting, lighting track, ceiling-sus-

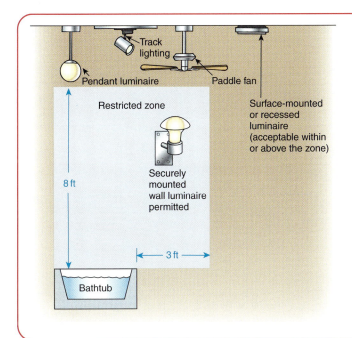

Track lighting

Pendant luminaire

Paddle fan

Restricted zone

Surface-mounted or recessed luminaire (acceptable within or above the zone)

Securely mounted wall luminaire permitted

8 ft

3 ft

Bathtub

FIGURE 8.7 Luminaires, lighting tracks, and ceiling-suspended (paddle) fans located near a bathtub must be out of reach of an individual standing on a bathtub rim. The restricted zone shown is defined in *NEC* 410.4(D).

pended (paddle) fans, and similar electrical items are prohibited within a "bathtub and shower zone" measured 8 ft vertically and 3 ft horizontally from the bathtub rim or shower stall threshold. The intent of this rule is to keep cord-connected, hanging, or pendant luminaires from coming into contact with a person standing on a bathtub rim. However, surface-mounted or recessed luminaires are permitted within the bathtub zone (Figure 8.7), because they don't present a significant shock hazard.

NOTE: Although the NEC has no special environmental rules for luminaires (lighting fixtures) installed in bathrooms, packaged bathtub or shower units with integral lighting do have luminaires listed for use in damp and wet locations. When surface-mounted or recessed luminaires within the bathtub or shower zone are installed in the field, it's a good idea to check with the authority having jurisdiction (AHJ) for any special local requirements.

• *Type IC Recessed Ceiling Luminaires.* Section 410.66(B) requires that recessed luminaires (lighting fixtures) installed in bathroom ceilings, with thermal insulation above them, be Type IC or that the installation be arranged so that no thermal insulation is installed within 3 in. of the fixture, wiring compartment, or ballast (Figure 8.8). Typically, this rule only applies to bathrooms on the top floor of a dwelling, directly below the attic.

• *Prohibited on Receptacle Circuit.* As explained earlier in this chapter, *NEC* rules generally don't permit lighting outlets to be supplied by the required 20-ampere branch circuit for bathroom receptacles. Locating bathroom lights and receptacles on separate circuits may actually contribute to safety: If a GFCI circuit breaker were to trip at night because a homeowner accidentally dropped an electric shaver into a bathroom basin filled with water, the lights would stay on, thus reducing the chances of slipping or falling in the dark.

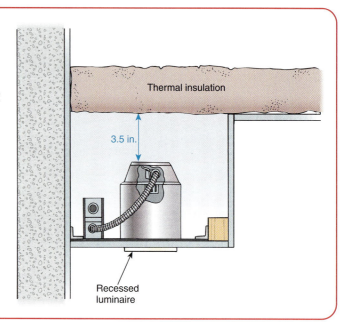

FIGURE 8.8 This recessed luminaire installed in a decorative soffit above a bathroom mirror does not have to be rated Type IC because it is at least 3 in. away from thermal insulation.

Thermal insulation

3.5 in.

Recessed luminaire

OTHER RULES FOR BATHROOMS

Switch Rules

There are few *National Electrical Code* rules that deal specifically with wall switches installed in bathrooms. A bathroom is required to have at least one wall switch-controlled lighting outlet [210.70(A)(1)]. Many bathrooms also have additional wall switches for controlling heat lamps and exhaust fans.

Prohibited in Tub and Shower Spaces. Switches are prohibited "within wet locations in tub or shower spaces unless installed as part of a listed tub or shower assembly" [404.4]. Although *NEC* 404.4 doesn't define the dimensions of this prohibited zone, a safe approach is to follow 410.4(D), which prohibits installation of several types of luminaires within a zone measured 8 ft vertically and 3 ft horizontally from the bathtub rim or shower stall threshold.

Exhaust Fans and Heat Lamps

Many bathrooms have ceiling-mounted heat lamps, exhaust fans to remove humid air, or both. Sometimes the heat lamp and fan are combined into a single unit (Figure 8.9). Typically, these appliances are controlled by conventional wall switches, but sometimes a timer switch is installed for the heat lamp. The primary reason for using a timer switch is to prevent this high-heat, high-load item from being left ON indefinitely.

Switches Controlling Motor Loads. Alternating-current snap switches are the type most often used for general applications today, particularly in dwellings. The *NEC* permits general-use snap switches to control motor loads that do not exceed 80 percent of their rating [404.14(A)(3)]. Thus, a 15-ampere switch operating at 120 volts can handle a motor load of 1440 volt-amperes (120 volts × 15 amperes ×

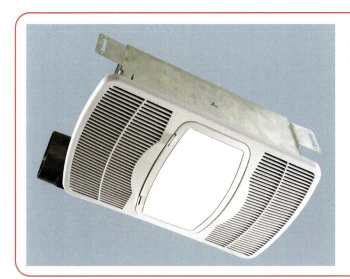

FIGURE 8.9 This ceiling-mounted unit combines a heat lamp with an exhaust fan. (Courtesy of Air King Ventilation Products)

80 percent). A typical small exhaust fan used in a bathroom is rated ⅛ to ¼ horsepower (approximately 100 to 200 watts), well within the rating of the snap switch.

Hydromassage Tubs

Increasing numbers of new homes, especially upscale homes, have indoor hydromassage tubs installed in bathrooms or bedrooms. Often these tubs are known as "whirlpool baths." Hydromassage tubs are similar in some ways to spas and hot tubs, since all are equipped with electric pumps to circulate water. But there is a significant difference as well. Spas and hot tubs, like swimming pools, are designed to remain filled with water for extended periods and have filtering and heating systems. Hydromassage tubs, like ordinary bathtubs, are intended to be filled, used, and then drained after each use. Electrical requirements for installing hydromassage bathtubs are as follows.

GFCI Protection. Hydromassage bathtubs and any associated equipment are required to have GFCI protection. Units for home use are typically rated ½ horsepower or less at 120 volts and require an individual branch circuit protected by a 15- or 20-ampere, single-pole GFCI circuit breaker installed according to the manufacturer's instructions. The electrical connection is made at a junction box or power panel located within an access panel (Figure 8.10).

All 125-volt, single-pole receptacles not exceeding 30 amperes must have GFCI protection if located within 5 ft horizontally of the inside walls of a hydromassage tub. (In bathrooms, of course, all receptacles are required to be GFCI protected.)

Bonding. Pump motors, metal piping, and metal parts of electrical equipment associated with hydromassage tubs must be bonded together using a minimum 8 AWG solid copper conductor. Normally, the hydromassage tub is furnished as a packaged unit with these bonding conductors already installed at the factory. During installation, however, the tub assembly must be bonded to the interior water supply piping system [680.74].

Luminaires, Switches, and Receptacles. Section 680.72 requires that other electrical equipment not directly associated with a hydromassage tub comply with normal

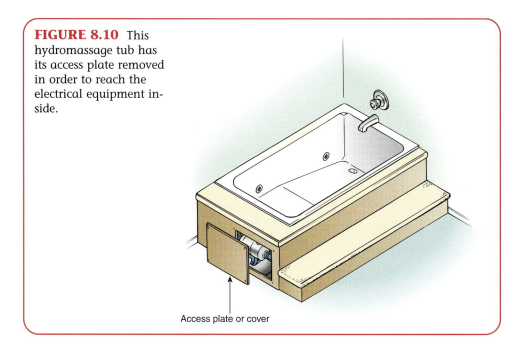

FIGURE 8.10 This hydromassage tub has its access plate removed in order to reach the electrical equipment inside.

Access plate or cover

Code rules for bathrooms. This requirement means that luminaires, switches, and receptacles cannot be installed within the hydromassage bathtub space.

ADDITIONAL SHOCK PROTECTION REQUIREMENTS

Most rules in the *National Electrical Code* deal with electrical construction. Requirements for GFCI protection, overcurrent protection, proper securing of wiring methods, and many other topics are intended to insure that permanently installed electrical products and systems operate safely. However, *Code* safety rules aren't limited to permanently installed equipment. Section 90.1 makes the following statement.

> **(A) Practical Safeguarding.** The purpose of this *Code* is the practical safeguarding of persons and property from hazards arising from the use of electricity.

Some *NEC* rules govern the construction and operation of consumer products as well as electrical safety at construction sites.

Product Construction Requirements

The *Code* contains many rules dealing with electrical products, such as the requirement in 422.41 that portable appliances likely to be used in bathrooms come equipped with immersion-detector circuit interrupters (IDCIs). These non-installation requirements in the *NEC* provide guidance to product manufacturers and are incorporated into UL listing standards for the products manufactured. Sections 90.7 and 110.3 of the *Code* provide detailed information about testing and listing of electrical equipment.

Immersion-Detector Circuit Interrupters (IDCIs)

Although receptacles in residential bathrooms have been required to have GFCI protection since the 1975 edition of the *National Electrical Code*, many bathroom receptacles in older houses aren't GFCI protected. For this reason, 422.41 requires that hand-held hair dryers and portable hydromassage units (which are used to circulate water in ordinary bathtubs) be equipped with "protection for personnel against electrocution when immersed." Some other types of personal grooming appliances, such as electric hair stylers and curling irons, are also manufactured with integral shock protection.

The most common type of integral shock protection device is an IDCI. IDCIs are normally built into the attachment plug cap of a hair dryer or other appliance. Like ground-fault circuit interrupters, IDCIs open the circuit when they detect leakage current to ground in the range of 4 to 6 milliamperes. Like GFCIs, IDCIs have TEST and RESET buttons.

IDCI Operation. IDCIs aren't identical to portable in-line GFCIs of the type often used with cord-and-plug-connected electrical tools. They work somewhat differently than GFCIs. Operation of an IDCI depends on a special wire in the cordset that is connected to a sensor inside the appliance. This sensor detects leakage current if conductive liquid enters the appliance and makes contact with live parts. When contact is made, the IDCI trips and interrupts the circuit.

No Panelboards in Bathrooms

Panelboards cannot be located in bathrooms of dwellings. *NEC* 230.70(A)(3) prohibits installing service disconnecting means in bathrooms, and 240.24(E) prohibits installation of overcurrent protection devices in dwelling unit bathrooms.

SPECIAL RULES FOR TWO-FAMILY DWELLINGS

There are no special *Code* rules that apply to bathrooms in two-family dwellings. Residential branch circuits are only permitted to supply loads located within, or associated with, each individual dwelling unit [210.25]. Also, occupants are required to have "ready access" to the overcurrent devices protecting conductors in their occupancy [240.24(B)]. Taken together, these rules mean that the bathrooms in a two-flat dwelling cannot be wired together on a single branch circuit to economize on GFCIs.

8 CHAPTER REVIEW

MULTIPLE CHOICE

1. The following method(s) can be used to provide GFCI protection for receptacle outlets installed in bathrooms:
   A. GFCI circuit breaker
   B. GFCI receptacle at each outlet
   C. GFCI receptacle providing feed-through protection
   D. All of the above

2. Ground-fault circuit interrupters of the type used to protect bathroom receptacle outlets are known as
   A. Class A
   B. Class B
   C. UL 943
   D. Toroidal

3. Table 210.21(B)(2) specifies that when a branch circuit supplies two or more receptacle outlets, a 15-ampere receptacle is not permitted to supply a total cord-and-plug-connected load exceeding
   A. 10 amperes
   B. 1440 volt-amperes
   C. 12 amperes
   D. 15 amperes

4. The minimum number of receptacle outlets required in a bathroom is
   A. One
   B. Two
   C. One for each basin (sink)
   D. None, but if provided, GFCI protection for personnel is required

5. The easiest GFCI protection approach for homeowners to understand is
   A. Using a GFCI circuit breaker to protect one or more receptacle outlets
   B. Installing a GFCI receptacle at each outlet required to have such protection
   C. Using a single GFCI receptacle to provide feed-through protection for multiple receptacles
   D. A "hybrid" wiring method that combines elements of A and C

6. Pump motors, metal piping, and metal parts of electrical equipment associated with hydromassage tubs must be bonded together using

A. A minimum 8 AWG solid copper conductor
B. A minimum 6 AWG aluminum or copper-clad aluminum conductor
C. Bonding jumpers 8 in. in length
D. Either A or B

7. GFCI-protected receptacle outlets in the following locations *only* are permitted to be connected to the 20-ampere bathroom branch circuit:
   A. Kitchen
   B. Unfinished basement or crawl space
   C. Outdoors
   D. None of the above

8. *National Electrical Code* rules that apply to residential construction cover only
   A. Permanently installed electrical products and systems
   B. Utilization equipment including cord-and-plug-connected appliances
   C. Temporary wiring used during construction
   D. All of the above

9. Article 422 states that the following types of utilization equipment must be constructed to provide protection for personnel against electrocution when immersed:
   A. Hand-held hair dryers
   B. Freestanding hydromassage units
   C. Any other portable appliance used within a bathroom
   D. Both A and B

10. GFCIs are available in the following type(s) of construction:
    A. Receptacle with integral GFCI
    B. Circuit breaker with integral GFCI
    C. Fuse with accessory GFCI sensing circuit
    D. Both A and B

FILL IN THE BLANKS

1. The required branch circuit for bathroom receptacle outlets must be rated _____.

2. Wiring requirements for hydromassage bathtubs are found in *NEC* Article _____.

3. A GFCI-protected receptacle is required to be located within _____ of the outside edge of each

210

basin on a wall or partition adjacent to the basin or countertop.

4. Installation of some types of luminaires is prohibited within a zone measured _____ vertically and _____ horizontally from the bathtub rim or shower stall threshold.

5. The official *Code* name for "whirlpool bath" is _____.

6. The only types of luminaires permitted to be installed within a zone measured 8 ft vertically and 3 ft horizontally from the bathtub rim or shower stall threshold are _____ and _____.

7. To protect construction workers from electric shock, all 125-volt, single-phase, 15-, 20-, and 30-ampere receptacles used for construction purposes, such as supplying power tools and portable work lights, are required to be protected by _____.

8. Timer switches are sometimes used to control _____ installed in bathrooms to insure that they aren't left ON accidentally.

9. According to the definition in Article 100, *a tub or shower space* in a bathroom must be considered a _____ location.

10. Although a ground-fault circuit interrupter opens a circuit quickly enough to prevent electrocution of a healthy adult, a GFCI cannot prevent the sensation of _____.

TRUE OR FALSE

1. GFCI-protected receptacle outlets are prohibited within a bathtub or shower space, unless equipped with enclosures that remain weatherproof when an attachment plug is inserted.

     ___ True   ___ False

2. Bathtub spaces and shower enclosures are considered to be wet locations.

     ___ True   ___ False

3. GFCI circuit breakers must be installed in accordance with the manufacturer's special instruction to provide only protection against faults between phase and grounded (neutral) conductors or between two different phases.

     ___ True   ___ False

4. Lighting outlets are permitted to be supplied by the 20-ampere bathroom branch circuit but must be connected ahead (upstream) of any ground-fault circuit-interrupter device.

     ___ True   ___ False

5. Only bathroom receptacle outlets located within 3 ft of the outside edge of the basin are required to have GFCI protection for personnel.

     ___ True   ___ False

6. Section 406.8(C) prohibits the installation of receptacles within a zone measured 8 ft vertically and 3 ft horizontally from the bathtub rim or shower stall threshold.

     ___ True   ___ False

7. It is common practice to circuit together all receptacle outlets in a dwelling that are required to have GFCI protection and supply them from a single GFCI circuit breaker.

     ___ True   ___ False

8. A bathroom is required to contain a basin, a toilet, and a bathtub or shower.

     ___ True   ___ False

9. A receptacle built into a luminaire (lighting fixture) is permitted to serve as the bathroom receptacle required by 210.52(D) if it is located within 3 ft of the outside edge of the basin and furnished with GFCI protection.

     ___ True   ___ False

10. GFCIs don't prevent the sensation of receiving an electric shock.

     ___ True   ___ False

CHALLENGE QUESTIONS

1. What is one possible reason the *Code* doesn't define a restricted bathtub or shower space for receptacles as it does for ceiling-suspended (paddle) fans and some types of luminaires (lighting fixtures)?

2. Why does the *NEC* require GFCI protection for all receptacle outlets in bathrooms, while only receptacles serving countertop areas in kitchens are required to be GFCI-protected?

3. Explain why ground-fault circuit-interrupter receptacles can be installed on branch circuits (such as those

in bathrooms of existing dwellings) that don't have separate equipment-grounding conductors.

4. The *Code* requires that the required 20-ampere bathroom branch circuit supply only receptacle outlets in bathrooms except in those cases when the required branch circuit supplies only a single bathroom (in this case, it is permitted to supply other outlets such as lighting, an exhaust fan, and a heater). Why might it be a good idea to always supply bathroom lighting on a different branch circuit than bathroom receptacles?

5. What is a primary difference in the way ground-fault circuit-interrupter circuit breakers, GFCI receptacles, and immersion detection circuit interrupters (IDCIs) operate?

The Bedroom

9

OBJECTIVES

After completing this chapter, the student will be able to understand the following:

- Requirements for arc-fault circuit-interrupter (AFCI) protection of bedroom branch circuits
- Circuiting considerations for bedrooms
- Installation of luminaires (lighting fixtures) inside closets
- Installation of ceiling-suspended (paddle) fans
- Required locations for receptacle outlets in habitable rooms
- Receptacle locations in rooms with electric baseboard heating

IMPORTANT NEC TERMS

Arc-Fault Circuit Interrupter
Branch Circuit
Branch Circuit, Multiwire
Circuit Breaker
Grounded Conductor
Outlet
Receptacle
Receptacle Outlet

INTRODUCTION

The chapter covers *National Electrical Code*® rules that apply to wiring methods in the bedroom, including the rules for arc-fault circuit interrupters, fans, and wiring in closets. The following major topics are included:

- Locating receptacle outlets in bedrooms
- Arc-fault circuit interrupters
- Luminaires and paddle fans
- Bedroom closet lighting
- Branch circuits in bedrooms
- Special rules for two-family dwellings

LOCATING RECEPTACLE OUTLETS IN BEDROOMS

Bedroom wiring must be installed in accordance with all standard *Code* rules governing living areas of residences.

Receptacle Spacing

Spacing for receptacle outlets installed in bedrooms must comply with the requirements of 210.52(A)(1). These rules, which are explained in detail in Chapter 6, The Living Room, can be summarized as follows:

- All 125-volt, 15- and 20-ampere receptacle outlets must be installed so that no point on a wall is more than 6 ft, measured horizontally along the floor, from another receptacle outlet in that same wall space.
- Any wall space 2 ft or wider must have its own receptacle outlet.
- The same receptacle spacing rules apply to fixed glass panels (floor-to-ceiling picture windows or the nonsliding portions of glass doors). Sliding panels are considered doors, so they aren't counted when determining the required spacing of receptacle outlets along bedroom walls.
- Additional receptacles can be provided for extra convenience or where required for a particular purpose. For example, receptacles are sometimes centered below windows for plugging in holiday lighting, or a quad receptacle in a single outlet box (often called a *four-plex*) might be provided for powering computer equipment when it's known that a spare bedroom will be used as a home office.
- Even when a bedroom has more than the minimum number of receptacle outlets required by the *NEC®,* they must still be located so that no point on a wall is more than 6 ft from a receptacle, and all required receptacles must be accessible. Receptacle outlets located more than 5½ ft above the floor, or installed within shelf alcoves or cabinets, can't be counted among the receptacle outlets required by 210.52(A)(1) and (2).

Other Rules for Bedrooms

Receptacle outlets aren't permitted to be located above electric baseboard heaters. To meet *Code* receptacle spacing rules, electric baseboard heaters are available with built-in receptacle modules (Figure 9.1). This subject is discussed at greater length in Chapter 6, The Living Room.

Receptacle outlets are permitted inside clothes closets, but they aren't required by the *Code*.

ARC-FAULT CIRCUIT INTERRUPTERS

Bedrooms in dwellings are required to have arc-fault circuit-interrupter (AFCI) protection.

AFCI *Code* Requirement

The 2005 *National Electrical Code* requirement for AFCI protection in bedrooms seems complex, at least on paper. (See Exhibit 9.1.)

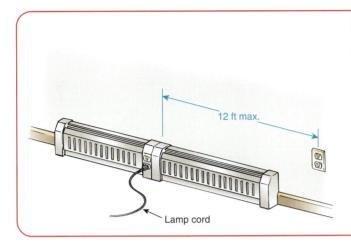

FIGURE 9.1 Permanently installed baseboard heaters equipped with receptacle outlets can be used to meet receptacle outlet spacing requirements.

12 ft max.

Lamp cord

EXHIBIT 9.1

NEC **210.12(B) Dwelling Unit Bedrooms**

All 120-volt, single phase, 15- and 20-ampere branch circuits supplying outlets installed in dwelling unit bedrooms shall be protected by a listed arc-fault circuit interrupter, combination type installed to provide protection of the branch circuit.

Branch/feeder AFCIs shall be permitted to be used to meet the requirements of 210.12(B) until January 1, 2008.

FPN: For information on types of arc-fault circuit interrupters, see UL 1699-1999, *Standard for Arc-Fault Circuit Interrupters*.

Exception: The location of the arc-fault circuit interrupter shall be permitted to be at other than the origination of the branch circuit in compliance with (a) and (b):

(a) The arc-fault circuit interrupter installed within 1.8 m (6 ft) of the branch circuit overcurrent device as measured along the branch circuit conductors.

(b) The circuit conductors between the branch circuit overcurrent device and the arc-fault circuit interrupter shall be installed in a metal raceway or a cable with a metallic sheath.

- *Branch/feeder AFCIs* are the circuit-breaker units currently sold by electrical suppliers (Figure 9.2). They protect feeder and branch-circuit wiring against the unwanted effect of arcing. These can be used to meet the requirements of 210.12(B) until January 1, 2008, any branch/feeder AFCIs installed up to that time can remain in place indefinitely.
- *Combination type AFCIs* protect branch-circuit wiring along with the power-supply cords attached to the outlets. They will be available in both circuit-breaker and receptacle designs. The reason for the January 1, 2008 effective date in the *Code* is to give manufacturers time to develop this new type of AFCI.

Exception. The exception to 210.12(B) permits arc-fault circuit interrupters to be installed not just in panelboards, but within 6 ft of the panel under certain circum-

FIGURE 9.2 AFCI circuit breakers are used to protect an entire branch circuit.

stances. The purpose of this exception is to permit development of a new AFCI product that can be used with fuse-type panelboards. This product doesn't exist yet.

What Are AFCIs?

AFCIs are designed to prevent a type of electrical hazard that conventional overcurrent protective devices don't guard against very well. An example of such a hazard is a frayed lamp cord that exposes the copper wire inside, creating a low-energy electrical arc to some other piece of nearby grounded metal. AFCIs protect against the following three types of arcing faults (Figure 9.3):

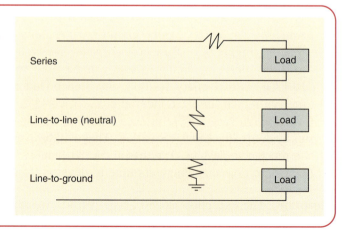

FIGURE 9.3 AFCIs protect against three different types of arcing faults.

1. *Series faults,* such as an arcing loose connection or damaged conductor

2. *Line-to-line (neutral) faults*, such as an arcing short in a two-wire branch circuit

3. *Line-to-ground faults,* such as the frayed lamp cord example

These types of arcing faults may not create an overload that will cause a circuit breaker to trip or a fuse to blow. Overcurrent protective devices are intentionally designed with built-in delays that allow a brief surge of inrush current when a motor or microwave oven starts up, without causing "nuisance trips."

Also, current in arcing faults tends to be low because of the impedance of the arc itself. Both circuit breakers and fuses react fairly slowly to overcurrents of low magnitude. However, an AFCI circuit breaker trips at an arcing current level of 75 amperes. On arcing faults to ground, listed AFCIs are required to trip at 5 amperes. However, most will actually trip at lower levels of around 50 milliamperes (mA).

Even low-energy arcs represent a fire hazard and a threat to life when flammable material is present. Since many fire-related fatalities are caused by smoke inhalation when people are sleeping, bedrooms were chosen as the place to start requiring this new protective technique.

AFCIs and Ground-Fault Circuit Interrupters (GFCIs)

AFCIs are technically similar in some ways to ground-fault circuit interrupters (GFCIs). Both types of circuit breakers have a white wire pigtail that is connected to the neutral bus in the panelboard and a TEST button. AFCIs are reset the same way as are GFCI circuit breakers: by moving the operating handle from the tripped or center position to the full OFF position and then back to the ON position (Figure 9.4).

However, AFCIs have a different purpose than GFCIs. The function of an AFCI is to protect against faults or damage to wiring and equipment that could initiate fire-causing arcs anywhere in a branch circuit. The function of a GFCI is to protect

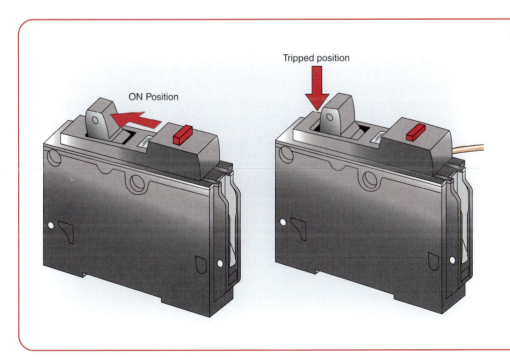

FIGURE 9.4 When the TEST button on a circuit breaker AFCI is pressed, the device will trip if working properly.

people from shock or electrocution if they come into contact with energized parts connected to receptacles.

Prohibited on Multiwire Branch Circuits

Arc-fault circuit interrupters cannot protect branch circuits that share a grounded (neutral) conductor. The reason is that current flowing out and returning on a two-wire circuit is monitored for the presence of arcing faults. Therefore, the circuit must have a distinct hot conductor and a distinct neutral conductor. Otherwise, the AFCI cannot detect arcing occurrences.

AFCI Protection in Bedroom Closets

Whether a bedroom closet is considered part of the bedroom itself is up to the authority having jurisdiction (AHJ). However, the intent of 210.12 is to provide protection against arcing faults for the entire bedroom. So when a lighting outlet or other wiring is installed in bedroom closets, it's a good idea for these circuits to have AFCI protection.

Smoke Detectors on AFCI-Protected Branch Circuits

Installation of smoke detectors is covered in Chapter 15, Special Systems. Smoke detectors must be located in all bedrooms and are required to be direct-connected. Branch circuits supplying smoke alarms in bedrooms must have AFCI protection for the same reason as other bedroom circuitry: to prevent a fault in the smoke detector wiring from causing an electrical arc that might ignite flammable furnishings or other materials.

Installing and Troubleshooting AFCIs

Good workmanship is particularly important when installing AFCIs, because there's no test equipment available for verifying that the units have been installed correctly. For safety reasons, a tester can't create an actual arcing fault in order to exercise the arc-detection circuitry. AFCI circuit breakers and receptacles have self-test buttons, like GFCI units. The steps for installing and testing an AFCI circuit breaker follow.

Step 1. *Turn* **OFF** *the panelboard main circuit breaker or other disconnect device.*

Step 2. *With the power off, install the AFCI circuit breaker in the panelboard.*

Step 3. *Turn* **OFF** *all loads on the protected branch circuit.*

Step 4. *Turn* **ON** *the power to the panelboard.*

Step 5. *Turn* **ON** *the circuit breaker handle.*

Step 6. *Push the* **TEST** *button on the AFCI circuit breaker. This step generates a signal that looks like an arcing fault to the AFCI sensor. If the unit is operating properly, the circuit breaker will trip and the handle will move to the tripped center position, as shown in Figure 9.4.*

Step 7. *To restore power to the branch circuit, move the circuit breaker handle to the full* **OFF** *position and then back to the* **ON** *position.*

Causes of AFCI Tripping. In addition to arcing faults, several other abnormal conditions may cause AFCIs to trip:

- *High-Voltage Surges.* These can be caused by lightning or utility switching transients.
- *Voltage or Frequency Fluctuations.* These can be due to poorly regulated backup generators.
- *Mechanical Shocks.* These could be caused by earthquakes, for example.
- *Neutral Grounding.* If the grounded conductor (neutral) of an AFCI-protected circuit touches grounded metal, the AFCI can trip if the path to ground has very low impedance.
- *Overcurrents.* AFCI circuit breakers trip if they sense currents above their normal rating, which can't be distinguished from tripping due to arc faults.

Troubleshooting AFCI Installations. If an AFCI circuit breaker trips, the following steps can be used to determine the cause.

Step 1. *Gather information from people who may have been in the bedroom at the time the AFCI tripped.* Buzzing noises, visible arcing or sparking, smoke, or strange odors may indicate defective appliances or other utilization equipment.

Step 2. *Unplug all appliances, table lamps, and extension cords plugged into the affected circuit.* If tripping occurs again when the AFCI is turned on, the fault is in the fixed wiring system.

Step 3. *Turn off all fixed appliances such as luminaires (lighting fixtures) and ceiling fans that have switches.* However, since this equipment can't be completely disconnected (that is, switches interrupt the hot circuit conductor but not the grounded/neutral or green/grounding wires), AFCI tripping with switches turned off doesn't rule out the possibility of faults in these appliances.

Step 4. *If the AFCI doesn't trip when all plugs are disconnected and all fixed appliances are turned off, begin reconnecting plugs one by one.* Then, fixed lights and appliances should be turned on one by one. This procedure may locate the faulted cord or appliance.

Rooms Not Requiring AFCI Protection

The *NEC* requires 120-volt, 15- and 20-ampere branch circuits that supply outlets in dwelling unit bedrooms to have AFCI protection [210.12(B)]. Rooms identified on construction plans as other than bedrooms (such as dens, home offices, sewing rooms, and storage spaces) don't require AFCI protection. However, a den that could also serve as a bedroom may need AFCI-protected branch circuits. In doubtful situations, the authority having jurisdiction should be consulted. The *NEC* doesn't prohibit AFCI protection on branch circuits in rooms other than bedrooms.

Bathrooms in Bedroom Suites

Dwellings are required to have at least one 20-ampere branch circuit that supplies only bathroom receptacle outlets. No other outlets are permitted on this required branch circuit, except that the required 20-ampere bathroom branch circuit can feed other outlets when it supplies only a single bathroom [210.11(C)(3), Exception].

Because no other outlets are permitted on the 20-ampere bathroom branch circuit, GFCI receptacles in bathrooms shouldn't be connected to an AFCI-protected circuit supplying bedroom outlets. In addition, AFCIs won't function properly on branch circuits that also have GFCIs installed on them.

The *NEC* does not prohibit supplying bathroom lighting from an AFCI-protected branch circuit that also supplies bedroom outlets. However, in case of a fault affecting the bedroom circuits, the bathroom lighting supplied by the same circuits will also be de-energized.

LUMINAIRES AND PADDLE FANS

General *Code* requirements for lighting are explained in Chapter 5, Lighting the Home. However, special attention is often necessary for unique lighting characteristics in bedrooms, such as those that follow.

- Most bedrooms have a lighting outlet located in the middle of the ceiling. In some cases, the luminaire (lighting fixture) is replaced by a ceiling-suspended (paddle) fan or a paddle fan with built-in lighting.
- Some bedrooms also have other types of lighting, such as wall-washers, lighting track, or wall-mounted decorative fixtures.

Wall-Mounted Luminaires

Accessories known as *hickeys* or *fixture studs* are used to attach luminaires to outlet boxes and help support their weight. Often the hickey or fixture stud is supplied with the luminaire itself.

Lightweight Luminaires. Wall-mounted luminaires weighing up to 6 lb are permitted to be mounted directly to outlet boxes, using no fewer than two screws, No. 6 or larger [314.27(A), Exception]. Using standard device boxes is an easy and inexpensive way to mount lightweight wall-mounted luminaires, such as sconces.

Ceiling-Mounted Luminaires

Standard ceiling outlet boxes can be used to support luminaires weighing up to 50 lb [314.27(B)]. Heavier luminaires must be supported by outlet boxes listed for the weight to be supported or must be supported independently of the box. The installation and support of boxes is covered in more detail in Chapter 4, Wiring Methods.

Boxes for Ceiling-Suspended (Paddle) Fans

Special boxes listed for use with paddle fans (Figure 9.5) are marked "Acceptable for Fan Support" or "Suitable for Fan Support." Standard outlet boxes intended for luminaires can't be used to support paddle fans [314.27(D)] because of torque and vibration problems. Typical paddle fans are also heavier than typical lighting fixtures and sometimes use larger attachment screws.

Installing Paddle Fan–Rated Ceiling Boxes in Bedrooms. When ceiling-suspended (paddle) fans are installed at the time of original house construction, installers use the correct type of box/support system for the weight of the fan to be supported. However, after the original construction has been completed, homeowners frequently replace existing luminaires with paddle fans in bedrooms and other areas.

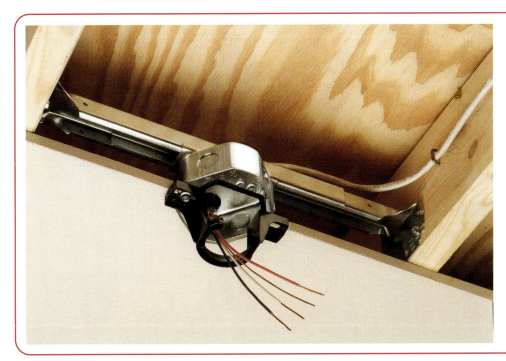

FIGURE 9.5 This fan-rated box has an adjustable bracket for positioning the unit between ceiling joists. (Courtesy of Arlington Industries, Inc.)

Because existing luminaires are often replaced, it's a good idea to install boxes identified for supporting paddle fans up to 35 lb at ceiling-mounted lighting outlets in bedrooms during original construction. The extra cost to provide these special outlet boxes during original construction is very low and can help avoid possible safety problems later on. A different type of ceiling fan box is shown in Figure 5.16 of Chapter 5, Lighting the Home.

Ceiling Fan/Light Wiring

There are many types of ceiling fan controls. Some have a pull-chain located on the fan itself, but pull-chains aren't used much in new construction. Most paddle fans installed in new dwellings have wall-mounted speed controls.

Fans with built-in lighting also have wall-mounted ON/OFF switches or dimmers. The fan motor and lighting controls for a combination unit are normally mounted in the same faceplate. Some types of fan/light controls require a double-gang box, while more modern designs fit both into a single device box (Figure 9.6). When conductors are installed between wall-mounted controls and a ceiling fan or fan/ light combination, the fan manufacturer's wiring diagram should be followed.

A typical wiring arrangement for a ceiling fan with integral lighting is shown in Figure 9.5. Three-wire-with-ground Type NM cable is run from the wall box for the fan control up to the ceiling box.

- The black conductor runs from the fan speed control to the fan motor.
- The red conductor runs from the wall switch to the light.
- The white conductor and bare equipment grounding conductor are common to both the fan and light controls.

FIGURE 9.6 This combination fan-speed control and light dimmer can be installed in a single-gang device box. (Courtesy of Pass & Seymour/ Legrand®)

BEDROOM CLOSET LIGHTING

As indicated in Chapter 5, Lighting the Home, the *Code* permits lighting to be installed inside closets but doesn't require it. In many cases, ceiling lighting outlets in bedrooms provide sufficient illumination for shallow closets without installing additional outlets.

However, when lighting is installed inside closets, the *NEC* has strict safety requirements [410.8]. Broken incandescent lamps with hot filaments present a fire danger in closets filled with flammable clothing and other stored materials. For this reason, only fluorescent luminaires and recessed or completely enclosed incandescent luminaires are permitted inside closets.

Prohibited Luminaires

The following types of luminaires are *never* permitted inside closets:

- Incandescent luminaires with open or partially enclosed lamps
- Pendant luminaires and lampholders of any kind

Clearances Inside Closets

Figure 9.7 illustrates the clearances required for different types of luminaires installed inside closets. The shaded areas indicate the *storage space,* as defined in 410.8, and the following requirements apply:

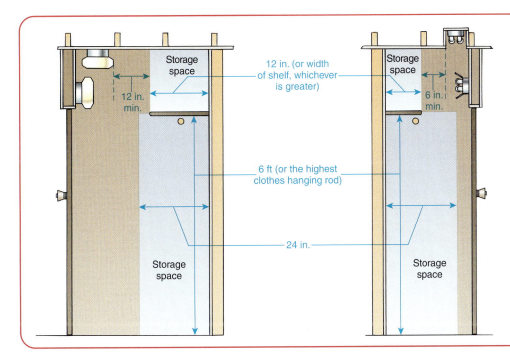

FIGURE 9.7 Different types of luminaires inside closets require different clearances from storage space (lights are optional in closets).

- A volume at least 6 ft high and 24 in. wide must be allowed for hanging clothes.
- A volume at least 12 in. wide must be allowed above the 6-ft point for shelf space.

NOTE: These are minimum dimensions that must be allowed for storage space when clothes rods and shelves haven't been installed inside closets at the time lighting outlets are roughed in. But if the highest rod for hanging clothes will be installed above 6 ft, or if the shelves used are deeper than 12 in., then these actual dimensions must be used in calculating storage space and required clearances from closet luminaires.

Walk-In Closets

Walk-in closets are more like small rooms. Typically, there is sufficient space for a surface-mounted ceiling luminaire with adequate clearances from storage space. A wall switch can be provided either inside or outside the door of the walk-in closet. A door-jamb switch can also be used to turn the light on automatically whenever the door is opened.

BRANCH CIRCUITS IN BEDROOMS

In addition to the *Code* requirements for residential branch circuits explained in Chapter 6, The Living Room, bedrooms have some special requirements and wiring considerations.

Laying Out Branch Circuits

The *Code* specifically permits lighting and other loads to be connected together on 15- or 20-ampere branch circuits in homes [210.23(A)], except in a few special cases [210.23(A), Exception]. Mixing together lighting and receptacle outlets in different rooms on the same branch circuit(s) is permitted by the *NEC* and is a normal practice in residential wiring.

One Room, Multiple Circuits. Many installers also feel it's a good idea to have more than one branch circuit supplying the lighting and receptacles in a room or other area. Then, if a circuit breaker trips or needs to be turned off to perform maintenance, energized receptacles are still available in the same room to supply portable lamps and power tools.

AFCI Protection. The requirement for AFCI protection of bedroom branch circuits also dictates certain wiring practices:

- All lighting, receptacles, smoke detectors, ceiling fans, and other loads in bedrooms must be connected to AFCI-protected branch circuits. A single outlet in a bedroom can't be tacked onto some other non-AFCI branch circuit just because it seems convenient to do so during original construction. (It will seem less convenient later, when the job is red-tagged.)
- All 120-volt, 15- or 20-ampere branch circuits serving other loads, such as permanently installed electric heaters, must also have AFCI protection.

Although not required by the *Code*, loads in rooms or areas other than bedrooms are permitted to be supplied by AFCI-protected branch circuits, as the following two examples indicate:

1. A smoke detector in a hall outside bedrooms can be connected to the same circuit as the bedroom detectors.
2. Receptacle or lighting outlets in a hall or other room can be connected to the same circuit as bedroom outlets.

SPECIAL RULES FOR TWO-FAMILY DWELLINGS

There are no special *Code* rules that apply to bedrooms in two-family dwellings. Residential branch circuits are only permitted to supply loads located within, or associated with, each individual dwelling unit [210.25]. Also, occupants are required to have "ready access" to the overcurrent devices protecting conductors in their occupancy [240.24(B)]. Taken together, these rules mean that the bedrooms in a two-family dwelling can't be wired together on a single branch circuit to economize on AFCIs.

CHAPTER REVIEW 9

MULTIPLE CHOICE

1. Arc-fault circuit-interrupter protection is required for
   A. Bedroom receptacle outlets only
   B. 15- and 20-ampere branch circuits in bedrooms
   C. 15- and 20-ampere branch circuits in bathrooms
   D. None of the above

2. All 125-volt, 15- and 20-ampere receptacle outlets must be installed so that no point along the floor in any wall space is more than how many feet, measured horizontally, from another receptacle outlet in that space?
   A. 2 ft
   B. 6 ft
   C. 12 ft
   D. 4 ft in residences that will be occupied by people with disabilities

3. The following type of room often found in close proximity to bedrooms requires AFCI protection for all 120-volt, 15- and 20-ampere branch circuits:
   A. Den
   B. Home office
   C. Sewing room
   D. None of the above

4. Circuits supplying the following items in bedrooms are required to have AFCI protection:
   A. Smoke detectors
   B. Ceiling fans
   C. Permanently installed electric heaters
   D. All of the above

5. A bedroom with two entrances is required to have how many wall switches (or equivalent controls such as dimmers and occupancy sensors) controlling the lighting outlet(s) installed in that room?
   A. One
   B. Two
   C. Two three-way switches
   D. None of the above

6. When listed electric baseboard heaters are installed in bedrooms
   A. Wall-mounted receptacles cannot be installed above them.
   B. Receptacle units are available for baseboard heaters to meet *NEC* spacing rules.
   C. Receptacles installed as part of electric baseboard heating units are not permitted to be supplied from the heater circuit.
   D. All of the above

7. AFCI protection is permitted for 120-volt, 15- and 20-ampere branch circuits supplying outlets in which of the following?
   A. Living rooms
   B. Bedrooms
   C. Unfinished basements
   D. Any of the above

8. A duplex receptacle has the following number(s) of contact devices on the same yoke:
   A. 1
   B. 2
   C. 3
   D. None of the above

9. Receptacles located at heights above the floor greater than which of the following are not permitted to be counted among the receptacle outlets required by 210.52(A)(1) and (2)?
   A. 2 ft
   B. 3½ ft
   C. 5½ ft
   D. 6 ft

10. The following types of fluorescent luminaires (fixtures) are permitted to be installed inside closets:
    A. Surface-mounted enclosed
    B. Recessed nonenclosed
    C. Pendant-mounted enclosed
    D. Both A and B

FILL IN THE BLANKS

1. Accessories known as _____ are used to attach luminaires to outlet boxes and help support their weight and often are supplied with the luminaire itself.

2. AFCIs will not perform properly if installed on _____ branch circuits.

3. Standard ceiling outlet boxes can be used to support luminaires weighing up to _____ lb.

4. Any wall space _____ or wider must have its own receptacle outlet.

5. Occupants in a two-family dwelling are required to have _____ to the overcurrent devices protecting conductors in their occupancy.

6. A minimum of _____ lighting outlet(s), controlled by a wall switch, must be installed in every habitable room of a dwelling.

7. Most 120-volt branch circuits in a typical dwelling unit are rated _____ amperes.

8. According to the definition of the term *receptacle* in Article 100, a duplex receptacle is considered to be the same as _____ receptacle(s).

9. Wall-mounted luminaires (lighting fixtures) weighing not more than 6 lb are permitted to be supported by standard outlet boxes using No. _____ or larger screws.

10. On arcing faults to ground, listed AFCIs are required to trip at _____ amperes.

TRUE OR FALSE

1. The *National Electrical Code* requires that closets have two independent sources of lighting, so that the failure of any single source does not leave the closet in darkness.

    _____ True   _____ False

2. Duplex receptacles mounted up to 6 ft above the floor can be counted among the receptacle outlets required by 210.52(A)(1) and (2).

    _____ True   _____ False

3. A 125-volt, 15-ampere duplex receptacle can be installed on either a 15-ampere or 20-ampere branch circuit.

    _____ True   _____ False

4. Special outlet boxes are required for ceiling-suspended (paddle) fans only when the fan weighs 35 lb or more.

    _____ True   _____ False

5. Bedroom lighting outlets are permitted to be of the type defined by 210.70(A).

    _____ True   _____ False

6. AFCI protection is required for all branch circuits in bedrooms, associated closets, and associated bathrooms.

    _____ True   _____ False

7. Receptacle outlets are permitted to be located above electric baseboard heaters in bedrooms only when supplied from a different branch circuit than the heater.

    _____ True   _____ False

8. Bedroom circuits and other outlets/loads can be mixed together on the same AFCI-protected branch circuits.

    _____ True   _____ False

9. Bedrooms in dwelling units are required to have lighting outlets with permanently installed luminaires (lighting fixtures).

    _____ True   _____ False

10. All types of incandescent luminaires (lighting fixtures) are prohibited in closets unless protected by an AFCI.

    _____ True   _____ False

CHALLENGE QUESTIONS

1. The *NEC* prohibits installation of receptacles above electric baseboard heaters but not above hot-water radiator-type baseboard heaters. What is the probable reason for singling out electric baseboard heaters in this way?

2. Arc-fault circuit-interrupter (AFCI) protection is required for all branch circuits supplying 15- and 20-ampere outlets installed in dwelling unit bedrooms. Some people find it strange that there is no exemption for smoke-detector circuits since these devices are also intended to provide protection against fire hazards. Discuss the reason for this.

3. The *Code* doesn't require lighting in closets. But when lighting is provided, incandescent luminaires with open or partially enclosed lamps are prohibited, as are pendant luminaires and lampholders of any kind. What are the reasons for these rules?

4. Should switch locations be classified as *outlets* as defined in *NEC* Article 100? Why or why not?

5. Why is it a good idea to install ceiling-fan-rated boxes for ceiling-mounted lighting outlets in bedrooms?

6. AFCI and GFCI circuit breakers are technically simi-

lar in some ways. Both have a white wire pigtail that is connected to the neutral bus in the panelboard and a TEST button. Describe one or more ways in which AFCI and GFCI circuit breakers are different from one another.

7. What is the general concept behind the receptacle outlet spacing rules that apply to bedrooms?

The Basement, Family Room, and Laundry

10

OBJECTIVES

After completing this chapter, the student will be able to understand the following:

- Requirements for GFCI protection in basements
- Installation of fluorescent luminaires (fixtures) in lay-in ceilings
- Rules for receptacle outlets within 6 ft of wet bars
- Special rules for basement receptacles that serve appliances in dedicated spaces
- *Code* requirements for basement lighting and switching

INTRODUCTION

This chapter covers the *National Electrical Code*® rules for electrical installations in unfinished and finished basement areas in dwellings. The following major topics are included:

- Unfinished basement areas
- Family rooms and wet bars
- Laundry rooms
- Bedrooms in basement areas
- Special rules for two-family dwellings

IMPORTANT NEC TERMS

Branch Circuit

Ground-Fault Circuit Interrupter

Identified (as applied to equipment)

Lighting Outlet

Luminaire

Outlet

Receptacle

Receptacle Outlet

UNFINISHED BASEMENT AREAS

Homes in many parts of the country have basements located completely or partially below ground level. Basements frequently contain unfinished areas for storage space, electric service equipment, mechanical equipment for heating and cooling, laundry equipment, and workshops. They may also contain finished areas such as family/recreation rooms, home offices, and bedrooms. In parts of the country where basements are uncommon, some of the electrical equipment may be located on other floors, but the same *NEC*® rules apply in any case.

Different rules apply to electrical installations in finished and unfinished parts of basements. General rules for unfinished areas in the basement apply to wiring methods for receptacles and lighting, while more specific rules apply to sump pumps and other equipment. Heating, ventilating, and air-conditioning equipment (HVAC), along with water heaters, is covered in Chapter 14, HVAC Equipment and Water Heaters. Electric service equipment is covered in Chapter 3, Services, Service Equipment, and Grounding.

Receptacle Outlet(s)

Each unfinished portion of a basement is required to have at least one receptacle outlet, in addition to any provided for laundry equipment [210.52(G)]. Often it's convenient to provide more than one receptacle outlet to minimize the use of extension cords and cube taps.

Separate Circuit Not Required. Receptacle outlets in the unfinished portion of a basement aren't required to have their own branch circuit. They can be wired to lighting outlets or receptacle outlets in finished portions of basements and other areas of the house. It is common to connect grade level ground-fault circuit-interrupter (GFCI) protected outdoor receptacle outlets at the front and back of a dwelling [210.52(E)] to basement receptacle circuits.

Grounding Receptacles to Boxes. Generally speaking, the *Code* requires that an equipment-grounding jumper be used to connect the grounding terminal of a receptacle to a grounded box [250.146]. However, when a receptacle is installed in a surface-mounted box, direct metal-to-metal contact between the device yoke and the box is permitted to ground the receptacle. To ensure an effective ground-fault current path, at least one of the insulating washers on the yoke screws must be removed (Figure 10.1) [250.146(A)].

Receptacles Mounted on Covers. Receptacles in unfinished areas of basements are commonly installed exposed in device boxes with faceplates or in utility boxes with covers that accommodate duplex receptacles. Receptacles mounted in these types of boxes must be held rigidly against the cover by more than one screw (Figure 10.2), unless the box cover is listed and identified for securing a device by means of a single screw [406.4(C)].

As described above, receptacles in surface-mounted boxes are generally permitted to be grounded to the box by direct metal-to-metal contact. However, this *NEC* provision does not apply to cover-mounted receptacles unless the box and cover combination are listed for this application [250.146(A)].

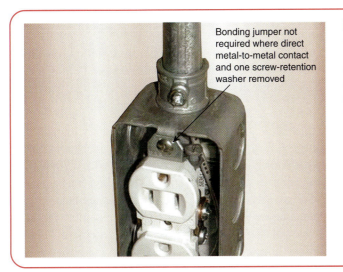

FIGURE 10.1 A receptacle installed in a surface-mounted box is not required to have a bonding jumper if at least one of the insulating washers is removed.

Bonding jumper not required where direct metal-to-metal contact and one screw-retention washer removed

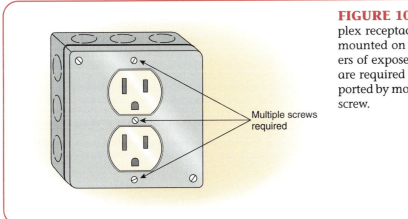

Multiple screws required

FIGURE 10.2 Duplex receptacles mounted on raised covers of exposed boxes are required to be supported by more than one screw.

GFCI Protection. The *Code* rules require that receptacles in unfinished basements have GFCI protection. However, 210.8(A)(5) allows several exceptions, which follow.

   1. *Not Readily Accessible.* Receptacle outlets installed more than 6½ ft above the floor aren't required to be GFCI protected. For example, some builders install duplex receptacles up in the joists supporting the first floor ceiling. Homeowners can then purchase fluorescent luminaires with short cordsets at the local home center and hang them in the joists, plugging them into these receptacles that aren't required to have GFCI protection [210.8(A)(2)], Exception No. 1].

   2. *Appliances in Dedicated Space.* A single receptacle outlet located within dedicated space for a cord-and-plug-connected appliance that isn't easily moved from one place to another (or a duplex receptacle for two appliances) isn't required to have GFCI protection [210.8(A)(2)], Exception No. 2]. The intent of this rule is to permit refrigerators and freezers located in unfinished basements to be supplied by non-GFCI receptacles so that "nuisance tripping" doesn't cause food to spoil.

(Receptacles for refrigerators in kitchens also aren't required to have GFCI protection.)

3. *Alarms.* A receptacle outlet supplying only a permanently installed fire alarm or burglar alarm system is not required to have GFCI protection [210.8(A)(2), Exception No. 3]. This is explained in greater detail in Chapter 15, Special Systems.

Unfinished Basement Lighting

An unfinished basement area used for storage or containing equipment must have a lighting outlet containing a switch or controlled by a wall switch [210.70(A)(3)]. The outlet must be located at or near any equipment requiring servicing. If a basement has a separate utility room containing HVAC equipment and a water heater, then two lighting outlets are required: one in the utility room and a second in the area devoted to storage and/or electric service equipment. Required basement lighting outlets must have a lampholder or luminaire installed. Additional lighting outlets provided at the option of the builder or homeowner can be receptacles, as described previously.

Basement Stairway Lighting. When a stairway leads from the first floor down to a basement, lighting must be provided to illuminate the stairs. More detail is provided about this subject in Chapter 11, Hallways, Stairways, and Attics.

Outdoor Entrance Lighting. Some basements are entered by stairs from the outside. In houses built on sloping lots, the basement may be partially below grade (often at the front) and partially at grade level with a door to the outside. A lighting outlet must be installed on the exterior side of any door entrance and controlled by a wall switch inside the basement [210.70(A)(2)(b)].

Switch Control of Basement Lighting. Basements used for storage or containing equipment that requires servicing must have at least one lighting outlet containing a switch or controlled by a wall switch [210.70(3)]. This rule allows pull-chains and other types of switches built into lampholders and luminaires, but this solution may not be right for every lighting situation.

General Guidelines for Planning and Installing Basement Lighting Controls

- ***Usual Point of Entry.*** At least one switch should be provided at the "usual point of entry" to a basement or utility room, as required by 210.70(A)(3). If there are two or more points of entry, multiple switches are needed.

- ***Basement Stairs.*** When a basement is entered by a stairway from the floor above, the stairway itself becomes the point of entry and requires at least one light controlled by a switch. In this case, the stairway lighting can be controlled from the top of the stairs, and the lighting within the basement can be controlled by its own switch. However, the preferable method is to install three-way switches that allow control from both the top and foot of the stairs (Figure 10.3).

- ***Multiway Switching.*** Using three- and four-way switches is often the most convenient way to control basement lighting when there's more than one entrance.

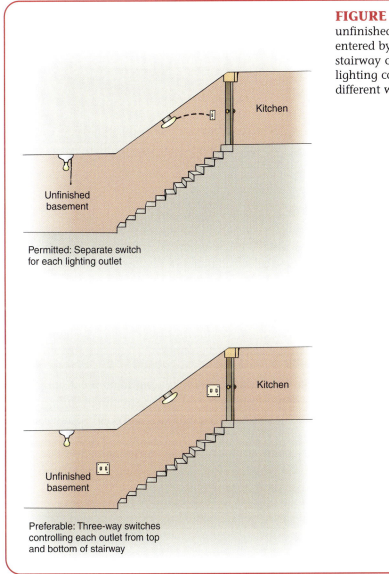

FIGURE 10.3 An unfinished basement entered by means of a stairway can have its lighting controlled in different ways.

Kitchen

Unfinished basement

Permitted: Separate switch for each lighting outlet

Kitchen

Unfinished basement

Preferable: Three-way switches controlling each outlet from top and bottom of stairway

Sump Pumps

Some unfinished basements have sump pumps installed to prevent water accumulation. Typically, these units are cord-and-plug connected.

GFCI Not Required. The *Code* does not require a single receptacle used to supply a sump pump to have GFCI protection. Likewise, a duplex receptacle used to supply a sump pump plus another appliance occupying dedicated space (that is, the appliance cannot easily be moved from one place to another) is not required to have GFCI protection [210.8(A)(5), Exception No. 2].

NOTE: Receptacles serving dedicated appliances don't satisfy the requirement for "at least one receptacle outlet" for each unfinished area of a basement [210.52(G)].

Sump Pump Lighting. The *Code* requires a lighting outlet at or near equipment requiring service in basements and utility rooms [210.70(A)(3)]. If a sump pump is located away from the mechanical or electric service equipment, a separate lighting outlet must be provided. The separate outlet is not required to be controlled by a wall switch.

FAMILY ROOMS AND WET BARS

Family/recreation rooms may be located in any area of a dwelling but are often located in basements. The *NEC* rules governing electrical installations in family rooms are the same no matter where they are located.

Lighting (Lay-In Fluorescent Luminaires)

Many family rooms and other finished spaces in basements are lighted with 2 × 4 fluorescent luminaires (fixtures) installed in a dropped ceiling with acoustic tiles. Listed lighting fixtures for this type of installation are marked as follows:

- *Recessed fluorescent luminaire (fixture)*, which means the luminaire can be installed either in cavities in ceilings and walls or in suspended (dropped) ceilings
- *Suspended-ceiling fluorescent luminaire (fixture)*, which means the luminaire is only permitted to be installed in suspended ceilings

Fixture Whips. The most common method of wiring lay-in fluorescent luminaires (sometimes called *troffers*) is to install outlet boxes on the structural ceiling above and then run unsupported lengths of cable or flexible raceway from the outlet boxes to each fixture (Figure 10.4). These cables are known as *fixture whips.*

A single outlet box typically has up to four fixture whips running to fluorescent fixtures supported by the dropped ceiling system. The following raceway and cable types are permitted to be installed as fixture whips:

Armored cable (AC)—6 ft
Nonmetallic-sheathed cable (NM)—4½ ft
Metal-clad cable (MC)—6 ft

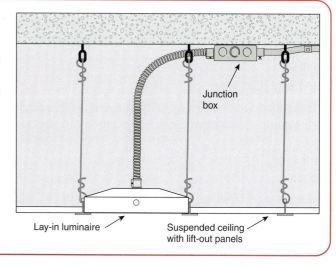

FIGURE 10.4 Fluorescent luminaires in suspended ceilings are often supplied by means of fixture whips connected to junction boxes mounted on structural elements above.

Junction box

Lay-in luminaire

Suspended ceiling with lift-out panels

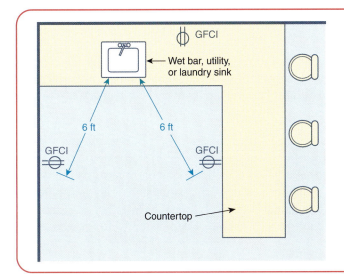

FIGURE 10.5 Receptacles located within 6 ft of a wet bar sink and installed to serve countertop surfaces must have GFCI protection.

Flexible metal conduit (FMC)—6 ft

Electrical nonmetallic tubing (ENT)—6 ft

Liquidtight flexible metal conduit (LFMC)—6 ft

Liquidtight flexible nonmetallic conduit (LFNC)—6 ft

Type AC and Type NM cable are most commonly used in residential construction. In general, fixture whips can be up to 6 ft long, but unsupported lengths of Type NM cable are limited to 4½ ft.

General-Purpose Receptacles

Receptacle outlets in family rooms must be located according to the "12-foot rule," and any wall at least 2 ft wide must have a receptacle outlet, such as spaces between doors and fireplaces. For more details on this subject, see Chapter 9, The Bedroom. Fixed room dividers such as freestanding bar-type counters or railings must also have receptacles spaced to meet the 12-foot rule [210.52(A)(2)(3)].

GFCI Not Required. Receptacle outlets in finished areas of basements such as family rooms aren't required to have GFCI protection, but special rules apply to countertop receptacles at wet bars.

Wet Bar Receptacles

GFCI Protection. Many family rooms include a wet bar with a sink. Receptacle outlets installed within 6 ft of the outside edge of a sink are required to have GFCI protection (Figure 10.5). This rule isn't limited to receptacles that serve appliances on countertop surfaces. Any receptacle outlet installed within 6 ft of the wet bar sink must be GFCI protected [210.8(A)(7)].

Other Considerations. Receptacle outlets cannot be installed face-up in countertops, to avoid the problem of dirt and liquid getting into the slots [406.4(E)]. Unlike kitchen countertop receptacles, wet bar counter receptacles aren't required to be supplied by a separate branch circuit that has no other outlets.

LAUNDRY ROOMS

Some dwellings have a separate laundry room in the basement. Others have a laundry area combined with a kitchen or bathroom or designed as a closet area near the bedrooms. The *Code* requirements are the same no matter where laundry equipment is installed.

Laundry Lighting

A separate laundry room must have its own wall switch-controlled lighting outlet [210.70(1)]. Often fluorescent luminaires, either surface-mounted or lay-in fixtures in suspended ceilings, are installed in laundry rooms. Luminaires installed in suspended ceilings are covered in the section of this chapter titled "Family Rooms and Wet Bars."

Laundry Branch Circuit

The laundry branch circuit is rated 20 amperes and supplies one or more laundry receptacle outlets [210.11(C)(2)]. It is one of four special circuits required in one- and two-family dwellings (the others are the bathroom branch circuit and a minimum of two small-appliance branch circuits).

At least one receptacle outlet for laundry equipment [210.52(F)] must be installed within 6 ft of the intended location of the appliance [210.50(C)]. The required laundry receptacle(s) supplies the clothes washer and sometimes a gas dryer (which requires electric power for its controls). Receptacle outlets in other rooms are not allowed on the laundry branch circuit.

General-Purpose Receptacles. The *NEC* doesn't require additional receptacles in the laundry room. When additional receptacle outlets are provided for ironing or to serve as a wall switch-controlled lighting outlet, they must be connected to a different branch circuit. Additional receptacle outlets aren't permitted to be on the 20-ampere laundry branch circuit.

GFCI Protection for Laundry Receptacles

Unfinished Areas. When laundry equipment is installed in an unfinished portion of a basement, 15- and 20-ampere receptacle outlets are generally required to have GFCI protection. However, receptacle outlets that supply appliances occupying dedicated space (those that aren't easily moved from one place to another) are exempt from this *NEC* requirement [210.8(A)(5)]. This exemption means that a single receptacle used to supply a clothes washer or gas dryer in an unfinished basement area is not required to have GFCI protection (Figure 10.6). A duplex receptacle used to supply both a washer and a dryer also isn't required to have GFCI protection [210.8(A)(5)].

Finished Areas. Receptacle outlets in finished laundry rooms generally don't need GFCI protection under *NEC* rules. (But see the next section, "Bedrooms in Basement Areas," as well.)

Near Laundry Sinks. Receptacle outlets installed within 6 ft of the outside edge of a laundry sink are required to have GFCI protection [210.8(A)(7)]. This rule

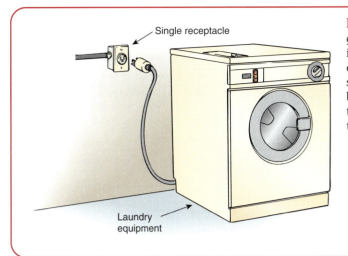

Single receptacle

Laundry
equipment

FIGURE 10.6 A single receptacle supplying an appliance that occupies dedicated space in an unfinished basement is not required to have GFCI protection.

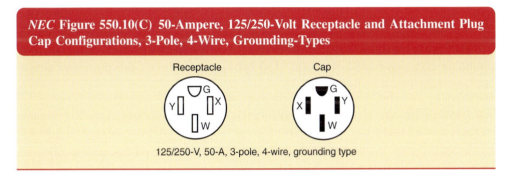

NEC **Figure 550.10(C) 50-Ampere, 125/250-Volt Receptacle and Attachment Plug Cap Configurations, 3-Pole, 4-Wire, Grounding-Types**

Receptacle   Cap

125/250-V, 50-A, 3-pole, 4-wire, grounding type

EXHIBIT 10.1

applies to both the required receptacle for laundry equipment, and to general-purpose receptacles, if they are located within 6 ft of the laundry sink.

Clothes Dryers

Electric Clothes Dryers. Nameplates on motorized appliances are marked to indicate the minimum circuit conductor ampacity and maximum overcurrent protection (circuit breakers or fuse) [422.62(B)]. A typical electric clothes dryer is rated 240 volts, 30 amperes, single-phase, 3-wire, and uses 10 AWG conductors. Some larger models are rated 50 amperes. (See Exhibit 10.1.)

Wiring Methods. Electric clothes dryers can be direct-connected (hard-wired) or supplied through a plug and receptacle. The second method is more common for the following two reasons:

1. It's easier to move the appliance for cleaning, repair, or replacement (Figure 10.7).
2. The attachment plug and receptacle can serve as the required appliance disconnecting means [422.33(A)].

Dryer receptacles are available in both surface and recessed models. The recessed type requires a 4-in. square box. Often the cordset is supplied with the appliance, but sometimes it must be purchased separately.

FIGURE 10.7 Supplying electric dryers by means of a 4-wire cordset and attachment plug is the most common and convenient method.

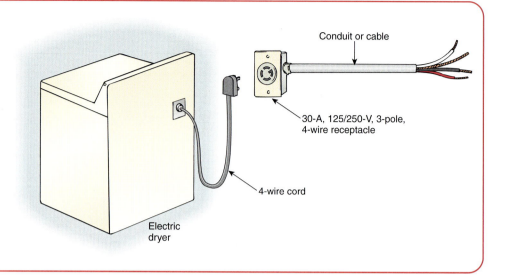

Conduit or cable

30-A, 125/250-V, 3-pole, 4-wire receptacle

4-wire cord

Electric dryer

All wiring methods for clothes dryers are required to include an equipment grounding conductor [250.114(3)(b), 250.140] in order to ground the metal frame and minimize shock hazard.

Gas Clothes Dryers. Gas clothes dryers require a small amount of electric power for their controls and gas igniters. They come supplied with a cordset and NEMA 5-15P plug, which is plugged into the laundry receptacle.

Exhaust Fans. Some laundry rooms have exhaust fans to help control heat and humidity. Installation of exhaust fans is covered in Chapter 8, The Bathroom.

BEDROOMS IN BASEMENT AREAS

The *National Electrical Code* doesn't contain a definition of the term *bedroom,* and rooms in dwellings are classified by building codes. Typically, a bedroom is required to have a window and a closet. If a room on the basement level of a dwelling is designated as a bedroom, special rules affect how its electrical systems are planned and installed. See Chapter 9, The Bedroom, for more information.

The rules for bedrooms in basement areas can be summarized as follows.

• *Smoke Detectors.* Hard-wired smoke detectors must be installed inside and outside the basement bedroom (see Chapter 15, Special Systems).

• *Receptacles.* Receptacle outlets in bedrooms must be located according to the 12-foot rule, and any wall at least 2 ft wide must have a receptacle outlet, such as those between doors and fireplaces. Fixed room dividers such as freestanding bar-type counters or railings must also have receptacles spaced to meet the 12-foot rule [210.52(A)(2)].

• *Arc-Fault Circuit Interrupter.* All branch circuits feeding outlets in a basement bedroom must have arc-fault circuit-interrupter (AFCI) protection (Figure 10.8).

GFCI Not Required. Receptacle outlets in finished areas of basements such as bedrooms and home offices aren't required to have GFCI protection.

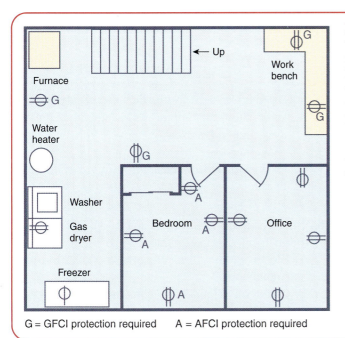

FIGURE 10.8 Receptacle outlets in a basement bedroom must have AFCI protection, and those in unfinished areas must be GFCI protected. Receptacles in a basement home office don't require any type of special protection.

G = GFCI protection required   A = AFCI protection required

SPECIAL RULES FOR TWO-FAMILY DWELLINGS

Some two-family dwellings have a basement with storage space, laundry facilities, or HVAC equipment that is used by both units. The branch circuits feeding lighting, receptacles, and other loads in these common areas must be supplied from a house loads panel [210.25]. Electrical equipment in common areas cannot be supplied from either dwelling unit.

10 CHAPTER REVIEW

MULTIPLE CHOICE

1. An unfinished basement area used for storage or equipment is required to have (a) lighting outlet(s)
   A. At or near equipment requiring servicing
   B. At the foot of each stairway
   C. In any area without a window
   D. Both A and B

2. The following areas of a basement are required to have receptacle outlets:
   A. Unfinished area
   B. Bedroom
   C. Laundry room
   D. All of the above

3. GFCI protection is required for receptacles installed in the following area(s) of a basement:
   A. Workshop in finished room
   B. Bedroom
   C. Within 6 ft of a wet bar sink
   D. All of the above

4. Wall switch-controlled lighting outlets are required in the following area(s) of a basement:
   A. Foot of a stairway
   B. Outside each exterior entrance
   C. At or near equipment requiring servicing
   D. None of the above

5. AFCI protection is required for basement receptacle outlets in the following area(s):
   A. Laundry room or area
   B. Exterior walls
   C. Bedroom
   D. All of the above

6. Three-way switches are required to control lighting outlets when a basement
   A. Has multiple entrances
   B. Is entered by a stairway from the floor above
   C. Has equipment requiring servicing located in two or more areas
   D. None of the above

7. The following receptacle outlets in a basement do not require GFCI protection:

A. Those on branch circuits that also supply lighting
B. Those supplying appliances occupying dedicated space that are not easily moved from one place to another
C. Those with an isolated equipment grounding conductor
D. Both B and C

8. The following *NEC* rule(s) can be applied to laundry rooms located in basements:
   A. Electric clothes dryers must have a cord-and-plug connection.
   B. One or more laundry receptacles must be supplied by a 20-ampere branch circuit that has no other outlets.
   C. The laundry room must have a wall switch-controlled lighting outlet with a permanently installed lampholder or luminaire (lighting fixture).
   D. All of the above

9. The following statement(s) about the electric supply to a gas clothes dryer located in an unfinished area of a basement is (are) true:
   A. It uses a cord-and-plug connection.
   B. The receptacle is not required to have GFCI protection.
   C. The wiring method used to supply the receptacle must include an equipment grounding conductor.
   D. All of the above

10. The following wiring method(s) can be used in the basement of a one-family dwelling:
    A. Type AC cable
    B. Type NM cable
    C. Conductors in electrical metallic tubing (EMT)
    D. All of the above

FILL IN THE BLANKS

1. At least one control for basement lighting is required to be located at _____.

2. The minimum circuit ampacity and maximum overcurrent protection for an electric clothes dryer is marked on the appliance _____.

3. The laundry branch circuit is required to supply at least _____ receptacle outlet(s).

4. Receptacles installed more than 6½ ft above floor level are considered to be _____.

5. Receptacle outlets installed in basement bedrooms are required to have _____ protection.

6. A wall switch-controlled lighting outlet is required on the _____ side of any outdoor entrance to a basement.

7. All clothes dryer wiring methods are required to include a(n) _____ conductor.

8. A single receptacle that supplies a freezer in an unfinished basement area is not required to have _____ protection.

9. Electric clothes dryer receptacles are typically rated 240 volts, _____ or _____ amperes.

10. Receptacles serving wet bar countertops located within _____ ft of the sink are required to have GFCI protection.

TRUE OR FALSE

1. Receptacle outlets in finished areas of basements are required to have GFCI protection.

_____ True   _____ False

2. Article 100 defines *bedroom* as "an area intended for sleeping and provided with a window and closet."

_____ True   _____ False

3. The *Code* requires a wall switch-controlled lighting outlet near equipment requiring servicing in a utility room.

_____ True   _____ False

4. A duplex receptacle supplying only a refrigerator located in an unfinished part of a basement requires GFCI protection.

_____ True   _____ False

5. All branch-circuit wiring installed in unfinished basement areas is required to be run in EMT or Schedule 40 PVC for physical protection.

_____ True   _____ False

6. A bedroom located in a basement is required to include a smoke detector.

_____ True   _____ False

7. Lampholders or luminaires (lighting fixtures) may be installed in all areas of basements.

_____ True   _____ False

8. Each unfinished area of a basement is required to have a minimum of one receptacle outlet in addition to any required for laundry equipment.

_____ True   _____ False

9. Receptacle outlets in unfinished areas of basements must be supplied by separate branch circuits from those in finished areas.

_____ True   _____ False

10. A bathroom located in a basement area below grade is required to have AFCI protection on all receptacle outlets.

_____ True   _____ False

CHALLENGE QUESTIONS

1. Receptacle outlets in unfinished areas of basements as well as those located outdoors and in garages are required to have GFCI protection for personnel. Explain why receptacle outlets located more than 6½ feet above the finished floor (AFF) or above grade are exempt from this GFCI requirement.

2. Why does the *Code* require that a single appliance located within a dedicated space be supplied from a single non–GFCI-protected receptacle rather than a standard duplex receptacle?

3. One reason the *Code* sometimes permits flexible wiring methods is to provide vibration isolation for motorized equipment. What is the reason for allowing unsupported lengths of cables and flexible raceways to supply luminaires or other electrical equipment located within accessible ceilings?

4. A bedroom in a basement must have arc-fault circuit-interrupter (AFCI) protection. A home office in a basement isn't required to. Can the same branch circuit(s) be used to supply receptacle outlets and lighting outlets in both rooms?

5. Receptacle outlets in unfinished areas of basements are required to have GFCI protection for personnel while those in finished areas aren't. What's the reason for these different requirements?

6. Does the *NEC* permit luminaires in basements, utility rooms, and similar spaces to have pull chains?

7. Does the "12-foot-rule" apply to receptacle outlet locations in basements? Explain your answer.

Hallways, Stairways, and Attics

11

OBJECTIVES

After completing this chapter, the student will be able to understand the following:

- Number of lighting outlets that must be provided in a hallway 10 ft long
- Switch requirements for stairway lighting
- When occupancy sensors can be used to control residential lighting
- Receptacle requirements for hallways
- Lighting requirements for attics

IMPORTANT NEC TERMS

Branch Circuit

Lighting Outlet

Luminaire

Outlet

Receptacle

Receptacle Outlet

Switch, General-Use Snap

INTRODUCTION

This chapter covers three areas of the home—hallways, stairways, and attics—that don't fall into the classification of habitable rooms and are thus subject to somewhat different *National Electrical Code*® requirements. The following major topics are included:

- Lighting and receptacles in hallways
- Lighting and receptacles in stairways
- Lighting, switching, and receptacles in attics
- Special rules for two-family dwellings

LIGHTING AND RECEPTACLES IN HALLWAYS

Lighting in Hallways

The *NEC*® requires that at least one wall switch-controlled lighting outlet be installed in every hallway [210.70(A)(2)(a)]. The *Code* doesn't have switching rules for hallway lighting, probably because so many different layouts are possible. Safety and convenience are the main criteria for deciding where to install switches.

General Guidelines for Planning and Installing Hallway Lighting Controls

- *Single Point of Entry.* A hallway with only one point of entry (such as one leading to a closet or powder room) should have a wall switch placed where a person enters the hallway. Only one switch is needed for this type of hallway.

- *Two or More Points of Entry.* A hallway that can be entered from more than one point should use multiway switching. The goal is to avoid forcing homeowners to walk through a dark space looking for a light switch—that's how accidents happen. As a rule of thumb, residents shouldn't have to walk more than 6 ft to find a hallway lighting switch.

- *Hallway Leading to a Habitable Room(s).* A hallway leading to habitable rooms should also use multiway switches (Figure 11.1).

Automatic Controls. The *NEC* permits automatic control of hallway lighting by devices such as occupancy sensors (Figure 11.2) and time clocks [210.70(A)(2)(a), Exception].

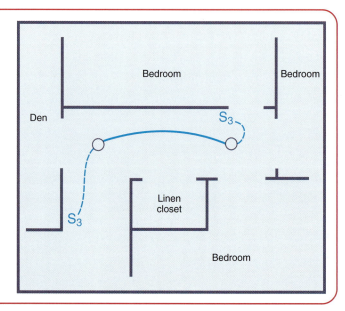

FIGURE 11.1 Three-way switches are often a convenient method for controlling hallway lighting.

FIGURE 11.2 Lighting in hallways and stairways of dwelling units is permitted to be controlled by automatic means such as occupancy sensors. (Courtesy of Pass & Seymour/Legrand®)

Receptacle Outlets in Hallways

Section 210.52(H) requires residential hallways 10 ft or more in length to have at least one receptacle outlet (Figure 11.3). The *Code* doesn't specify a required location for this outlet. In longer hallways, providing additional receptacle outlets improves

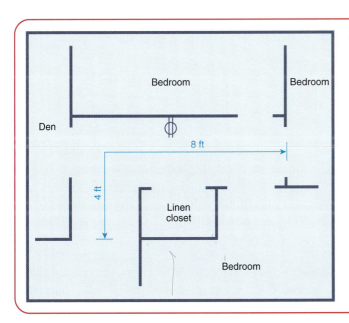

FIGURE 11.3 In dwellings, hallways 10 ft or more in length must have at least one receptacle outlet.

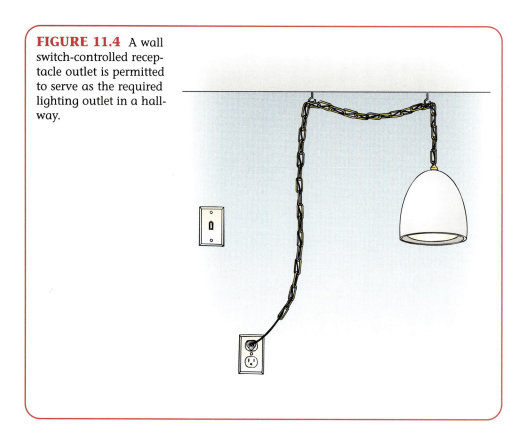

FIGURE 11.4 A wall switch-controlled receptacle outlet is permitted to serve as the required lighting outlet in a hallway.

convenience, since they are frequently used for supplying table lamps or picture-frame lights.

Lighting Outlets. The receptacle outlet required in a hallway 10 ft or longer is also permitted to serve as the required hallway lighting outlet. For this type of installation, the receptacle could be split-wired, as shown in Figure 11.4. The top half is switched to control a lamp, while the lower half is "always on" for use by a vacuum cleaner or other appliance.

LIGHTING AND RECEPTACLES IN STAIRWAYS

Lighting in Stairways

The *Code* requires at least one wall switch-controlled lighting outlet in each interior stairway of a dwelling [210.70(A)(2)]. Because there are so many designs and locations for stairways, this rule is sometimes subject to interpretation, and the authority having jurisdiction (AHJ) should be consulted. The following are examples of different stairway designs:

- *Open Design.* In houses having a two-story entrance foyer or "great room" with an open stairway to the second floor, a single chandelier or other luminaire (lighting fixture) may provide the required lighting for both the foyer or great room and the stairway.

- *Split-Level Design.* In dwellings, such as a split level, that have short stairways between levels, the stairs may be adequately illuminated by the required lighting outlets on each floor.
- *Circular Stairway Design.* It is often impractical to install a lighting outlet "in" a circular stairway itself, but again, as in the case of a split-level design, the treads may be adequately illuminated by other required lighting outlets on each floor.

Switching in Stairways. Section 210.70(A)(2)(c) requires that a wall switch be installed at each floor level and at each landing that includes an entryway, where the stairway between the floor levels has six risers or more (Figure 11.5). Using three- and four-way switches is usually a practical approach.

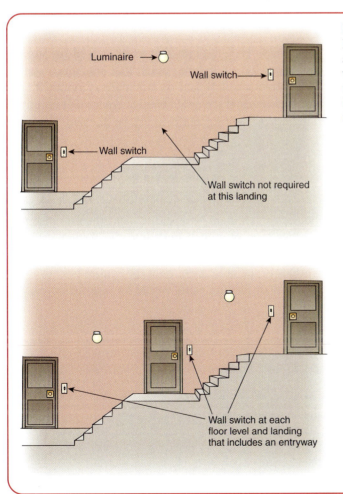

FIGURE 11.5   Where lighting outlets are installed for interior stairways, a wall switch is required at each floor level and landing that includes an entryway.

Automatic Controls. The *NEC* permits automatic control of interior stairway lighting by devices such as occupancy sensors and time clocks [210.70(A)(2)(c), Exception]. (See Figure 11.2.)

Receptacle Outlets in Stairways

The *National Electrical Code* doesn't require receptacles in stairways, but, as previously noted, architectural designs vary greatly. Some stairways have landings large enough for side tables with lamps or to display artwork that needs supplemental lighting. In such cases, it may make sense to provide a receptacle outlet. However, receptacles should not be installed in stairways where appliance or extension cords running over risers could create a tripping hazard.

LIGHTING, SWITCHING, AND RECEPTACLES IN ATTICS

The installation of branch-circuit wiring in attics is covered in Chapter 4, Wiring Methods. Other rules and considerations are covered in this section.

Lighting in Attics

Section 210.70(A)(3), shown in Exhibit 11.1, requires that an attic used for storage or that contains mechanical equipment have a lighting outlet that contains a switch or is controlled by a wall switch. Often a lampholder or luminaire is mounted near the scuttle hole or other entrance. However, if the attic contains equipment requiring servicing, such as a ventilation fan, a lighting outlet must be located near that equipment (Figure 11.6).

Switching in Attics

Usual Point of Entry. At least one switch must be located at the "usual point of entry" to the attic. If a lampholder or luminaire is installed near the entrance, the switch can be a pull-chain or other type of switch built into the fixture itself. A lighting outlet located away from the attic entrance must be controlled by a wall switch, which can be installed either in the attic itself, near the scuttle hole or door, or on the floor below where a ladder or folding stairs would be used to access the attic.

EXHIBIT 11.1

NEC **210.70(A)(3) Storage or Equipment Spaces**

For attics, underfloor spaces, utility rooms, and basements, at least one lighting outlet containing a switch or controlled by a wall switch shall be installed where these spaces are used for storage or contain equipment requiring servicing. At least one point of control shall be at the usual point of entry to these spaces. The lighting outlet shall be provided at or near the equipment requiring servicing.

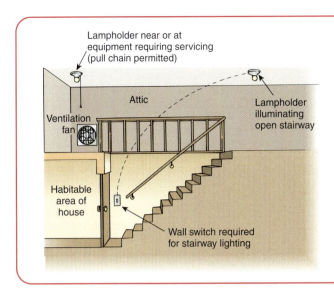

Lampholder near or at equipment requiring servicing (pull chain permitted)

Attic

Lampholder illuminating open stairway

Ventilation fan

Habitable area of house

Wall switch required for stairway lighting

FIGURE 11.6 Lighting in an attic accessed by a permanent stairway must comply with *NEC* 210.70(A)(2)(a) and 210.70(A)(3).

Pilot Light Switch. When the attic lighting control is located on the floor below, installing a wall switch with an integral pilot light is a good way to prevent accidentally leaving the lights on in an untended area.

Receptacles in Attics

The *National Electrical Code* doesn't require receptacle outlets in attics under most circumstances. However, providing a receptacle can improve convenience, particularly if the attic contains equipment such as a ventilation fan that requires servicing. Both lampholders and luminaires are also available with built-in receptacles.

Heating, Air-Conditioning, and Refrigeration Equipment. If an attic contains heating, air-conditioning, or refrigeration (HACR) equipment, a 15- or 20-ampere, 125-volt receptacle must be installed at an accessible location on the same level and within 25 ft of the equipment. This receptacle cannot be connected to the load side of the equipment disconnecting means [210.63].

Stairways in Attics

When permanent stairs lead to an attic, they must be lighted according to *Code* rules.

Lighting Outlet. A lampholder or luminaire mounted at the top of the attic stairs is normally sufficient. (See Figure 11.6.)

Switching. The *NEC* requires wall switches to control stairway lighting [210.70(A)(2)(c)]. Pull-chains are not permitted on attic stairways. Since an unfinished attic isn't a habitable room where people spend long periods of time, the use of three-way switches to control a single lighting outlet may not be necessary for the stairway. However, unlike lighting in unfinished basements, attic lighting is not permitted to be controlled by occupancy sensors.

SPECIAL RULES FOR TWO-FAMILY DWELLINGS

Some two-family dwellings have a hallway, interior stairway, or attic with storage space used by both units. The branch circuit feeding these common areas must be supplied from a house loads panel [210.25].

Lighting outlets for interior stairways at two-family dwellings are often controlled by a time clock or occupancy sensor, as permitted by the Exception to 210.70(A)(2)(c). The use of an automatic control avoids the need for a wall switch at each landing with a point of entry, when landings are separated by six or more risers.

MULTIPLE CHOICE

1. The following hallway(s) is (are) required to have a receptacle outlet:
   A. Hallway with two or more points of entry
   B. Exit hallway
   C. Hallway 10 ft or more in length
   D. Both B and C

2. Attics are required to have more than one lighting outlet if they
   A. Are accessible
   B. Contain equipment requiring servicing
   C. Contain habitable space
   D. None of the above

3. Interior stairway lighting is permitted to be controlled by the following:
   A. Automatic control
   B. Occupancy sensor
   C. Wall switch at each level that includes a point of entry
   D. All of the above

4. The following area(s) is (are) required to have a receptacle outlet:
   A. Attic with equipment requiring maintenance
   B. Stairway landing with floor area exceeding 20 ft^2
   C. Hallway 12 ft in length
   D. Both A and C

5. An attic luminaire (lighting fixture) mounted near a scuttle hole can be controlled by a
   A. Pull-chain switch
   B. Snap switch that is part of the fixture
   C. Wall switch
   D. All of the above

6. If an attic contains heating, air-conditioning, or refrigeration equipment, a 15- or 20-ampere, 125-volt receptacle must be installed
   A. On the floor below and within 50 ft of the equipment
   B. On the same level and within 25 ft of the equipment
   C. On the same level and within 10 ft of the equipment
   D. All of the above

7. A hallway more than 10 ft long, with three points of entry, is required to have a lighting outlet controlled by the following number of switches:
   A. One
   B. Two
   C. Three
   D. None

8. The following types of interior stairways are required to have wall switch-controlled lighting:
   A. Permanent stairs to an attic
   B. Open stairways
   C. Circular stairways
   D. All of the above

9. A lighting outlet located away from an attic entrance can be controlled by the following type(s) of switch:
   A. Pull-chain switch
   B. Wall switch located in the attic near the entrance
   C. Wall switch located on the floor below, near the attic entrance
   D. Both B and C

10. Receptacle outlets are required to be installed in attics when they
    A. Are used for storage
    B. Contain heating, air-conditioning, or ventilating equipment requiring servicing
    C. May be converted to habitable space in the future
    D. Both A and B

FILL IN THE BLANKS

1. An attic used for storage is required to have at least _____ lighting outlet(s).

2. A hallway with three entrances must have a minimum of _____ wall switch(es).

3. A wall switch to control interior stairway lighting is required at each floor or landing level with a point of entry when these levels are separated by _____ or more risers.

4. The *Code* requires a minimum of _____ lighting outlet(s) for an interior stairway that changes direction at a landing.

5. Branch circuits serving common areas in a two-family dwelling must be supplied from a _____.

6. The required receptacle outlet in a hallway 10 ft or more in length is also permitted to serve as the required _____ outlet.

7. Lighting outlets in hallways of dwellings are permitted to be controlled by wall switches, by remote, central, or automatic means, or by _____ sensors.

8. When an attic contains heating, air-conditioning, or refrigeration equipment, a receptacle rated _____ must be installed at an accessible location on the same level and within 25 ft of the equipment.

9. When a lighting outlet is installed in an interior stairway, a wall switch must be installed at each floor level and at each landing that includes a point of entry where the stairway between the floor levels has six or more _____.

10. The receptacle outlet required in attics with mechanical equipment requiring servicing must be located on the same level and within _____ ft of the equipment.

TRUE OR FALSE

1. Wall switch-controlled lighting outlets are required at the top and bottom of an interior stairway.

    _____ True   _____ False

2. A minimum of one receptacle outlet is required in an attic containing equipment that requires servicing.

    _____ True   _____ False

3. A split-wired receptacle controlled by three-way switches can serve as the lighting outlet in a hallway 14 ft long.

    _____ True   _____ False

4. A stairway between two floors with a single landing requires three wall switches (two three-way and one four-way) to control lighting.

    _____ True   _____ False

5. Occupancy sensors are permitted to be used to control hallway and interior stairway lighting.

    _____ True   _____ False

6. An attic that contains mechanical equipment must have a separate lighting outlet for each piece of equipment requiring maintenance.

    _____ True   _____ False

7. Lighting outlets in non-habitable spaces such as attics used for storage purposes are required by *Code* to have switches with pilot lights.

    _____ True   _____ False

8. General-use snap switches are constructed so that they can be installed in device boxes or on box covers.

    _____ True   _____ False

9. *NEC* 210.70(A)(2)(c), Exception, permitting automatic control of lighting in hallways, in stairways, and at outdoor entrances, applies only to common areas of two-family dwellings.

    _____ True   _____ False

10. Attic lighting is permitted to be controlled by an occupancy sensor equipped with a manual override that allows the sensor to function as a wall switch.

    _____ True   _____ False

CHALLENGE QUESTIONS

1. What are some reasons that the *Code* probably permits automatic control lighting in hallways and stairways without requiring that a wall switch or its equivalent be provided as is the case in habitable rooms?

2. In stairways, a wall switch is required at each floor level and at each landing that includes an entryway where the stairway between the floor levels has six risers or more. Why do you think the *Code* doesn't require the use of three- and four-way switches for convenient control of stairway lighting?

3. Why doesn't the *Code* require receptacle outlets in stairways as it does in habitable rooms and outdoors?

4. Why does *NEC* 210.63 require that a 125-volt, 15- or 20-ampere receptacle provided for servicing HVAC equipment not be connected to the load side of the equipment disconnecting means?

Outdoor Areas

12

OBJECTIVES

After completing this chapter, the student will be able to understand the following:

- Ground-fault circuit-interrupter (GFCI) protection requirements for outdoor receptacles
- Underground wiring practices
- *National Electrical Code®* rules for wiring and luminaires (lighting fixtures) installed in trees
- Types of weatherproof covers required on outdoor receptacles
- Lighting rules for attached and detached garages

INTRODUCTION

This chapter covers installation practices for most electrical equipment used in residential outdoor areas or in accessory buildings. The following major topics are included:

- Outdoor electrical installations
- Outdoor receptacle outlets
- Outdoor lighting
- Underground wiring
- Garages and accessory buildings
- Special rules for two-family dwellings

IMPORTANT NEC TERMS

Bonding (Bonded)

Branch Circuit

Branch Circuit, Appliance

Branch Circuit, General-Purpose

Branch Circuit, Individual

Equipment

Ground

Grounded

Ground-Fault Circuit Interrupter

Grounding Conductor, Equipment

Labeled

Listed

Location, Damp

Location, Wet

Luminaire

Receptacle

Receptacle Outlet

Weatherproof

OUTDOOR ELECTRICAL INSTALLATIONS

All dwellings have some electrical equipment installed outdoors or in accessory buildings such as garages and storage sheds. Some of these items, such as GFCI-protected receptacle outlets and outdoor lighting at entrances, are mandatory. Others, such as decorative landscape lighting, are optional.

The *NEC®* has special rules for equipment located outdoors. Generally, these rules deal with protecting the equipment itself from the elements (rain, snow, and direct sunlight) or with providing extra shock protection for people using electrical equipment outdoors under potentially damp conditions.

This chapter describes installation practices for most electrical equipment installed outdoors or in accessory buildings. The *Code* rules for swimming pools, spas, and similar installations are covered in Chapter 13, Swimming Pools, Hot Tubs, and Spas. Air-conditioning and heat pump equipment located outdoors is covered in Chapter 14, HVAC Equipment and Water Heaters.

This chapter doesn't cover electrical installations at docks and boathouses associated with one- and two-family dwellings. (See Article 555 of the *National Electrical Code*.)

OUTDOOR RECEPTACLE OUTLETS

Required Receptacle Outlet Locations

For a one-family dwelling and for each unit of a two-family dwelling located at grade level, at least one receptacle outlet accessible from grade and not more than 6½ ft above it must be installed outdoors at the front and back of the dwelling (Figure 12.1) [210.52(E)].

Heating, Air-Conditioning, and Refrigeration (HACR) Maintenance Outlet. When a dwelling has heating, air-conditioning, or refrigeration equipment located outdoors, a 125-volt, 15- or 20-ampere receptacle outlet must be located at an

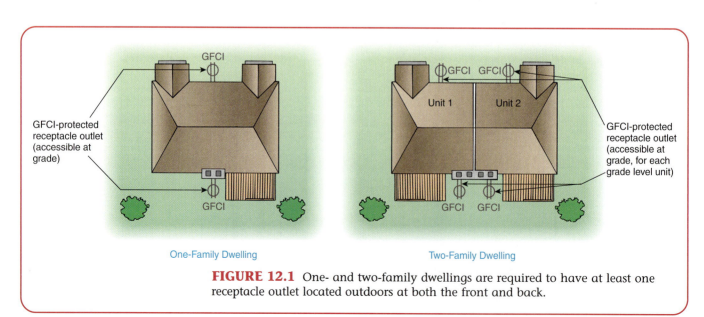

FIGURE 12.1 One- and two-family dwellings are required to have at least one receptacle outlet located outdoors at both the front and back.

accessible location (on the same level and within 25 ft of the equipment) for servicing the equipment. This receptacle outlet must not be connected to the load side of the equipment disconnecting means [210.63]. However, it isn't required to be on an individual branch circuit.

The required outdoor receptacle can also serve as the HACR maintenance outlet if it is located within 25 ft of the heat pump, air-conditioning compressor, or other equipment. Similarly, the same receptacle outlet can serve more than one item of HACR equipment if properly located.

GFCI Protection of Outdoor Receptacles

Outdoor receptacle outlets installed at dwellings are required to have ground-fault circuit-interrupter protection for personnel [210.8(A)(3)]. That rule covers the following receptacles:

- Required outlets at the front and back of the dwelling
- Receptacles on outdoor porches or balconies (any height above grade)
- Accessible receptacles on rooftops [see 210.8(A)(3), Exception]
- Receptacles in a crawl space at or below grade level
- Receptacles mounted in yard areas
- Receptacles mounted on the outside walls of garages or other accessory buildings
- Maintenance receptacles for heating, air-conditioning, or refrigeration equipment

Protection can be provided by using a GFCI circuit breaker for the entire branch circuit, by installing a GFCI receptacle at each outdoor outlet, or by using a single GFCI receptacle to provide "downstream protection" for multiple other receptacles.

If a GFCI receptacle with a visible test button is *not* installed at each outdoor receptacle outlet, a label that reads "GFCI-Protected Outlet" or equivalent wording should be applied to the face of the device. These labels are supplied with GFCI circuit breakers and receptacles.

Outdoor Receptacle Outlets Prohibited on Bathroom or Small-Appliance Circuits. The bathroom branch circuit and small-appliance branch circuits in a dwelling aren't permitted to supply any other outlets [210.11(C)(3), 210.52(B)(2)]. GFCI-protected outdoor receptacles must be supplied from different branch circuits and are permitted to be on the same general-purpose branch circuits with other lighting and receptacle outlets.

For example, a required outdoor receptacle outlet on the front wall of a house can be connected to an indoor receptacle outlet in the living room, and a receptacle-type GFCI can be installed at that location. However, an outdoor receptacle outlet *cannot* be circuited to an indoor outlet in a dining room, because dining room receptacles must be supplied by a small-appliance branch circuit, which is not permitted to have any other outlets.

Outdoor Receptacles Not Required to Have GFCI Protection. Receptacle outlets for snow-melting or de-icing equipment that aren't readily accessible and that are supplied by a dedicated branch circuit don't require GFCI protection [210.8(A)(3), Exception]. These include receptacles mounted at house eaves for anti-freezing cables for gutters and those located in crawl spaces to power heat tapes for water supply piping (Figure 12.2).

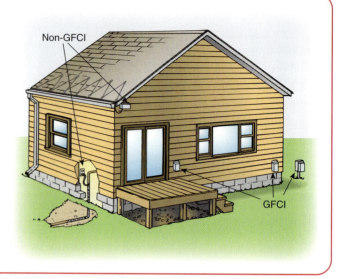

FIGURE 12.2 Most outdoor receptacles at dwellings are required to have GFCI protection but may be exempt if not in a readily accessible location.

Installing Receptacle Outlets in Damp and Wet Locations

Weatherproof Covers. Receptacles in outdoor damp locations, such as those located beneath a porch roof where they are protected from the weather, must have covers that are weatherproof only when a receptacle is not inserted (that is, when the receptacle cover is closed) [406.8]. All 15- and 20-ampere receptacles in outdoor wet locations must have covers that are weatherproof, even when an attachment plug is inserted (Figure 12.3) [406.8(B)(1)].

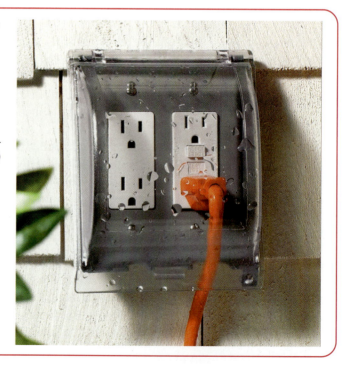

FIGURE 12.3 All 15- and 20-ampere receptacles installed outdoors in wet locations must have covers that are weatherproof, regardless of whether an attachment plug is inserted. (Courtesy of Carlon® Lamson & Sessions)

FIGURE 12.4 A conventional outlet box installed on the outside wall of a building is protected by a weatherproof receptacle cover. (Courtesy of Thomas & Betts Corp.)

Standard Device Boxes. When receptacle outlets are installed on exterior walls of houses and other buildings or under a roof or overhang, a standard device box recessed into the wall construction can be used. The weatherproof receptacle cover installed on the box keeps moisture out of the box (Figure 12.4).

Boxes in Wet Locations. When receptacle outlets are installed at other outdoor locations, such as freestanding receptacles for pool equipment or landscape lighting, enclosures listed for wet locations must be used [314.15(A)]. These boxes typically have threaded hubs or entries.

Raceway-Supported Enclosures. When freestanding boxes containing wiring devices are supported by conduits (Figure 12.5), they must have threaded entries or hubs identified for the purpose. They must be supported by at least two conduits threaded into the enclosure and secured within 18 in. of the enclosure [314.23(F)].

Boxes aren't permitted to be supported by raceways attached with locknuts and bushings. This means that boxes aren't permitted to be supported by electrical metallic tubing (EMT) [358.12(5)].

Wiring Methods

When outdoor receptacles are installed on device boxes recessed into exterior walls and under roofs, Type NM or Type AC cables are the most common wiring methods used in residential construction.

Raceway Wiring Methods. When receptacles are installed in freestanding outdoor boxes, wires in conduits are typically installed. Insulated conductors used in damp and wet locations must be of the types listed in *NEC* 310.8(C). These conductor types typically have a "W" ("wet") in their designations, such as RHW, THW, THHW, and THWN-2.

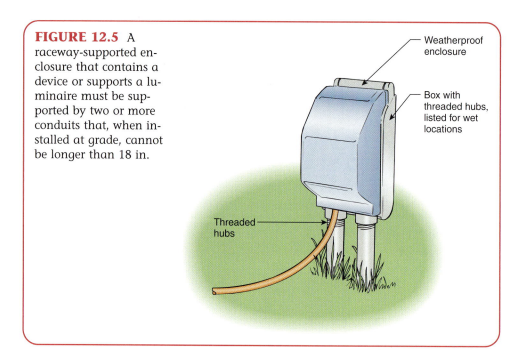

FIGURE 12.5 A raceway-supported enclosure that contains a device or supports a luminaire must be supported by two or more conduits that, when installed at grade, cannot be longer than 18 in.

Weatherproof enclosure

Box with threaded hubs, listed for wet locations

Threaded hubs

Cable Wiring Methods. Certain cable wiring methods can also be used for outdoor receptacles. Types AC and NM cable aren't permitted in damp or wet locations [320.12(2), 334.12(B)(4)]. However, Types UF, SE, and USE cables are permitted to be used for outdoor branch-circuit wiring [338.10(B)(4)(b), 340.10(3)]. Cables installed in locations exposed to the direct rays of the sun must be listed as sunlight resistant or covered with insulating material such as tape or sleeving that's listed as being sunlight resistant [310.8(D), 340.12(9)].

NOTE: Direct-burial wiring methods are described in the section of this chapter titled "Underground Wiring."

OUTDOOR LIGHTING

Outdoor lighting at dwellings must be installed to resist the effects of moisture. Doing so involves both the proper equipment (luminaires, outlet boxes, wiring) and the proper installation techniques.

Luminaires for Wet and Damp Locations

NEC 410.4(A) addresses luminaires for wet and damp locations. Lighting fixtures installed in wet locations must be marked "Suitable for Wet Locations." Such luminaires include fixtures mounted on outside walls without a roof or overhang to protect them, pole-mounted luminaires, bollards, and luminaires mounted in trees.

Fixtures mounted in damp locations protected from being saturated with water, such as those under porch or carport roofs, must be marked "Suitable for Damp Locations." A fixture marked for wet locations can also be used in damp locations.

Installing Outdoor Luminaires

Luminaires in wet and damp locations must be installed so that water cannot accumulate in wiring compartments, lampholders, or other electrical parts [410.4(A)]. Boxes used with outdoor luminaires are the same types as those used with outdoor receptacles.

Standard Device Boxes. When luminaires are installed on exterior walls of houses and other buildings, or under a roof or overhang, a standard device box can be used, but it must be recessed into the wall construction. The wet- or damp-location luminaire installed on the box will have a gasket that keeps moisture out of the box.

Boxes in Wet Locations. When luminaires are installed at other outdoor locations, enclosures listed for wet locations must be used [314.15(A)]. These boxes typically have threaded hubs or entries.

Raceway-Supported Enclosures. When freestanding boxes with luminaires are supported by conduits, they must have threaded entries or hubs identified for the purpose. They must be supported by at least two conduits threaded into the enclosure and secured within 18 in. of the enclosure [314.23(F)].

An enclosure supporting a luminaire or lampholder isn't permitted to be supported by a single conduit, except under certain special conditions described in 314.23(F), Exception No. 2.

Boxes also aren't permitted to be supported by raceways attached with locknuts and bushings. This means that boxes aren't permitted to be supported by electrical metallic tubing (EMT) [358.12(5)].

Wiring Methods

Wiring methods for outdoor luminaires are the same as those for outdoor receptacles, described earlier in this chapter

Luminaires Mounted in Trees

Outdoor luminaires and associated equipment, such as wiring and photocell switches, are permitted to be supported by trees [410.16(H)]. However, trees and other vegetation are not permitted to support overhead conductor spans [225.26]. This prohibition means that if several luminaires are mounted in trees in a yard or other outdoor area, branch-circuit wiring can't be run between them from tree to tree. Instead, wiring must run to each tree using an underground wiring method, and then conductors can run up each tree to the fixture.

Protection. When direct-buried cables emerge from grade, they must be protected from physical damage by a raceway extending from the minimum cover distance below grade (18 in. under most circumstances) to at least 8 ft above grade [300.5(D)(1)]. A bushing, terminal fitting, or seal must be used on the direct-burial end of the raceway (Figure 12.6) [300.5(H)].

Support. When Type UF or Type USE cables are installed on trees, they must be supported in the same manner as Type NM cables used indoors. They must be

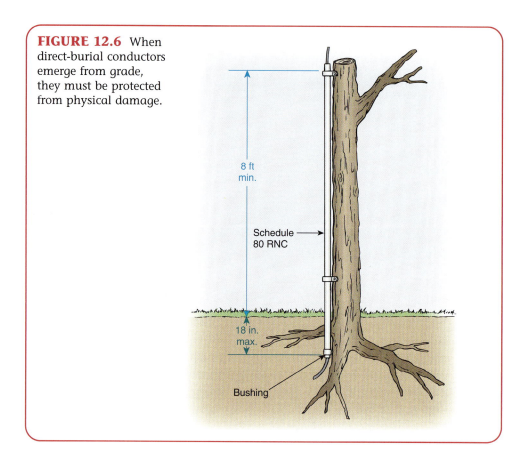

FIGURE 12.6 When direct-burial conductors emerge from grade, they must be protected from physical damage.

8 ft min.

Schedule 80 RNC

18 in. max.

Bushing

secured at least every 4½ ft and within 12 in. of every box, cabinet, or fitting [338.10(B)(4), 340.10(4)].

NOTE: Direct-burial wiring methods are described in greater detail in the section of this chapter titled "Underground Wiring."

Low-Voltage Landscape Lighting

Landscape lighting operating at 30 volts or less (usually 12 volts) is often used outdoors at residences (Figure 12.7). Low-voltage landscape lighting is sold as a listed assembly that includes an isolation transformer that steps 120 volts down to the operating voltage, a weatherproof cable permanently connected to the transformer, and a number of low-voltage luminaires. These systems are listed for use in damp or wet locations.

Power Supply. Low-voltage lighting systems must be supplied from a branch circuit rated 20 amperes maximum. Typically, the 120-volt to 12-volt isolation transformer is plugged into an outdoor receptacle. All 15- and 20-ampere receptacles installed outdoors in wet locations are required to have a cover that is weatherproof, regardless of whether an attachment plug is inserted [406.8(B)(1)]. (See Figure 12.3.)

A receptacle in a damp location, such as one located beneath a porch roof where it is protected from the weather, is permitted to have an enclosure that is weatherproof

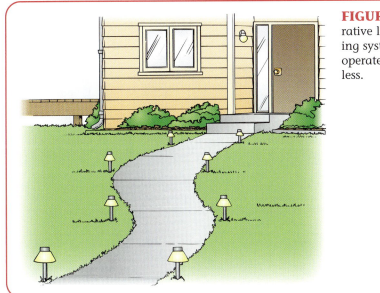

FIGURE 12.7 Decorative landscape lighting systems typically operate at 30 volts or less.

only when a receptacle is not inserted (that is, when the receptacle cover is closed) [406.8(A)].

Installing Low-Voltage Landscape Lighting. Low-voltage luminaires typically have stakes that are inserted into the ground. They are connected together and to the supplying isolation transformer using weatherproof cable. The luminaires are installed on the continuous cable wherever indicated by the lighting layout or as desired by the lighting installer. The fixtures clamp onto the cable using insulation-displacement connectors (pointed prongs that pierce the cable insulation, making contact with the 12-volt conductors inside).

Low-voltage lighting normally can't be installed within 10 ft of a swimming pool, spa, fountain, or similar body of water [411.4]. However, GFCI-protected lighting can be installed as close as 5 ft horizontally from pools and similar bodies of water [680.22(B)(4)].

Low-voltage luminaires can be mounted in trees, the same as 120-volt luminaires. However, separate cables must be installed up each tree trunk to the fixtures, since they aren't permitted to be strung through the air between trees [225.26].

Wiring Methods. The cable supplied as part of prewired, listed low-voltage lighting systems can be installed exposed. Larger custom landscape lighting installations use more conventional wiring methods, such as direct-buried Type UF cable or conductors in raceways. These types of installations must comply with the minimum burial depths specified in Column 5 of *NEC* Table 300.5 (shown in Exhibit 12.1) [300.5(A)] and must be grounded and bonded [250.86].

Voltage Drop. Listed low-voltage lighting systems consist of a transformer, a fixed length of weatherproof cable, and multiple luminaires. These systems will operate properly and safely when installed according to the manufacturer's instructions [110.3(B)]. However, they typically can't be extended using additional cable and

EXHIBIT 12.1

NEC TABLE 300.5 Minimum Cover Requirements, 0 to 600 Volts, Nominal, Burial in Millimeters (Inches)										
	Type of Wiring Method or Circuit									
Location of Wiring Method or Circuit	Column 1 Direct Burial Cables or Conductors		Column 2 Rigid Metal Conduit or Intermediate Metal Conduit		Column 3 Nonmetallic Raceways Listed for Direct Burial Without Concrete Encasement or Other Approved Raceways		Column 4 Residential Branch Circuits Rated 120 Volts or Less with GFCI Protection and Maximum Overcurrent Protection of 20 Amperes		Column 5 Circuits for Control of Irrigation and Landscape Lighting Limited to Not More Than 30 Volts and Installed with Type UF or in Other Identified Cable or Raceway	
	mm	in.	mm	in.	mm	in.	mm	in.	mm	in.
All locations not specified below	600	24	150	6	450	18	300	12	150	6
In trench below 50-mm (2-in.) thick concrete or equivalent	450	18	150	6	300	12	150	6	150	6
Under a building	0 (in raceway only)	0	0	0	0	0	0 (in raceway only)	0	0 (in raceway only)	0
Under minimum of 102-mm (4-in.) thick concrete exterior slab with no vehicular traffic and the slab extending not less than 152 mm (6 in.) beyond the underground installation	450	18	100	4	100	4	150 (direct burial) 100 (in raceway)	6 4	150	6
Under streets, highways, roads, alleys, driveways, and parking lots	600	24	600	24	600	24	600	24	600	24
One- and two-family dwelling driveways and outdoor parking areas, and used only for dwelling-related purposes	450	18	450	18	450	18	300	12	450	18
In or under airport runways, including adjacent areas where trespassing prohibited	450	18	450	18	450	18	450	18	450	18

Notes:
1. Cover is defined as the shortest distance in millimeters (inches) measured between a point on the top surface of any direct-buried conductor, cable, conduit, or other raceway and the top surface of finished grade, concrete, or similar cover.
2. Raceways approved for burial only where concrete encased shall require concrete envelope not less than 50 mm (2 in.) thick.
3. Lesser depths shall be permitted where cables and conductors rise for terminations or splices or where access is otherwise required.
4. Where one of the wiring method types listed in Columns 1–3 is used for one of the circuit types in Columns 4 and 5, the shallowest depth of burial shall be permitted.
5. Where solid rock prevents compliance with the cover depths specified in this table, the wiring shall be installed in metal or nonmetallic raceway permitted for direct burial. The raceways shall be covered by a minimum of 50 mm (2 in.) of concrete extending down to rock.

luminaires without either causing voltage drop problems or overloading the transformer.

UNDERGROUND WIRING

Underground wiring is installed at dwellings to supply outdoor receptacles, outdoor lighting, pool pumps, irrigation systems, snow-melting cables, or accessory buildings such as garages, greenhouses, and pool-equipment structures. Circuits can be run as either direct-buried cables or conductors in raceways. All conductors and cables installed in underground enclosures or raceways must be listed for use in wet locations [300.5(B)].

Wiring Methods

The following raceway and cable types can be installed underground:

- Rigid metal conduit (Type RMC)
- Intermediate metal conduit (Type IMC)
- Rigid nonmetallic conduit (Type RNC)
- Liquidtight flexible metal conduit (Type LFMC) (where marked for underground use)
- Liquidtight flexible nonmetallic conduit (Type LFNC)
- Underground feeder and branch-circuit cable (Type UF)
- Service entrance cable (Type USE) (Type SE is not permitted to be installed underground)
- Mineral-insulated, metal-sheathed cable (Type MI) (not normally used in residential construction)

Minimum Depths

Underground wiring must be installed at the minimum depths specified in *NEC* Table 300.5, shown in Exhibit 12.1. Two special provisions that apply to underground wiring installed at one- and two-family dwellings follow.

1. *Vehicle Areas.* Direct-burial cables or raceways can be installed at a minimum depth of 18 in. under dwelling driveways and parking areas. (The required minimum depth is 24 in. beneath paved vehicle areas at structures other than dwellings.)
2. *GFCI Protection.* All 15- and 20-ampere, 120-volt residential branch circuits with GFCI protection can also be installed at lesser minimum depths than circuits at other occupancies (Figure 12.8). The required depth depends on the type of wiring method installed, as specified in Column 4 of *NEC* Table 300.5, which is shown in Exhibit 12.1.

NOTE: The entire branch circuit, not just individual outlets, must be GFCI protected. This requirement means a GFCI circuit breaker must be used.

Communications and Signaling Conductors. The minimum burial depths for underground wiring shown in Table 300.5 apply only to power conductors. The *Code* doesn't specify burial depths for communications and signaling conductors [90.3].

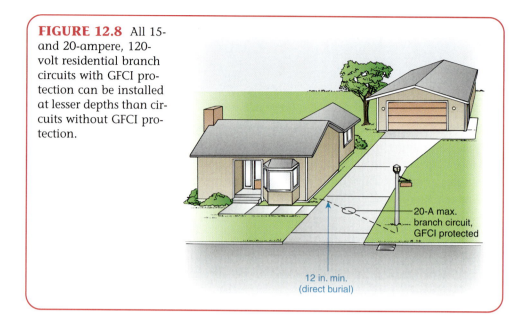

FIGURE 12.8 All 15- and 20-ampere, 120-volt residential branch circuits with GFCI protection can be installed at lesser depths than circuits without GFCI protection.

20-A max. branch circuit, GFCI protected

12 in. min. (direct burial)

However, Table 300.5 provides useful guidance; following its guidelines when installing direct-buried telephone cables or other communications and signal conductors will help prevent damage to these circuits from people digging in their yards or similar causes.

Grounding

All underground installations are required to be grounded and bonded [250.86]. The rules for grounding can be summarized as follows.

- *Direct-Buried Cables.* Type UF or Type USE cables, which are manufactured with integral equipment grounding conductors, must be used (Figure 12.9).
- *RMC and IMC.* Section 250.118 permits both RMC and IMC to function as equipment grounding conductors for the branch-circuit or feeder conductors installed in them. Some installers also choose to run a separate equipment grounding conductor along with the circuit conductors.
- *LFMC.* Since LFMC is only approved to function as an equipment grounding conductor in lengths up to 6 ft, longer runs must have a separate equipment grounding conductor installed with the branch-circuit conductors.
- *Nonmetallic Raceways.* When RNC or LFNC is used for underground wiring, a separate equipment grounding conductor must be run with the circuit conductors.

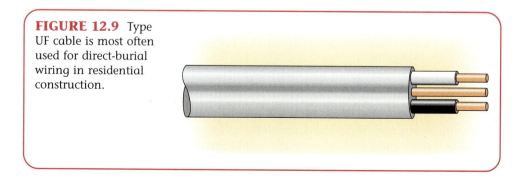

FIGURE 12.9 Type UF cable is most often used for direct-burial wiring in residential construction.

- *Metal Elbows Used with RNC.* Metal elbows are often used with RNC installed below grade (since RNC elbows can be damaged by friction during the process of pulling in conductors) and aren't required to be grounded [250.80, Exception; 250.86, Exception No. 3]. However, the elbow must be isolated from possible human contact by at least 18 in. of cover (Figure 12.10). The RNC must still contain an equipment grounding conductor, but this green ground wire doesn't have to be connected to the metal conduit elbow.

GARAGES AND ACCESSORY BUILDINGS

Attached Garages

Lighting. Garages attached to houses are required to contain at least one wall switch-controlled lighting outlet [210.70(A)(2)(a)]. They also must have one wall switch-controlled lighting outlet installed on the exterior side of outdoor entrances with grade-level access. A vehicle door isn't considered an outdoor entrance [210.70(A)(2)(b)].

Receptacles. Attached garages are required to have at least one receptacle outlet in addition to any provided for laundry equipment [210.52(G)]. Receptacle outlets must have GFCI protection for personnel, unless they meet one of two conditions (Figure 12.11). The rules for receptacles in garages can be summarized as follows.

- *Receptacle Not Readily Accessible.* A receptacle located more than 6½ ft above the floor, such as a ceiling-mounted outlet that supplies a garage door opener, isn't required to have GFCI protection [210.8(A)(2), Exception No. 1].
- *Appliance(s) in Dedicated Space.* When a single receptacle (or a duplex receptacle for two appliances) is located within dedicated space for a cord-and-plug-connected appliance that can't be easily moved from one place to another, the outlet isn't required to have GFCI protection [210.8(A)(2), Exception No. 2]. The intent of this rule is to permit refrigerators and freezers located in garages to be supplied by non-GFCI receptacles so that "nuisance tripping" doesn't cause food to spoil. (Receptacles serving refrigerators in kitchens also aren't required to have GFCI protection.)

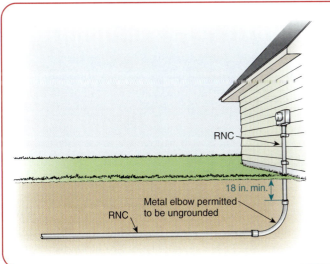

FIGURE 12.10 A metal elbow used in an underground run of RNC does not have to be grounded when it is buried at least 18 in. below grade.

RNC

18 in. min.

Metal elbow permitted to be ungrounded

RNC

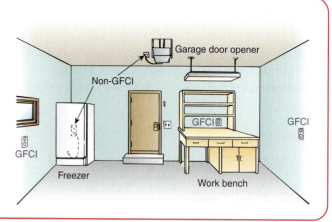

FIGURE 12.11 Most receptacle outlets in garages and outbuildings are required to have GFCI protection unless they are not readily accessible or they supply appliances in dedicated spaces.

Detached Garages

Lighting. Garages separated from a dwelling aren't required to be wired for electricity. When they do have electric service, they are subject to the same lighting rules as attached garages [210.70(A)(2)(a) and (b)].

Receptacles. Detached garages with electric service are subject to the same receptacle rules as attached garages [210.52(G)].

Other Accessory Buildings

Other non-dwelling buildings, such as storage sheds, greenhouses, and garden structures, aren't required to be wired for electricity. But certain rules apply when they are.

Lighting. When an accessory building is used for storage or contains electrical equipment such as a water pump, providing a lighting outlet improves safety and usability. Open lampholders are a low-cost solution for locations where illumination is needed only on an occasional basis.

NOTE: Some AHJs may interpret an accessory building with electricity as being a "storage or equipment space" that requires a wall switch-controlled lighting outlet in accordance with 210.70(A)(3).

Receptacles. When an accessory building with electric service has floors at or below grade, the same rules that apply to garages and unfinished basements should be followed. At least one receptacle outlet should be installed, and provided with GFCI protection, unless it isn't readily accessible or is located within dedicated space for a cord-and-plug-connected appliance that isn't easily moved.

Branch Circuit or Feeder to an Accessory Building

Garages, storage sheds, greenhouses, and similar structures are often supplied by a single 15- or 20-ampere branch circuit from the main dwelling. Structures containing more electric equipment, such as workshops or pool buildings, may have a remote panelboard (sub-panel) supplied by a feeder from the service equipment.

Wiring Methods. Any of the wiring methods previously described in this chapter can be used for an underground branch circuit or feeder. Several special considerations apply to underground cables entering buildings (Figure 12.12):

1. *Protection.* Underground cables installed under a building must be in a raceway that extends beyond the outside walls of the building [300.5(C)].
2. *Type USE Cable.* When Type USE cable emerges from the ground at terminations, it must be protected in accordance with 300.5(D).
3. *Type SE Cable.* After Type SE cable enters a building, it can be installed indoors following the same rules as Type NM cable [338.10(4)(a)].
4. *Type UF Cable.* After Type UF cable enters a building, it can be installed indoors following the same rules as Type NM cable [340.10(4)].

Disconnecting Means. Disconnecting means must be provided for a feeder or branch circuit that supplies an accessory building [225.31]. The disconnecting means can be installed either inside or outside the building in a readily accessible location near the point of entrance of the conductors [225.32].

- *Feeder.* When a remote panelboard is located in the accessory building, the main circuit breaker (MCB) in the panel can serve as the required disconnecting means. Otherwise, a disconnect switch must be provided close to the point where the feeder enters the building.

NOTE: For this reason, the feeder circuit breaker in the service equipment panelboard cannot serve as the required disconnecting means.

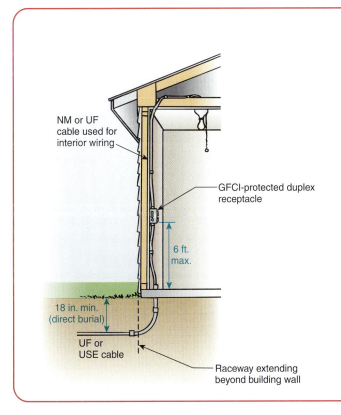

NM or UF cable used for interior wiring

GFCI-protected duplex receptacle

6 ft. max.

18 in. min. (direct burial)

UF or USE cable

Raceway extending beyond building wall

FIGURE 12.12 A single branch circuit can be used to provide electric service to an accessory building, such as a storage shed. Structures other than garages don't require a wall switch-controlled lighting outlet.

- *Branch Circuit.* For garages and outbuildings on residential property, a snap switch or set of three-way or four-way switches is permitted to serve as the required disconnecting means [225.36, Exception].

Grounding Requirements

Branch Circuit. When a single branch circuit supplies an accessory building and it includes an equipment grounding conductor, a separate grounding electrode isn't required at the accessory building [250.32(A), Exception]. Instead, the grounding electrode conductor from the main dwelling service equipment connects to the non-current-carrying metal parts of electrical equipment such as metal boxes, faceplates, and luminaire housings (Figure 12.13).

Feeder. When a feeder from the main dwelling service equipment supplies a panelboard in an accessory building, the structure must have a separate grounding electrode installed [250.32(A)]. The equipment grounding conductor run with the feeder connects to the grounding terminal bar (bus) in the accessory building panelboard. The grounding electrode conductor is sized in accordance with 250.32(E), and with *NEC* Table 250.66, which is shown in Exhibit 12.2.

Ground-Neutral Connection Prohibited. No direct connection is allowed between the grounding bus and neutral bus in an accessory building panelboard [250.32(B)(1)]. Instead, the feeder grounded (neutral) conductor connects to the neutral bus, and the feeder equipment grounding conductor connects to the grounding bus. Both of these conductors also connect to the appropriate buses in the service equipment (Figure 12.14).

The equipment grounding conductor and grounded (neutral) conductors are connected together only once, inside the service equipment enclosure, by means of the main bonding jumper [250.28]. This subject is explained in greater detail in Chapter 3, Services, Service Equipment, and Grounding.

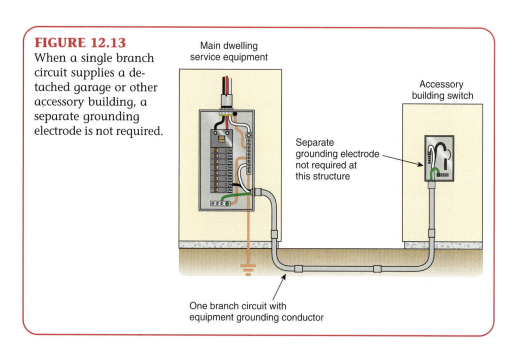

FIGURE 12.13
When a single branch circuit supplies a detached garage or other accessory building, a separate grounding electrode is not required.

Main dwelling service equipment

Accessory building switch

Separate grounding electrode not required at this structure

One branch circuit with equipment grounding conductor

EXHIBIT 12.2

NEC Table 250.66 Grounding Electrode Conductor for Alternating-Current Systems

Size of Largest Ungrounded Service-Entrance Conductor or Equivalent Area for Parallel Conductors[a] (AWG/kcmil)		Size of Grounding Electrode Conductor (AWG/kcmil)	
Copper	Aluminum or Copper-Clad Aluminum	Copper	Aluminum or Copper-Clad Aluminum[b]
2 or smaller	1/0 or smaller	8	6
1 or 1/0	2/0 or 3/0	6	4
2/0 or 3/0	4/0 or 250	4	2
Over 3/0 through 350	Over 250 through 500	2	1/0
Over 350 through 600	Over 500 through 900	1/0	3/0
Over 600 through 1100	Over 900 through 1750	2/0	4/0
Over 1100	Over 1750	3/0	250

Notes:
1. Where multiple sets of service-entrance conductors are used as permitted in 230.40, Exception No. 2, the equivalent size of the largest service-entrance conductor shall be determined by the largest sum of the areas of the corresponding conductors of each set.
2. Where there are no service-entrance conductors, the grounding electrode conductor size shall be determined by the equivalent size of the largest service-entrance conductor required for the load to be served.
[a]This table also applies to the derived conductors of separately derived ac systems.
[b]See installation restrictions in 250.64(A).

SPECIAL RULES FOR TWO-FAMILY DWELLINGS

- *Required Outdoor Receptacles.* Each unit of a two-family dwelling that is located at grade level must have at least one GFCI-protected receptacle outlet installed at the front and back of the dwelling not more than 6½ ft above grade level [210.52(E)].
- *Optional Outdoor Receptacles.* No outdoor receptacles are required for a residence located on the second floor of a two-family dwelling. However, if a receptacle outlet is installed for convenience on an outdoor porch or balcony, it must have GFCI protection for personnel [210.8(A)(3)].
- *Common Area Equipment.* Section 210.25 permits residential branch circuits to supply only those loads located within, or associated with, each individual dwelling unit. Thus, outdoor lighting for a common space such as a yard or carport must be supplied from a house loads panel.

(Continued)

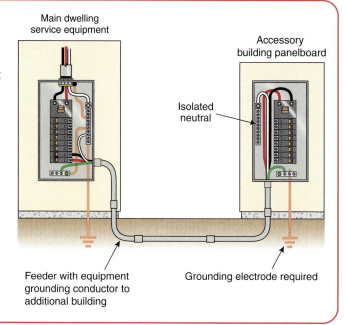

FIGURE 12.14
When a feeder supplies a panelboard in a detached garage or other accessory building, that structure must have a separate grounding electrode.

Main dwelling service equipment

Accessory building panelboard

Isolated neutral

Feeder with equipment grounding conductor to additional building

Grounding electrode required

- *Maintenance Receptacles.* If a single maintenance receptacle is provided for HACR equipment serving both units of a two-family dwelling, that receptacle must be supplied from a house loads panel (Figure 12.15). It cannot be fed from either of the dwelling unit panelboards.

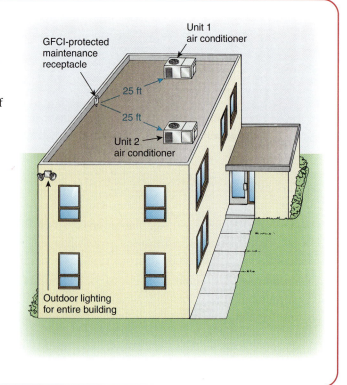

FIGURE 12.15
Electrical equipment serving common areas of two-family dwellings must be supplied from a common house loads panel, not from either of the dwelling units.

GFCI-protected maintenance receptacle

Unit 1 air conditioner

25 ft

25 ft

Unit 2 air conditioner

Outdoor lighting for entire building

MULTIPLE CHOICE

1. Receptacle outlets in the following locations are required to have GFCI protection:
   A. Required outlets at the front and back of the dwelling
   B. Readily accessible receptacles in accessory buildings
   C. Receptacles in yard areas
   D. All of the above

2. The following structures are required to have lighting outlets:
   A. All attached garages
   B. All detached garages
   C. All other accessory buildings at residences
   D. Both A and B

3. The following cable types are permitted to be used for branch-circuit wiring inside accessory buildings:
   A. Type UF
   B. Type NM-B
   C. Type USE
   D. Both A and B

4. When free-standing boxes containing wiring devices are supported by two or more raceways, each raceway must be secured
   A. By locknuts
   B. Within 18 in. of the enclosure
   C. With underground bracing of concrete or metal
   D. None of the above

5. Direct-burial cables or raceways can be installed under dwelling driveways and parking areas at a minimum depth of
   A. 12 in. when the circuit is GFCI protected
   B. 18 in.
   C. 24 in.
   D. Both A and B

6. Low-voltage luminaires generally cannot be located closer than which of the following to swimming pools, spas, fountains, or similar bodies of water?
   A. 5 ft
   B. 5 ft, if provided with GFCI protection
   C. 10 ft
   D. 12 ft

7. The following raceway(s) is (are) permitted to be installed below grade:
   A. Electrical metallic tubing (Type EMT)
   B. Flexible metallic tubing (Type FMT)
   C. Electrical nonmetallic tubing (Type ENT)
   D. Rigid metal conduit (Type RMC)

8. Receptacles in accessory buildings other than garages must have GFCI protection, except when they
   A. Aren't readily accessible
   B. Are located within dedicated space for a cord-and-plug-connected appliance
   C. Supply outdoor lighting
   D. Both A and B

9. The following cable type(s) is (are) rarely used in residential construction:
   A. Underground feeder and branch-circuit cable (Type UF)
   B. Mineral-insulated, metal-sheathed cable (Type MI)
   C. Armored cable (Type AC)
   D. None of the above

10. Type UF cable used for interior wiring must be secured as follows:
    A. It must be secured at intervals no greater than 4½ ft.
    B. It must be secured within 12 in. from every box, cabinet, or termination.
    C. It can be installed unsupported in whips up to 6 ft long for connecting to luminaires (lighting fixtures) within accessible ceilings.
    D. Both A and B.

FILL IN THE BLANKS

1. All 15- and 20-ampere receptacles installed outdoors in wet locations are required to have covers that are weatherproof, regardless of whether an attachment plug is _____.

2. Residential branch circuits rated 20 amperes or less with GFCI protection can be direct-buried at a minimum depth of _____ below driveways and parking areas.

3. Type SE cable can be installed as interior wiring, following the *Code* rules for Type _____ cable.

4. Garages attached to houses are required to contain at least _____ wall switch-controlled lighting outlet(s).

5. GFCI-protected outdoor receptacles cannot be supplied from _____ branch circuits serving dining rooms and kitchens.

6. Both Types RMC and IMC can function as equipment grounding conductors for the branch-circuit or feeder conductors installed in them. Some installers also choose to run a separate _____ along with the circuit conductors.

7. Most outdoor receptacle outlets installed at dwellings are required to have _____ protection for personnel.

8. When freestanding weatherproof boxes are supported by conduits, at least two of them must be threaded into the enclosure and be secured within _____ in. of the enclosure.

9. When Type UF cable is installed indoors, it can be secured and supported in the same way as Type _____ cable.

10. A(n) _____ door in a garage is not considered an outdoor entrance, and isn't required to have a wall switch-controlled lighting outlet installed on its exterior side.

TRUE OR FALSE

1. A two-family dwelling is required to have two HACR maintenance receptacle outlets.

   _____ True   _____ False

2. Luminaires (lighting fixtures) are permitted to be mounted in trees.

   _____ True   _____ False

3. Receptacles of 15 and 20 amperes installed outdoors in damp locations are required to have a cover that is weatherproof only when an attachment plug is inserted.

   _____ True   _____ False

4. When a duplex receptacle is located within dedicated space for a cord-and-plug-connected appliance that

can't be easily moved from one place to another, the outlet isn't required to have GFCI protection.

   _____ True   _____ False

5. Nonmetallic elbows are often used with Type RMC installed below grade, because they make it easier to pull conductors into the raceway.

   _____ True   _____ False

6. Lighting systems operating at 30 volts or less must be supplied from a branch circuit rated either 15 or 20 amperes.

   _____ True   _____ False

7. When receptacles or luminaires (lighting fixtures) are installed on device boxes recessed into exterior walls and under roofs, Types NM and AC cables are the most common wiring methods used in residential construction.

   _____ True   _____ False

8. Types NM, NMC, and NMS are all types of nonmetallic-sheathed cable defined in *NEC* Article 334.

   _____ True   _____ False

9. Pole-mounted luminaires (lighting fixtures), bollards, luminaires mounted in trees, and luminaires mounted on outside walls without a roof or overhang to protect them must be listed for use in wet locations.

   _____ True   _____ False

10. All 15- and 20-ampere, 125-volt receptacles installed outdoors in wet locations are required to have covers that are weatherproof even when an attachment plug is inserted.

   _____ True   _____ False

CHALLENGE QUESTIONS

1. How many GFCI-protected receptacle(s) is (are) required for a two-family dwelling with both units located at grade level?

2. Why does the *NEC* require attached garages to be wired for electricity but not detached garages?

3. Explain why service-lateral conductors emerging from grade to terminate at watt-hour meter cans often aren't provided with physical protection as required by *NEC* 300.5(D)(1) for direct-buried conductors.

4. Why does the *Code* permit 15- and 20-ampere, 120-volt residential branch circuits with GFCI protection

to be installed underground at lesser depths than circuits of other occupancies?

5. Explain the reason for requiring outlet boxes sup-ported by conduits to have at least two conduits. Why isn't a single conduit acceptable?

Swimming Pools, Hot Tubs, and Spas

13

OBJECTIVES

After completing this chapter, the student will be able to understand the following:

- Why electrical equipment must be kept at specified minimum distances from swimming pools
- What wiring methods are permitted for swimming pools
- Grounding requirements for pools, hot tubs, spas, and similar installations
- Differences among three methods of installing underwater pool lighting
- Ground-fault circuit-interrupter (GFCI) protection requirements for electrical equipment installed near pools and similar installations

INTRODUCTION

This chapter covers new installations of swimming pools, spas, and hot tubs, but not modifications to existing installations. The following major topics are included:

- Overview of swimming pool, hot tub, and spa installations
- Required clearances for outdoor pools
- Underwater pool lighting: 120 volts
- Low-voltage pool lighting: 12 volts
- Power distribution: general considerations
- Grounding, bonding, and GFCI protection
- Other pool equipment
- Spas and hot tubs
- Special rules for two-family dwellings

IMPORTANT NEC TERMS

Bonding (Bonded)
Conductor, Bare
Conductor, Covered
Conductor, Insulated
Cord-and-Plug-Connected Lighting Assembly
Dry-Niche Luminaire (Lighting Fixture)
Forming Shell
Fountain
Ground-Fault Circuit Interrupter
Hydromassage Bathtub
In Sight from (Within Sight from, Within Sight)
Maximum Water Level
No-Niche Luminaire (Lighting Fixture)
Packaged Spa or Hot Tub Equipment Assembly
Pool Cover, Electrically Operated
Self-Contained Spa or Hot Tub
Spa or Hot Tub
Storable Swimming or Wading Pool
Through-Wall Lighting Assembly
Wet-Niche Luminaire (Lighting Fixture)

OVERVIEW OF SWIMMING POOL, HOT TUB, AND SPA INSTALLATIONS

Installations of swimming pools, hot tubs, fountains, and decorative pools must follow many special *National Electrical Code®* rules intended to minimize the chance of electric shock around water. The effects of electric shock on people when *immersed* in water are much more severe than those received from simply *touching* energized metal out of water.

Because immersion in water increases the severity of an electric shock, *NEC®* Article 680 has many special safety rules that apply to installations of equipment that contains water and uses electricity. These rules are especially concerned with grounding, GFCI protection for personnel, and adequate clearances between water and electrical wiring and equipment. Article 680 also applies to hydromassage tubs (whirlpool baths), which are covered in Chapter 8, The Bathroom.

REQUIRED CLEARANCES FOR OUTDOOR POOLS

Overhead Conductors

Power. Service conductors and open overhead wiring must not be installed less than 22½ ft in any direction above the water level or deck area immediately surrounding the swimming pool, less than 10 ft horizontally from the inside wall, or less than 14½ ft from a diving board. These requirements are specified in *NEC* Figure 680.8, shown in Exhibit 13.1, and *NEC* Table 680.8, shown in Exhibit 13.2. These same limitations apply to cables for network-powered broadband communications systems [680.8(C)].

Communications. Coaxial cables and telephone wiring must be kept at a minimum vertical clearance of 10 ft above swimming pools and diving boards [680.8(B)].

Underground Wiring

Underground wiring isn't permitted under swimming pools or within 5 ft horizontally of the pool, unless the wiring supplies pool equipment. If this minimum spacing can't be maintained, then the underground wiring must be installed in rigid metal

EXHIBIT 13.1

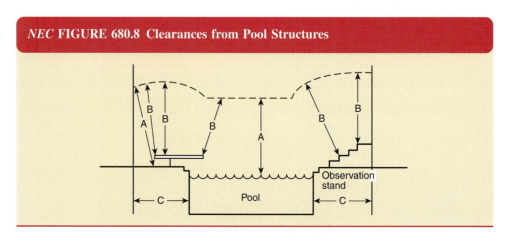

NEC FIGURE 680.8 Clearances from Pool Structures

EXHIBIT 13.2

NEC TABLE 680.8 Overhead Conductor Clearances

| Clearance Parameters | Insulated Cables, 0–750 Volts to Ground, Supported on and Cabled Together with an Effectively Grounded Bare Messenger or Effectively Grounded Neutral Conductor | | All Other Conductors Voltage to Ground | | | |
| | | | 0 through 15 kV | | Over 15 through 50 kV | |
	m	ft	m	ft	m	ft
A. Clearance in any direction to the water level, edge of water surface, base of diving platform, or permanently anchored raft	6.9	22.5	7.5	25	8.0	27
B. Clearance in any direction to the observation stand, tower, or diving platform	4.4	14.5	5.2	17	5.5	18
C. Horizontal limit of clearance measured from inside wall of the pool	This limit shall extend to the outer edge of the structures listed in A and B of this table but not to less than 3 m (10 ft).					

conduit (RMC), intermediate metal conduit (IMC), or a nonmetallic raceway system. Minimum burial depths are specified in *NEC* Table 680.10, shown in Exhibit 13.3.

Receptacle Outlets

Each permanently installed pool at a dwelling is required to have at least one 125-volt, 15- or 20-ampere receptacle on a general-purpose branch circuit that is located 10 ft to 20 ft from the inside wall of the pool, measured horizontally (Figure 13.1).

NEC TABLE 680.10 Minimum Burial Depths

EXHIBIT 13.3

| Wiring Method | Minimum Burial | |
	mm	in.
Rigid metal conduit	150	6
Intermediate metal conduit	150	6
Nonmetallic raceways listed for direct burial without concrete encasement	450	18
Other approved raceways*	450	18

*Raceways approved for burial only where concrete encased shall require a concrete envelope not less than 50 mm (2 in.) thick.

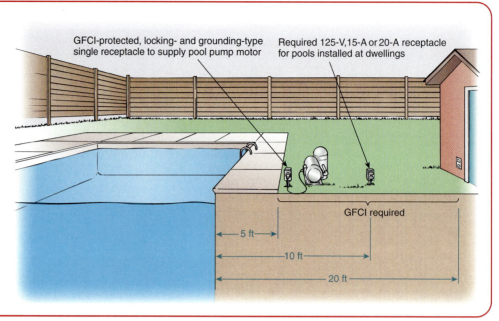

FIGURE 13.1 Receptacles for pumps or other pool equipment must be located at least 10 ft from the inside walls of a pool, unless GFCI protected and of a locking configuration.

GFCI-protected, locking- and grounding-type single receptacle to supply pool pump motor

Required 125-V,15-A or 20-A receptacle for pools installed at dwellings

GFCI required

5 ft

10 ft

20 ft

The receptacle(s) must be located no higher than 6½ ft above floor, platform, or grade level and is required to have GFCI protection [680.22(A)(3)].

Pool Equipment. Receptacles that provide power for pumps and other pool equipment are normally located at least 10 ft from the inside wall of the pool. Single receptacles for this purpose are permitted to be installed as close as 5 ft if they are grounding-type single receptacles of a locking configuration (twist-lock) and have GFCI protection [680.22(A)(1)].

Restricted Space. When a pool is installed within 10 ft of a dwelling and the size of the lot makes it impossible to maintain these required distances, 680.22(A)(4) permits one receptacle outlet to be installed a minimum of 5 ft horizontally from the inside wall (Figure 13.2). Receptacle outlets cannot be installed closer than 5 ft from a pool under any circumstances.

Switches

Switching devices such as general-use snap switches, circuit breakers, and automatic timers must be located at least 5 ft from an inside wall of a pool, unless separated from it by a solid wall, a fence, or other permanent barrier [680.22(C)]. Switches listed for the application are permitted to be installed closer to pools than 5 ft. These switching devices use pneumatic or other actuating means to prevent contact between users and live electrical parts.

NOTE: All of the minimum distances previously specified apply only to receptacles and switching devices located in the pool area. Receptacles and switching devices located inside buildings (for outdoor pools) or in other rooms (for indoor pools) are considered to be separated from the swimming pool by a permanent barrier (Figure 13.3).

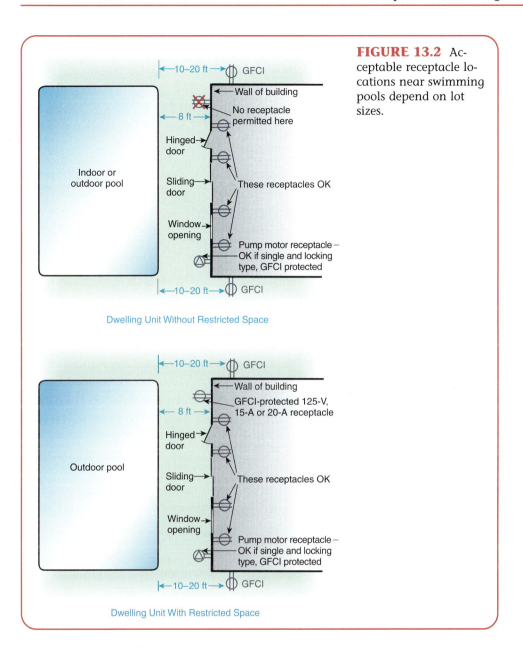

FIGURE 13.2 Acceptable receptacle locations near swimming pools depend on lot sizes.

REQUIRED CLEARANCES FOR LUMINAIRES AND CEILING-SUSPENDED (PADDLE) FANS

Outdoor Pools

In newly constructed outdoor pools, luminaires (lighting fixtures) and ceiling-suspended (paddle) fans can't be installed within 12 ft vertically above the maximum water level or within 5 ft horizontally from an inside wall of an outdoor pool (Figure 13.4) [680.22(B)(1)].

Indoor Pools

Luminaires and paddle fans normally cannot be installed within 12 ft vertically above the maximum water level or within 5 ft horizontally from the inside wall of

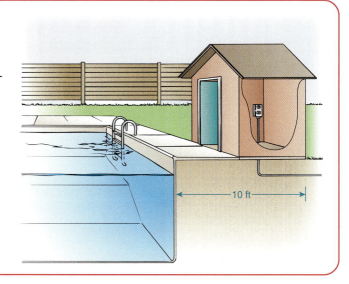

FIGURE 13.3 The minimum distance required by *NEC* 680.22(A) does not apply to a receptacle located in a structure.

FIGURE 13.4 Luminaires, lighting outlets, and ceiling-suspended (paddle) fans cannot be installed in the areas shown that surround outdoor and indoor pools.

Outdoor Pools

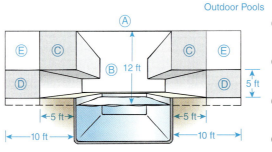

Ⓐ Luminaires, lighting outlets, and ceiling-suspended (paddle) fans permitted above 12 ft.

Ⓑ Luminaires, lighting outlets, and ceiling-suspended (paddle) fans not permitted below 12 ft.

Ⓒ Existing luminaires and lighting outlets permitted in this space if rigidly attached to existing structure (GFCI required).

Ⓓ Luminaires and lighting outlets permitted if protected by a GFCI.

Ⓔ Luminaires and lighting outlets permitted if rigidly attached.

Indoor Pools

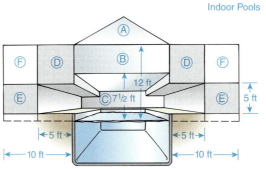

Ⓐ Luminaires, lighting outlets, and ceiling-suspended (paddle) fans permitted above 12 ft.

Ⓑ Totally enclosed luminaires protected by a GFCI and ceiling-suspended (paddle) fans protected by a GFCI permitted above 7½ ft.

Ⓒ Luminaires, lighting outlets, and ceiling-suspended (paddle) fans not permitted below 5 ft.

Ⓓ Existing luminaires and lighting outlets permitted in this space if rigidly attached to existing structure (GFCI required).

Ⓔ Luminaires and lighting outlets permitted if protected by a GFCI.

Ⓕ Luminaires and lighting outlets permitted if rigidly attached.

an indoor pool [680.22(B)(2)]. However, if the branch circuit supplying the equipment is GFCI protected, then totally enclosed luminaires and paddle fans can be mounted lower—at 7½ ft above the maximum water level (Figure 13.4).

Existing Structures

When a new pool is built near or inside an existing structure, existing luminaires and lighting outlets can be closer than 5 ft horizontally from the inside wall of the pool if they are located not less than 5 ft above the maximum water level, have GFCI protection, and are rigidly attached to the building structure [680.22(B)(3)]. These relaxed clearances are illustrated in Figure 13.4.

However, hanging or pendant luminaires aren't permitted to be located less than 7½ ft above the maximum pool water level under any circumstances, in either new or existing construction.

UNDERWATER POOL LIGHTING: 120 VOLTS

Pool lighting operates at either 120 volts or 12 volts. The rules applying to 120-volt luminaires are described in this section, and those applying only to 12-volt luminaires are covered in the next section, "Low-Voltage Pool Lighting: 12 Volts."

Most installation requirements are the same for all pool lighting, regardless of operating voltage. Because they are cooled by water, all underwater pool luminaires must be installed at least 18 in. below the normal water level, unless listed and identified for use at lesser depths [680.23(A)(5)]. The 18-in. depth ensures that, even if the water level goes down due to normal evaporation, the lighting fixtures won't be exposed to air and heat to temperatures that could burn people. Pool luminaires are designed so that the fixtures themselves won't fail in a hazardous manner through overheating [680.23(A)(7)].

Luminaire Types. Underwater pool luminaires come in three general types: *wet niche, dry niche,* and *no niche.* A fourth type of underwater pool lighting installation, known as a *through-wall lighting assembly,* is also permitted by 680.23(E). But this type of lighting method isn't normally used in pools at one- and two-family dwellings.

Pool Junction Boxes. Junction boxes that supply underwater pool lighting must be listed as swimming pool junction boxes, corrosion-resistant, and equipped with threaded hubs or conduit entries. Typically, pool junction boxes are manufactured of brass, plastic, or copper and have extra grounding terminals.

Pool lighting junction boxes are frequently installed above grade using conduits for support. A junction box isn't permitted to be supported by a single conduit, but two are acceptable [314.23]. Sometimes an additional support method is provided, as shown in Figure 13.5.

Grounding. An insulated copper equipment grounding conductor, 12 AWG or larger, must be run with the branch-circuit conductors for all three types of underwater lighting. This equipment grounding conductor must be continuous and cannot be spliced at junction boxes or enclosures, except as specifically permitted in 680.23(F) for control devices or multiple underwater luminaires supplied by the same branch circuit.

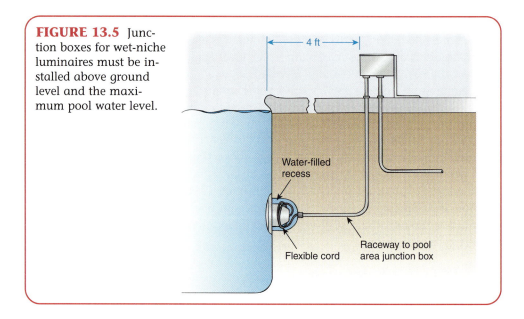

FIGURE 13.5 Junction boxes for wet-niche luminaires must be installed above ground level and the maximum pool water level.

Wet-Niche Luminaires

Wet-niche luminaires are completely immersed in water. They fit into a water-filled recess (niche) in the pool wall and are supplied by a flexible cord [680.23(B)]. When re-lamping is necessary, the wet-niche fixture is removed and allowed to float on the surface of the pool, where its lamp is replaced. Then the fixture is re-installed inside the wet niche (Figure 13.6).

Wiring Methods. Wiring between the wet-niche forming shell and a junction box or other enclosure must be installed in brass conduit, other approved corrosion-resistant metal, rigid nonmetallic conduit (RNC), or liquidtight flexible nonmetallic conduit (LFNC) [680.23(B)(2)]. Normally this wiring is a flexible cord or cable

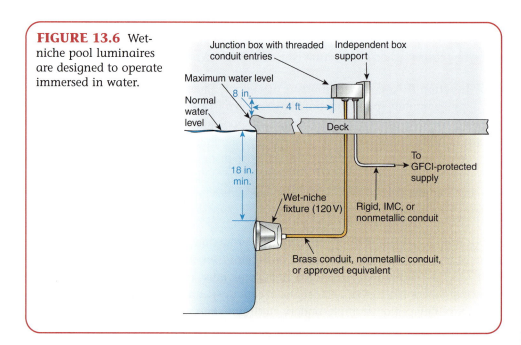

FIGURE 13.6 Wet-niche pool luminaires are designed to operate immersed in water.

supplied with the luminaire and includes an insulated copper equipment grounding conductor.

Branch-circuit wiring between the junction box and the source of supply must be installed in RMC, IMC, RNC, or LFNC [680.23(F)(1)]. Within buildings, electrical metallic tubing (EMT), electrical nonmetallic tubing (ENT), and Type MC cable are permitted. Where installed on buildings, EMT is permitted.

Grounding. When nonmetallic conduit is used for wiring to the wet-niche forming shell, an 8 AWG insulated copper equipment grounding conductor must be installed. This requirement means that the raceway must be sized large enough for both the flexible cord and the additional 8 AWG insulated conductor.

The terminations of the supply cord and the 8 AWG equipment ground (when used with nonmetallic raceway) at the wet-niche forming shell must be covered with a listed potting compound to protect them from corrosion [680.23(B)(2)(b)].

Junction Boxes. A junction box connected to a conduit from the forming shell of a wet-niche luminaire must be installed either 4 in. above ground level or 8 in. above the maximum pool water level, whichever provides the higher elevation. The reason for this requirement is that electrical raceways supplying wet-niche luminaires normally aren't sealed against the entry of water. The junction box must be located higher than the pool water level to prevent flooding.

Wet-niche luminaire junction boxes must also be located at least 4 ft from the inside wall of the pool, unless separated from the pool by a solid wall, a fence, or other permanent barrier [680.24(A)(2)]. The purpose of this rule is to prevent a person in the pool from being able to reach out and touch an electrical junction box.

Dry-Niche Luminaires

Dry-niche luminaires are installed in a dry recess (niche). They can be accessed for re-lamping or service from the top or rear by means of a hatch or handhole. The *Code* doesn't require a separate junction box. If a separate junction box is installed, it isn't required to be elevated above ground level, because raceways supplying dry-niche pool lighting don't contain water [680.23(C)(2)].

Wiring Methods. Branch-circuit conductors to the field-wiring compartment of a dry-niche luminaire must be installed in RMC, IMC, RNC, or LFNC [680.23(F)]. Within buildings, electrical metallic tubing (EMT), electrical nonmetallic tubing (ENT), and Type MC cable are permitted. Where installed on buildings, EMT is permitted.

Grounding. When nonmetallic conduit is used with dry-niche luminaires, a supplemental 8 AWG insulated copper equipment grounding conductor isn't required.

No-Niche Luminaires

No-niche luminaires are permanently installed in a mounting bracket recessed into the wall of a pool and are designed to be re-lamped from the front while underwater.

Wiring Methods. Wiring between the no-niche mounting bracket and a junction box or other enclosure must be installed in brass conduit, other approved corrosion-resistant metal, rigid nonmetallic conduit (RNC), or liquidtight flexible nonmetallic conduit (LFNC).

Branch-circuit conductors to the junction box must be installed in RMC, IMC, RNC, or LFNC [680.23(F)]. Within buildings, electrical metallic tubing (EMT), electrical nonmetallic tubing (ENT), and Type MC cable are permitted. Where installed on buildings, EMT is permitted.

Grounding. When nonmetallic conduit is used for wiring to the no-niche mounting bracket, an 8 AWG insulated copper equipment grounding conductor must be installed. This requirement means that the raceway must be sized large enough for both the flexible cord and the additional 8 AWG insulated conductor.

The terminations of the supply cord and the 8 AWG equipment ground (when used with nonmetallic raceway) at the wet-niche forming shell must be covered with a listed potting compound to protect them from corrosion [680.23(B)(2)(b)].

Junction Boxes. A junction box connected to a conduit from the mounting bracket of a no-niche luminaire must be installed either 4 in. above ground level or 8 in. above the maximum pool water level, whichever provides the higher elevation. The reason for this requirement is that electrical raceways supplying no-niche luminaires normally aren't sealed against the entry of water. The junction box must be located higher than the pool water level to prevent flooding.

No-niche luminaire junction boxes must also be located at least 4 ft from the inside wall of the pool, unless separated from the pool by a solid wall, a fence, or other permanent barrier [680.24(A)(2)]. The purpose of this rule is to prevent a person in the pool from being able to reach out and touch an electrical junction box.

Branch-Circuit Wiring Methods for 120-Volt Pool Lighting

According to *NEC* 680.23(F), branch-circuit wiring on the supply side of enclosures and junction boxes connected to conduits from wet-niche luminaire forming shells, no-niche luminaire mounting brackets, and the field-wiring compartments of dry-niche luminaires must be installed in

- Rigid metal conduit (RMC)
- Intermediate metal conduit (IMC)
- Rigid nonmetallic conduit (RNC)
- Liquidtight flexible nonmetallic conduit (LFNC)

Buildings. Within buildings, wiring for underwater pool lighting can be run in electrical metallic tubing (EMT), electrical nonmetallic tubing (ENT), or Type MC cable. EMT is permitted to be installed on buildings.

GFCI Protection. Branch circuits supplying 120-volt pool lighting luminaires must have ground-fault circuit-interrupter protection. Typically, this protection is provided by GFCI circuit breakers in the panelboard [680.23(A)(3)].

LOW-VOLTAGE POOL LIGHTING: 12 VOLTS

Transformers

Isolation transformers with an ungrounded secondary and a grounded metal barrier between the primary and secondary windings are used to supply 12-volt underwater

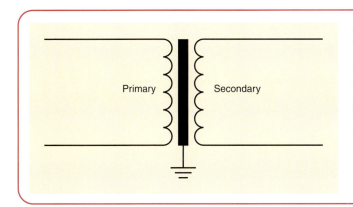

FIGURE 13.7 Low-voltage pool lighting must be supplied through an isolation transformer with a grounded metal barrier between the primary and secondary windings.

pool lighting [680.23(A)(2)]. This requirement reduces the chance of an accidental short circuit that could energize the secondary wiring to the underwater luminaires at higher voltage (Figure 13.7).

Underwater transformers are listed as "Swimming Pool and Spa Transformer." Transformers suitable for installation outdoors are marked "For Outdoor Use" or equivalent wording.

Branch-Circuit Wiring Methods for 12-Volt Pool Lighting

In general, low-voltage pool lighting uses the same wiring methods as those for line voltage, with the exception of the rules for junction boxes.

Junction Boxes. Lighting systems operating at 15 volts or less can use flush-mounted deck boxes rather than raised junction boxes (Figure 13.8). They must be located not less than 4 ft from the inside wall of the pool and be filled with an approved potting compound to keep out moisture [680.24(A)(2)(c)].

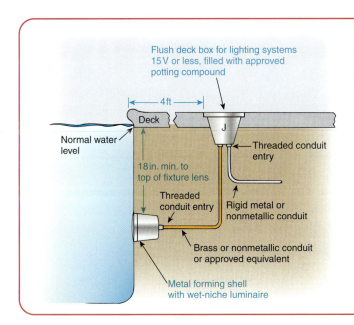

FIGURE 13.8 A flush deck (junction) box can be used with underwater pool lighting operating at 15 volts or less.

POWER DISTRIBUTION: GENERAL CONSIDERATIONS

Pool Equipment Panelboards

In new construction, a separate panelboard sub-fed from the service equipment is often provided to supply pool equipment. However, pool motors, lighting, and other related items can also be supplied from the service panelboard or from an existing remote panel originally installed for some other purpose.

Maintenance Disconnecting Means

All pool equipment other than lighting must have a means to disconnect all ungrounded conductors that is located in an accessible location in sight from the equipment [680.12]. In new construction, where the pool equipment is typically located in the same building or utility space with the panelboard that supplies it, the branch overcurrent protective devices can serve as maintenance disconnects.

Wiring Methods for New Pool Equipment Panelboards. Feeders for new pool equipment panels must be installed in RMC, IMC, RNC, or LFNC. EMT is permitted to be installed on buildings or other structures. Either EMT or ENT can be installed inside buildings [680.25(A)].

NOTE: An existing remote panelboard used to supply new pool equipment is permitted by the Exception to 680.25(A) to have a feeder that uses flexible metal conduit (FMC) or an approved cable assembly, as long as there is an equipment grounding conductor within the outer sheath (Figure 13.9).

Grounding. When the pool equipment panelboard is located within the dwelling, an insulated equipment grounding conductor sized in accordance with 250.122 must

FIGURE 13.9 When an existing remote panelboard supplies new pool equipment, a raceway is not required for the feeder if the existing wiring method contains an insulated equipment grounding conductor.

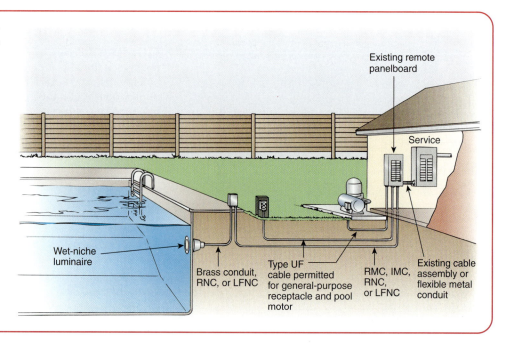

be installed in the same raceway as the feeder conductors [680.25(B)]. *NEC* Table 250.122 is shown as Exhibit 13.4.

Remote Panelboard (Sub-Panel). When the pool equipment panelboard is located in a separate structure, such as a pool equipment building, 250.32 requires that a separate grounding electrode be installed at this structure (Figure 13.10). The equipment grounding bus in the panelboard is connected to the grounding electrode using a conductor sized in accordance with 250.32(E) and 250.66. In this type of remote-panel installation, the grounded conductor (neutral) and the equipment grounding conductor are *not connected together* at the pool panelboard. Instead, the copper or aluminum equipment grounding conductor from the service equipment is connected directly to the pool building equipment grounding bus in the remote panelboard. The connection between grounded and grounding conductors occurs only once, at the service equipment (Figure 13.11).

EXHIBIT 13.4

NEC **Table 250.122 Minimum Size Equipment Grounding Conductors for Grounding Raceway and Equipment**

Rating or Setting of Automatic Overcurrent Device in Circuit Ahead of Equipment, Conduit, etc., Not Exceeding (Amperes)	Size (AWG or kcmil)	
	Copper	Aluminum or Copper-Clad Aluminum*
15	14	12
20	12	10
30	10	8
40	10	8
60	10	8
100	8	6
200	6	4
300	4	2
400	3	1
500	2	1/0
600	1	2/0
800	1/0	3/0
1000	2/0	4/0
1200	3/0	250
1600	4/0	350
2000	250	400
2500	350	600
3000	400	600
4000	500	800
5000	700	1200
6000	800	1200

Note: Where necessary to comply with 250.4(A)(5) or (B)(4), the equipment grounding conductor shall be sized larger than given in this table.
*See installation restrictions in 250.120.

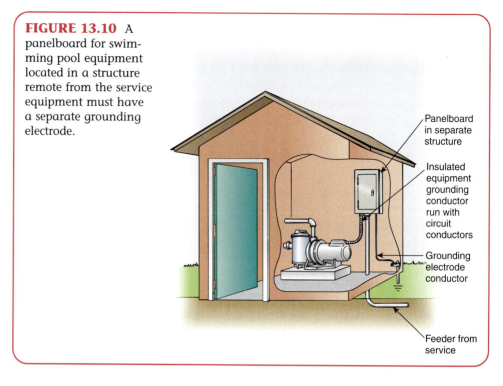

FIGURE 13.10 A panelboard for swimming pool equipment located in a structure remote from the service equipment must have a separate grounding electrode.

Panelboard in separate structure

Insulated equipment grounding conductor run with circuit conductors

Grounding electrode conductor

Feeder from service

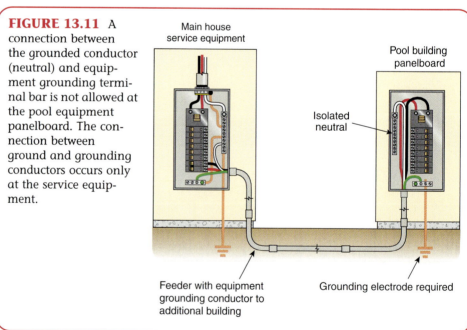

FIGURE 13.11 A connection between the grounded conductor (neutral) and equipment grounding terminal bar is not allowed at the pool equipment panelboard. The connection between ground and grounding conductors occurs only at the service equipment.

Main house service equipment

Pool building panelboard

Isolated neutral

Feeder with equipment grounding conductor to additional building

Grounding electrode required

Lighting and Receptacles in Pool Areas

This chapter has described the special wiring requirements of Article 680 for underwater pool lighting and associated items such as pool lighting junction boxes.

Other pool equipment wiring is permitted to use other wiring methods described in Chapter 3 of the *NEC,* such as underground feeder and branch-circuit cable (Type UF) for direct-buried branch circuits, as shown in Figure 13.9.

GROUNDING, BONDING, AND GFCI PROTECTION

Proper grounding and bonding are critical to safely installing pools, fountains, hot tubs, and similar installations. Proper grounding helps prevent electric shock and establishes an effective ground-fault current path that ensures the rapid operation of overcurrent protective devices in case of a fault.

NEC Article 250.2 defines *effective ground-fault current path* as follows.

> **Effective Ground-Fault Current Path:** An intentionally constructed, permanent, low-impedance electrically conductive path designed and intended to carry current under ground-fault conditions from the point of a ground fault on a wiring system to the electrical supply source and that facilitates the operation of the overcurrent protective device or ground fault detectors on high-impedance grounded systems.

Proper bonding reduces differences in potential (also called "voltage gradients") between metal parts. People perceive these differences as "touch voltage" (i.e., an unpleasant tingling sensation that may be dangerous when a person is immersed in water). Bonding also reduces the chance of shock in the event that non-current-carrying metal parts accidentally become energized.

The use of GFCIs is generally required by the *NEC* to provide protection for personnel in areas where electricity is used around moisture. GFCI protection, then, is particularly important around swimming pools and other bodies of water.

Some requirements for grounding, bonding, and GFCI protection of pool equipment have been explained in earlier parts of this chapter. They are repeated in the sections that follow for emphasis.

Grounding

Listed pool equipment—including luminaires, junction boxes, and transformer enclosures—is generally equipped with special terminals for connecting 8 AWG equipment grounding conductors.

The following equipment is required to be grounded by *NEC* 680.6, using insulated equipment grounding conductors:

- Underwater pool luminaires (lighting fixtures), except low-voltage systems listed for use without a grounding conductor
- Electrical equipment located within 5 ft of the inside wall of a pool or other body of water
- Circulating pumps and associated electrical equipment
- Junction boxes
- Transformer enclosures
- Ground-fault circuit interrupters
- Panelboards that are sub-fed from a dwelling's service equipment and that supply pool or fountain equipment

Bonding

Bonding is not the same as grounding. This is made clear by the FPN to 680.26(A), which states "The 8 AWG or larger solid copper bonding conductor shall not be required to be extended or attached to any remote panelboard, service equipment, or any electrode."

Establishing Continuity. Based on the FPN to 680.26(A), bonding does not involve establishing an effective fault-current path to ground. Bonding is the joining together of pieces of metal, often with wires called *jumpers,* to establish electrical and mechanical continuity. Pool equipment bonding wires must be minimum 8 AWG solid copper. The wires can have green insulation or a green coating (which is not recognized as insulation), or can be bare copper. The *Code* requires that specific parts of swimming pool installations be bonded [680.26(B)(1) through (5)]. These rules can be summarized as follows.

• *Metallic Parts of the Pool Structure.* Metallic parts of the pool structure, including reinforcing steel used in building concrete pools, metal shells of bolted or welded pools, ladders, diving board supports, drains, water inlets and outlets, and miscellaneous metal fittings, must be bonded. Small, isolated metal parts of the type described in 680.26(B)(3) don't have to be bonded.

• *Underwater Lighting Shells and Brackets.* Forming shells and mounting brackets of no-niche luminaires (lighting fixtures) must be bonded, unless listed low-voltage lighting with nonmetallic forming shells is used.

• *Electrical Equipment.* Motors and other equipment associated with pool covers must be bonded. Pool water heaters must be bonded in accordance with 680.26(E). Pool pump motors and associated equipment must be bonded, unless they are double-insulated [680.21(B)].

CAUTION: Even when a double-insulated water pump motor is installed, NEC 680.26(B)(4) requires that an 8 AWG solid copper conductor be installed from the bonding grid to an accessible point in the motor vicinity. The reason for this requirement is to provide a future bonding connection in case the double-insulated motor burns out and is replaced by a motor of standard construction.

• *Metal Wiring Methods and Equipment.* Metal raceways, metal-sheathed cables, metal piping, metal fences, and other fixed metal parts must be bonded when located within 5 ft of the inside walls of the pool or 12 ft vertically above the maximum water level, deck areas, and diving structure. These fixed metal parts are not required to be bonded when separated from the pool by a permanent barrier. This requirement means that metal equipment and parts located *inside* a house, pool-equipment building, garage, or other structure built within 5 ft of the inside wall of a pool do not have to be bonded.

Equipotential Bonding Grid. The metal equipment and parts specified in the previous list must be connected to an *equipotential bonding grid* using minimum 8 AWG solid copper conductors, brass rigid metal conduit (RMC), or other corrosion-resistant conduit [680.26(C)]. Most often, the structural steel reinforcing bars (rebar) of a concrete swimming pool, decorative fountain, or other body of water serve as the common bonding grid (Figures 13.12 and 13.13), although other types of pool construction require different solutions.

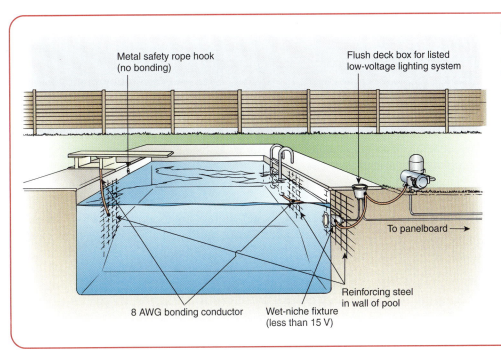

FIGURE 13.12 A poured-concrete pool has structural reinforcing steel that serves as the common bonding grid.

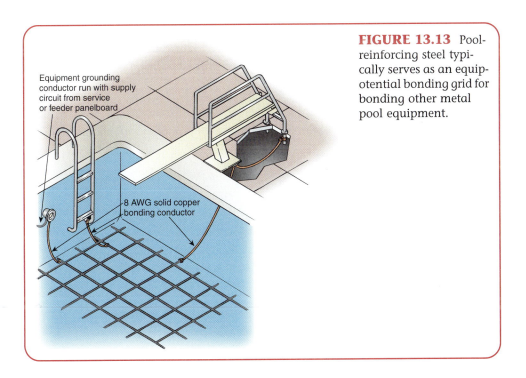

FIGURE 13.13 Pool-reinforcing steel typically serves as an equipotential bonding grid for bonding other metal pool equipment.

When reinforcing is encapsulated with a nonconductive compound (i.e., coated with epoxy to reduce corrosion), it isn't required to be bonded and can't serve as the equipotential bonding grid (Figure 13.14). When steel rebar isn't available, the wall of a metal pool (Figure 13.15), or several other methods described in 680.26(C)(3), can be used to establish an equipotential bonding grid.

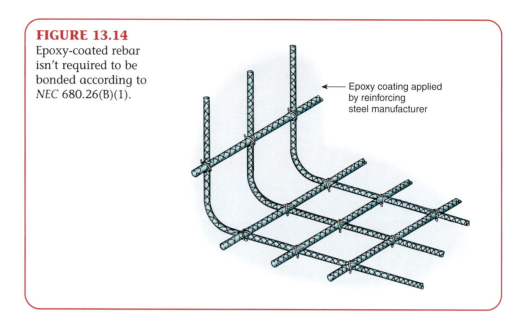

FIGURE 13.14 Epoxy-coated rebar isn't required to be bonded according to *NEC* 680.26(B)(1).

Epoxy coating applied by reinforcing steel manufacturer

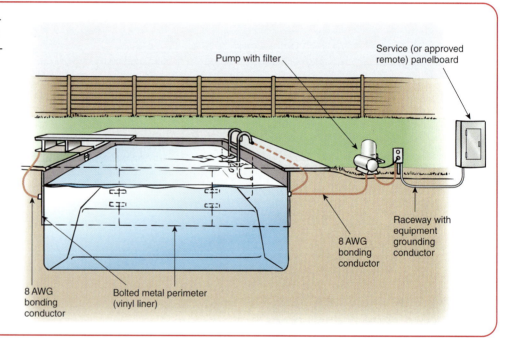

FIGURE 13.15 In a metal-perimeter pool with bolted or welded sections, the metal perimeter serves as the common bonding grid to which the metal ladder, metal diving board, and pump motor are connected.

Pump with filter

Service (or approved remote) panelboard

Raceway with equipment grounding conductor

8 AWG bonding conductor

8 AWG bonding conductor

Bolted metal perimeter (vinyl liner)

GFCI Protection

GFCI protection for personnel is required for the following equipment installed within pool areas:

- All 120-volt underwater pool lighting luminaires
- All 15- and 20-ampere, 125-volt, single-phase receptacles located within 20 ft of the inside wall of a pool or fountain
- All 15- and 20-ampere, 125- through 250-volt, single-phase receptacles that supply pool pump motors

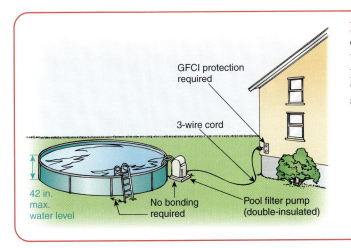

FIGURE 13.16 All electrical equipment used with a storable pool is required to have GFCI protection for personnel.

- Luminaires (lighting fixtures) and suspended-ceiling (paddle) fans installed in the area from 5 to 10 ft horizontally and 7½ to 12 ft vertically above indoor pools
- Electrically operated pool covers (see "Other Pool Equipment," which follows)

STORABLE POOLS

All electrical equipment of storable pools must be protected by ground-fault circuit interrupters [680.32] (Figure 13.16). Storable pools can have either 120-volt or low-voltage underwater lighting [680.33] and must be located so that no receptacle is located less than 10 feet from the inside walls [680.34]. They do not have special bonding requirements.

OTHER POOL EQUIPMENT

Electrically Operated Pool Covers

Section 680.27(B) addresses electrically operated pool covers. The rules for installing them can be summarized as follows.

- *Electric motors, controllers, and wiring* can't be located closer than 5 ft from an inside wall of a pool, unless separated from it by a wall, a cover, or other permanent barrier. The control for the pool cover motor must be located within sight of the pool.
- *Below-grade electric motors* must be the totally enclosed type.
- *Receptacles* that provide power for cord-and-plug-connected pool cover motors must be located at least 10 ft from an inside wall of a pool. They can be located as close as 5 ft if they are grounded single receptacles of a locking configuration.
- *GFCI protection* must be provided for the electric motor and controller for a pool cover.

Electric Heaters for Deck Areas

Section 680.27(C) addresses electric heaters for use in deck areas. Electric heaters cannot be mounted over pools. Radiant heating cables of the sort used for snow-melting purposes cannot be embedded in pool decks or installed below them for the following two reasons.

1. Electric heating equipment is subject to high leakage currents in normal operation.
2. Deteriorating insulation presents a possible shock hazard to people walking barefoot on wet pool decks.

The rules for installing electric heaters in pool deck areas can be summarized as follows.

- *Radiant Heaters.* These heaters must be located at least 5 ft horizontally from the inside walls and 12 ft vertically above the pool deck, unless otherwise approved. They must be permanently wired.

- *Unit Heaters.* These electric heaters with fans must be of the totally enclosed or guarded type and be located at least 5 ft horizontally from the inside walls of a pool. There is no minimum vertical mounting height requirement for unit heaters. Unit heaters are permitted to be cord-and-plug connected. Receptacles located within 20 ft horizontally from the inside walls of the pool are required to have GFCI protection [680.22(A)(5)].

SPAS AND HOT TUBS

Spas and hot tubs are similar to permanently installed swimming pools in that they are designed to remain filled with water for extended periods. However, while in-ground swimming pools are normally constructed in place, most spas and hot tubs are packaged systems that are simply set in place. The *Code* rules for installing them are different and, in general, simpler.

Installation of Spas and Hot Tubs

This section describes installation rules for spas and hot tubs that may be installed either indoors or outdoors. Installation considerations for hydromassage tubs of the type often called *whirlpool baths* and normally installed indoors are covered in Chapter 8, The Bathroom.

Emergency Switch. In general, spas and hot tubs are required to have a local disconnect that can be used to turn off the motor in an emergency, such as when long hair becomes caught and is being sucked into a drain. Section 680.41 requires that the switch be located within 5 ft of the spa or hot tub, be readily accessible to users, and be within sight from the spa or hot tub (Figure 13.17).

NOTE: While an emergency switch is not required for a spa or hot tub installed at a single-family dwelling, providing such a switch is an important safety improvement for users.

Wiring Methods. Spas and hot tubs can be connected using any regular wiring method specified in Chapter 3 of the *NEC*. Listed or self-contained units installed

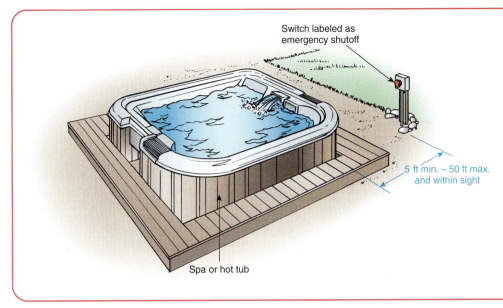

Switch labeled as emergency shutoff

5 ft min. – 50 ft max. and within sight

Spa or hot tub

FIGURE 13.17 A within-sight emergency switch is required for outdoor and indoor hot tub installations.

outdoors can use cord-and-plug connections if they have GFCI protection [680.42]. Listed spas and hot tub units installed indoors can be cord-and-plug connected if rated 20 amperes or less [680.43, Exception].

Receptacles. Each spa or hot tub installed indoors is required to have at least one 125-volt, 15- or 20-ampere receptacle on a general-purpose branch circuit located between 5 and 10 ft from an inside wall of the spa or hot tub. All receptacles rated 125 volts and 30 amperes or less, and located within 10 ft from an inside wall of the spa or hot tub, are required to have GFCI protection. Receptacles that provide power for a spa or hot tub are required to have GFCI protection [680.43(A)].

Clearances. Luminaires (lighting fixtures) and suspended-ceiling (paddle) fans normally cannot be installed within 12 ft vertically above the spa or hot tub or within 5 ft horizontally of the inside walls. However, where GFCI protection is provided, luminaires and paddle fans can be mounted 7½ ft above the spa or hot tub [680.43(B)]. Wall switches must be located at least 5 ft horizontally from the inside walls of the spa or hot tub [680.43(C)].

Luminaires (Lighting Fixtures). Permanently installed luminaires for spas and hot tubs are subject to the same rules as those for permanently installed pools. Listed lighting assemblies, with cord-and-plug connections of the same sort used with storable pools, can also be used [680.43(B)(2)].

Grounding and Bonding. All electric equipment located within 5 ft of an inside wall of a spa or hot tub must be grounded, along with all electric equipment associated with the circulating system [680.43(F)]. All metal parts within the spa or hot tub and within 5 ft of the inside walls must be bonded together, through either direct metal-to-metal contact or an 8 AWG copper bonding jumper (insulated, covered, or bare) [680.43(D), (E)].

SPECIAL RULES FOR TWO-FAMILY DWELLINGS

- *Wiring Methods.* Pool motor branch-circuit wiring run within a two-family dwelling is required to be installed in RMC, IMC, RNC, EMT, or Type MC cable [680.21(A)(1), (2)]. This wiring cannot be installed using any wiring method described in Chapter 3 of the *NEC,* as is permitted by 680.21(A)(4) for one-family dwellings.

- *Hot Tub Emergency Switch.* A spa or hot tub installed at a two-family dwelling must have an emergency shutoff switch. It must be clearly labeled, be located within 5 ft of the spa or hot tub, be readily accessible to users, and be within sight from the spa or hot tub [680.41].

MULTIPLE CHOICE

1. The following types of underwater lighting are used in permanently installed swimming pools:
   A. Wet-niche luminaires
   B. Dry-niche luminaires
   C. Through-wall lighting assemblies
   D. All of the above

2. Wiring to a dry-niche luminaire (lighting fixture) is required to be installed in
   A. RMC
   B. RNC
   C. ENT-C
   D. Both A and B

3. The power supply to pool cover motors is permitted to
   A. Use a cord-and-plug connection
   B. Include an equipment grounding conductor smaller than the ungrounded branch-circuit conductors
   C. Be either Type AC or Type MC cable
   D. None of the above

4. When nonmetallic conduit is used to enclose branch-circuit wiring to a wet-niche luminaire, the equipment grounding conductor must be
   A. Minimum 12 AWG uninsulated copper
   B. Minimum 8 AWG insulated copper
   C. The same size as the ungrounded circuit conductors and of either copper, aluminum, or copper-clad aluminum
   D. As specified in 250.134(A)(2)

5. All underwater pool luminaires are required to be installed at least 18 in. below the normal water level, unless listed and identified for use at lesser depths, because
   A. They are cooled by water.
   B. They are required to be readily accessible.
   C. Water contact is required for successful equipment grounding.
   D. Water contact is required for proper illumination diffusion.

6. Pool pump motors installed outdoors are required to have
   A. A weathershield, unless listed for outdoor use
   B. Cathodic corrosion protection
   C. A maintenance disconnecting means
   D. Supplementary overload protection

7. Underground wiring that does not supply pool equipment is not permitted under swimming pools or within 5 ft horizontally from an inside wall of the pool, unless installed in
   A. RMC
   B. Type UF cable with ground
   C. EMT with asphaltic corrosion protection
   D. All of the above

8. Each permanently installed pool at a dwelling is required to have at least how many 125-volt, 15- or 20-ampere receptacles on a general-purpose branch circuit located between 10 and 20 ft from the inside walls of the pool, measured horizontally, and no more than 6½ ft above floor, platform, or grade level?
   A. One
   B. Two
   C. Three
   D. As specified in Table 680.16, depending on lot size

9. Pool motor branch-circuit wiring is permitted to be installed in
   A. RMC
   B. EMT when run on buildings
   C. Any wiring method specified in Chapter 3 of the *NEC* when installed inside a one-family dwelling or associated accessory building (such as a pool equipment shed)
   D. All of the above

10. Power conductors and cables for network-powered broadband communications systems are required to be installed at the following minimum distance above the water level or deck area immediately surrounding the swimming pool:
    A. 7½ ft
    B. 12 ft
    C. 17½ ft
    D. 22½ ft

FILL IN THE BLANKS

1. When the branch circuit supplying equipment is GFCI protected, then totally enclosed luminaires and ceiling-suspended (paddle) fans are permitted to have a minimum vertical clearance of _____ above the maximum pool water level.

2. Pool lighting that operates at 15 volts or less is required to be supplied through a(n) _____ to reduce the chance of an accidental short circuit that could energize the secondary wiring to the underwater luminaires (lighting fixtures) at higher voltage.

3. A junction box connected to a conduit from the forming shell of a wet-niche luminaire (lighting fixture) must be installed either _____ above ground level or _____ above the maximum pool water level, whichever provides the higher elevation.

4. Switches and receptacle outlets must be located at least 5 ft from the inside walls of a swimming pool, unless separated from them by a _____.

5. Pool area junction boxes are frequently installed above grade using conduits for support. A junction box is not permitted to be supported by a _____ conduit.

6. Lighting systems operating at 15 volts or less are permitted to use _____ rather than junction boxes raised above grade level.

7. Branch-circuit breakers supplying 120-volt pool lighting luminaires (lighting fixtures) must have _____ protection.

8. No receptacle is permitted to be installed closer than _____ from an outdoor pool.

9. Underwater pool lighting luminaires used in residential construction typically are of three types: *wet niche, dry niche,* and _____.

10. Receptacles that supply pool pump motors and are rated _____ are required to have GFCI protection.

TRUE OR FALSE

1. Typically, pool area junction boxes are manufactured of aluminum, are equipped with threaded hubs or conduit entries, and have extra grounding terminals.

   _____ True   _____ False

2. An insulated copper equipment grounding conductor, 12 AWG or larger, must be run with the branch-circuit conductors for dry-niche underwater pool lighting.

   _____ True   _____ False

3. Each permanently installed pool at a dwelling is required to have at least one 125-volt, 15- or 20-ampere receptacle on a general-purpose branch circuit located between 5 and 10 ft from the inside walls of the pool, measured horizontally.

   _____ True   _____ False

4. No receptacle outlet is permitted to be installed closer than 6 ft from an outdoor pool, fountain, or other installation, unless GFCI protected.

   _____ True   _____ False

5. Switching devices such as general-use snap switches, circuit breakers, and automatic timers must be located at least 5 ft from an inside wall of a pool, unless separated from it by a solid wall, a fence, or other permanent barrier.

   _____ True   _____ False

6. All branch circuits supplying underwater pool lighting operating at more than 15 volts are required to be GFCI protected.

   _____ True   _____ False

7. Luminaires (lighting fixtures) and ceiling-suspended (paddle) fans cannot be installed within 12 ft vertically above the maximum water level or within 5 ft horizontally from the pool's inside walls.

   _____ True   _____ False

8. Reinforcing steel used for a pool bonding grid is required to have a corrosion-resistant nonmetallic coating.

   _____ True   _____ False

9. Receptacles and switching devices located within buildings are considered to be separated from the swimming pool by a permanent barrier.

   _____ True   _____ False

10. The operating mechanism for an electrically operated pool cover is required to be located within sight from the machinery controlled.

    _____ True   _____ False

CHALLENGE QUESTIONS

1. Do the overhead conductor clearance rules of *NEC* 680.8 prohibit utility-owned power and communications conductors from being installed over existing swimming pools, fountains, and similar installation? Why or why not?

2. Why are switching devices such as circuit breakers, automatic timers, and general-use snap switches prohibited within 5 ft horizontally from the inside wall of a pool unless separated from it by a permanent barrier?

3. The *NEC* requires that luminaires, lighting outlets, and ceiling-suspended (paddle) fans be located not less than 12 ft vertically above the maximum water level and not closer than 5 ft horizontally from the inside walls of a pool. What is the rationale for permitting lesser clearances for existing luminaires and lighting outlets protected by ground-fault circuit interrupters?

4. Why are junction boxes for supply conductors to wet-niche pool luminaires required to be installed above the maximum pool water level?

5. Explain why most metal objects permanently installed near pools are required to be bonded with minimum 8 AWG solid copper conductors, which are larger than bonding conductors specified in the *Code* for many other applications.

6. Why is a remote panelboard installed to serve pool equipment required to have a separate grounding electrode installed but not to have a connection between the grounding terminal bar and grounded (neutral) conductor terminal bar?

7. Describe the purpose of establishing an equipotential bonding grid that interconnects the metal parts of pools and associated equipment.

HVAC Equipment and Water Heaters

14

OBJECTIVES

After completing this chapter, the student will be able to understand the following:

- Thermostat controls for heating and air-conditioning systems
- Where to locate disconnecting means for outdoor units
- What is meant by a noncoincident load
- The circumstances under which heating, ventilating, and air-conditioning (HVAC) equipment can be protected only by fuses
- Special requirements for circuiting receptacles built into baseboard heaters

INTRODUCTION

This chapter covers *National Electrical Code*® rules for central heating and air-conditioning equipment and water heaters. The following major topics are included:

- Central heating equipment
- Central air-conditioning equipment
- Heat pumps
- Thermostats and Class 2 wiring
- Room air conditioners
- Electric baseboard heaters
- Water heaters
- Special rules for two-family dwellings

IMPORTANT NEC TERMS

Accessible (as applied to equipment)

Appliance

Attachment Plug (Plug Cap, Plug)

Automatic

Bonding (Bonded)

Branch Circuit

Branch Circuit, Appliance

Branch Circuit, General-Purpose

Branch Circuit, Individual

Continuous Load

Disconnecting Means

Equipment

Exposed (as applied to live parts)

Exposed (as applied to wiring methods)

Hermetic Refrigerant Motor Compressor

In Sight from (Within Sight from, Within Sight)

Labeled

Leakage Current Detection and Interruption (LCDI)

Plenum

Receptacle Outlet

Switch, General-Use

Weatherproof

CENTRAL HEATING EQUIPMENT

Electric heating equipment, air-conditioning equipment, electric water heaters, electric clothes dryers, and electric cooking appliances account for the largest loads in a dwelling. Electric furnaces consist of electric heating elements with blower motors for circulating hot air through ductwork to the rest of the house.

Sometimes electric furnaces are stand-alone units, but often they are combined with central air-conditioning. In this case, the indoor unit (usually installed in a basement or utility room) also includes a condensate coil and connections for refrigerant lines to an outdoor unit. Gas- and oil-fired furnaces may also be stand-alone units or be combined with central air-conditioning.

All types of residential furnaces are sold as listed, packaged units that include such items as internal controls and thermal overload protection. The electrician's primary concerns are to run the branch circuit, install the required disconnecting means, and install the thermostat.

Non-Coincident Loads

Heating and air-conditioning loads don't normally operate at the same time. For this reason, only the larger of the two loads must be considered when calculating service load [220.60]. This subject is discussed at greater length in Chapter 2, Planning the Installation—Required Branch Circuits and Load Calculations. However, two separate branch circuits are normally required for HVAC equipment: one for the inside unit (furnace) and one for the outside unit (air conditioner).

Branch Circuits for Central Heating Equipment

Electric Furnace. Fixed electric space-heating equipment can be supplied by an individual branch circuit of any rating [424.3(A)]. See "Branch Circuit Rating," which follows.

Gas Furnace. Subsection 8.6.4 of NFPA 54-2002, *National Fuel Gas Code,* also requires an individual branch circuit for fixed gas-fired heating equipment. Gas furnaces typically have a small electric load consisting of a gas ignition system and a fan rated ½ horsepower or less. This same individual branch circuit can also supply associated equipment such as a pump, humidifier, or electrostatic air cleaner.

Combination Unit. When central heating and air-conditioning are combined in the same system, a separate branch circuit is provided for the inside unit (furnace with air-handling fan) and outside unit (air-conditioning compressor). (See Figure 14.1.)

Permanent Wiring Methods. Fixed electric space-heating equipment must be installed using permanent wiring methods. Flexible cords with attachment plugs are not permitted, because central heating equipment isn't designed to permit ready removal for maintenance or repair and is not identified for flexible cord connection [400.7(A)(8), 422.16(A)]. In addition, ANSI Z21.47-2003, *Gas-Fired Central Furnaces,* does not allow cord-and-plug connections for gas furnaces.

Branch Circuit Rating. The manufacturer's nameplate normally specifies the overcurrent protection and conductor size needed for the central heating unit. Branch

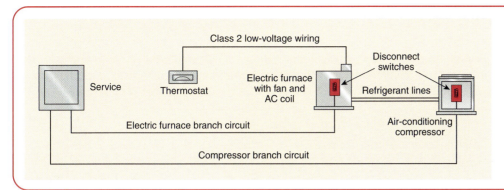

FIGURE 14.1 Central heating/air-conditioning systems have two separate branch circuits, one for the inside unit (furnace) and another for the outside unit (compressor).

circuit ratings of 60 to 80 amperes, 240 volts are common for electric furnaces. Gas furnaces typically have branch circuits rated 120 volts, either 15 or 20 amperes.

Disconnecting Means

Fixed electric space-heating equipment is required to have a disconnecting means located within sight from the unit or capable of being locked in the open position [424.19(A)]. Sometimes disconnect switches are mounted directly onto furnaces, although this arrangement becomes awkward if the furnace needs to be repaired. When a disconnect other than the circuit breaker is needed, the best solution is usually to mount it on a wall in an accessible location near the furnace.

Unit Switch. The required disconnecting means is permitted to be a unit switch with a marked OFF position built into the fixed heating equipment itself [424.19(C)].

Thermostatic Controls. A thermostatic control that directly interrupts all ungrounded conductors, that has a marked OFF position, and that can't be overridden automatically is permitted to serve as a required disconnecting means [424.20(A)]. Line-voltage thermostatic controls of this type are sometimes used on electric baseboard heaters, wall-mounted heaters, and unit heaters. See "Electric Baseboard Heaters" later in this chapter.

CAUTION: According to 424.20(B), a thermostat that does not directly interrupt all ungrounded conductors is not considered a disconnecting means. This rule includes all low-voltage thermostats and any type of thermostat that operates a remote-control circuit rather than actually interrupting the ungrounded supply conductors to the heating equipment.

Wiring Methods for Gas Furnaces

Gas furnace nameplates normally specify a 120-volt, 15- or 20-ampere branch circuit to supply components such as controls and a gas igniter. The required disconnecting means for gas furnaces is often a general-use switch built into the furnace itself [424.19(C)]. In these cases, the installer isn't required to provide a separate disconnect. However, the gas furnace must be located so that the integral ON/OFF switch is readily accessible (i.e., not blocked by ductwork or piping or mounted close to a facing wall).

CENTRAL AIR-CONDITIONING EQUIPMENT

Air-conditioning equipment with hermetic refrigerant motor compressors has different operating characteristics than other types of motor loads. In particular, it has a higher momentary starting load (inrush current) than other types of motorized appliances, because the compressor motor is liquid-cooled, operating inside the refrigerant itself. This special cooling feature allows the motor to work harder than a conventional electric motor of the same size.

Minimum Circuit Ampacity

Nameplates on central air-conditioning equipment specify the minimum circuit ampacity needed. This value is used to select the branch-circuit conductors and disconnect switch without performing any additional calculations. For example, the central air-conditioner nameplate shown in Figure 14.2 specifies a minimum circuit ampacity of 25.4 amperes. By referring to *NEC*® Table 310.16 and using the columns for conductors rated 60°C (140°F), a branch-circuit conductor size of 10 AWG is determined. The corresponding size for a non-fused disconnect switch would be 30 amperes. However, the minimum circuit ampacity is *not* used to select the proper circuit breaker or fuse size, which is based on another value provided in the nameplate.

Maximum Overcurrent Protection

The manufacturer's label or nameplate also specifies the maximum overcurrent protection required on central air-conditioning equipment. A typical nameplate might read "Maximum Fuse or HACR-Type Breaker Rating: 40 A."

Heating, Air-Conditioning, and Refrigeration (HACR) Circuit Breakers. HACR is a special rating for circuit breakers that are intended to serve heating, air-conditioning, and refrigeration equipment. HACR breakers can withstand high inrush currents—caused when equipment such as air-conditioning compressors start to operate—without tripping (Figure 14.3).

All 240-volt circuit breakers rated up to 100 amperes are now marked "HACR." So, in practical terms, any modern 2-pole breaker of the correct ampere rating can be used to provide overcurrent protection for central heating and air-conditioning

FIGURE 14.2 This is the information shown on a typical nameplate for an outdoor air compressor.

FRESHAIR, INC. Model: 345A-OU						
	VAC	Ø	HZ	RLC	LRC	FLA
Compressor	230	1	60	16.8	102	–
Fan 1/4 HP	230	1	60	–	–	1.5
Br ckt current:		19.3 A				
Min ckt ampacity:		25.4 A				
Max fuse / HACR bkr:		40 A				
Voltage range:		207–253 V				

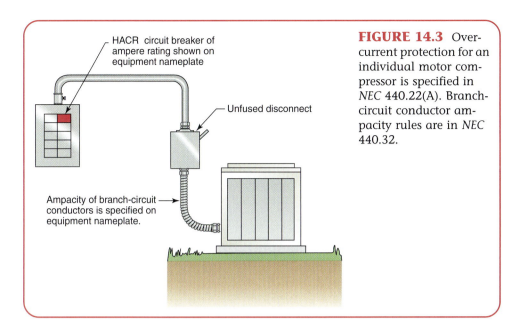

HACR circuit breaker of ampere rating shown on equipment nameplate

Unfused disconnect

Ampacity of branch-circuit conductors is specified on equipment nameplate.

FIGURE 14.3 Overcurrent protection for an individual motor compressor is specified in *NEC* 440.22(A). Branch-circuit conductor ampacity rules are in *NEC* 440.32.

equipment in dwellings, provided that the equipment label permits the use of circuit breakers. Typical ratings for supply branch circuits are 240 volts at 30, 40, or 50 amperes.

CAUTION: Older circuit breakers are not HACR rated, which means that circuit breakers in existing houses may not be suitable for supplying new heating, air-conditioning, and refrigeration equipment [110.3(B)].

Nearly all modern heating and air-conditioning equipment permits the use of either fuses or HACR circuit breakers. However, if a manufacturer's label or nameplate specifies maximum fuse size only, then 110.3(B) requires that fuses be used to provide the required overcurrent protection (Figure 14.4).

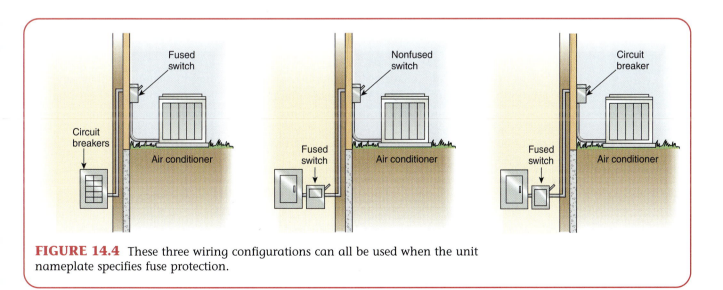

Fused switch

Circuit breakers

Air conditioner

Nonfused switch

Fused switch

Air conditioner

Circuit breaker

Fused switch

Air conditioner

FIGURE 14.4 These three wiring configurations can all be used when the unit nameplate specifies fuse protection.

Disconnecting Means

An outdoor air-conditioning unit is required to have a disconnecting means located within sight from the equipment and readily accessible [440.14]. Although disconnect switches are permitted to be mounted directly onto the air-conditioning equipment itself, this design is awkward if the unit needs to be replaced. Also, the *Code* doesn't permit a disconnecting means to be installed on any part of a panel of HVAC equipment that must be removed to service the unit [440.14].

Disconnect Location. A better practice is to install the disconnect switch on the wall of the house, as shown in Figure 14.4. However, mounting the disconnect behind an air conditioner or heat pump wouldn't be considered readily accessible by many electrical inspectors, and squeezing behind or leaning across other equipment to work on the disconnect isn't a safe practice.

The best approach is to install the required disconnect on the wall of the house, locating it to one side of the air-conditioning unit or heat pump. Disconnect switches mounted outdoors must be weatherproof [110.11] and are normally rated NEMA/UL Type 3R.

HEAT PUMPS

A heat pump is essentially a reversible air conditioner. In warm weather, it absorbs heat from inside air and exhausts it to the outdoors. In cold weather, it absorbs heat from outdoor air (there is some heat available even when it seems cold) and distributes it through air-supply ducts inside the home.

Heat pumps have two components: an outdoor compressor and an indoor air handler (fan unit). In colder climates, the indoor air handler often has supplemental electric-resistance heating elements. These heating elements turn on when the heat pump, acting by itself, can't deliver the temperature set at the thermostat.

Branch Circuits for Heat Pumps

Heat pumps must be installed according to all standard *NEC* rules governing air-conditioning and central heating equipment. These rules can be summarized as follows.

- The equipment must be supplied by an individual branch circuit that serves no other loads except associated equipment [424.3(A)]. In the case of heat pumps, each unit (outdoor and indoor) may be served by a separate circuit.
- Heat pumps must be installed using permanent wiring methods [400.7(A)(8), 422.16(A)].
- The manufacturer's nameplate or label normally specifies the overcurrent protection and conductor size needed for the heat pump unit(s).

THERMOSTATS AND CLASS 2 WIRING

A thermostat controls HVAC equipment based on sensing temperature inside the dwelling. For this reason, it should be mounted on an interior wall in a location where it won't be exposed to drafts, direct sunlight, heat from fireplaces or appliances, and hot-air heating registers. These variables can interfere with a thermostat's ability to properly regulate temperature throughout the home.

Thermostats are normally mounted at 60 in. above the finished floor. (See Table 6.1, Recommended Heights for Wall-Mounted Outlets, in Chapter 6 of this book.) Some thermostats simply turn HVAC equipment on and off in response to temperature. This type of thermostat often has a dial that twists to set the desired temperature. It may also have a selector switch that indicates operating modes, such as COOL, HEAT, AUTOMATIC, and FAN. Other thermostats are programmable and offer features such as separate day and night temperature settings, humidity control, and a digital clock on the thermostat itself.

Thermostat Wiring. Different types of thermostat central heating and air-conditioning systems use different wiring hookups between the thermostat and furnace/air conditioner/heat pump. Basic thermostats may use only two wires, while more advanced models with additional features may require four or five. The thermostat and HVAC manufacturers' instructions will include the wiring requirements.

Most residential thermostats operate at 24 volts and use Class 2 wiring. *Code* rules for installing Class 2 wiring for thermostats, door chimes, and other low-voltage circuits in the home can be summarized as follows.

- Class 2 wiring doesn't have to be installed in raceways. However, it must be installed in a neat and workmanlike manner and properly supported using straps, staples, hangers, and similar fittings [725.8].
- Class 2 wiring can't be installed in the same raceway with power wiring [725.55(A)]. It can enter the same enclosure as power wiring only when the two wiring systems are connected to the same piece of equipment. However, in such a case, the Class 2 wiring and power wiring must be separated by a barrier or kept ⅛ in. apart within the outlet or terminal box [725.55(D)].

ROOM AIR CONDITIONERS

Although central air-conditioning is the most common approach in new residential construction, window or through-wall-type air conditioners are also used. *NEC* rules for installing 120- and 240-volt, single-phase room air conditioners equipped with attachment cords and plugs can be summarized as follows.

- *Single Receptacle.* A single receptacle installed on an individual branch circuit must be rated the same as the branch circuit [210.21(B)(1)]. This rule means that a 15-ampere single receptacle can't be installed on a 20-ampere individual branch circuit.

- *Multiple Receptacles.* When a branch circuit supplies two or more receptacles and one or more of them is intended to serve a room air conditioner, the load of the appliance cannot exceed 80 percent of the branch-circuit rating [210.21(B)(2)]. Maximum cord-and-plug-connected loads are listed in *NEC* Table 210.21(B)(2), shown in Exhibit 14.1.

NOTE: A duplex receptacle is considered to be two receptacles [Article 100 definition, Receptacle].

EXHIBIT 14.1

NEC TABLE 210.21(B)(2) Maximum Cord-and-Plug-Connected Load to Receptacle		
Circuit Rating (Amperes)	Receptacle Rating (Amperes)	Maximum Load (Amperes)
15 or 20	15	12
20	20	16
30	30	24

- *Disconnecting Means.* The attachment plug can serve as the required disconnecting means, under certain conditions [440.63].

- *LCDI or AFCI.* A 120- or 240-volt, single-phase room air conditioner must have factory-installed leakage current detection and interruption (LCDI) or arc-fault circuit-interrupter (AFCI) protection built into the power supply cord or attachment plug [440.65].

- *Device Types.* Several types of receptacles are used to supply room air conditioners (Figure 14.5). When a dual-voltage duplex device is installed with one 120-volt and one 240-volt receptacle on the same yoke (strap), it must be supplied by a 2-pole circuit breaker that simultaneously opens all ungrounded conductors (Figure 14.6) [210.4(C), Exception No. 2].

ELECTRIC BASEBOARD HEATERS

Many electric baseboard heaters can operate at either 120 volts or 208/240 volts. As discussed in earlier chapters, some very large or all-electric homes have services

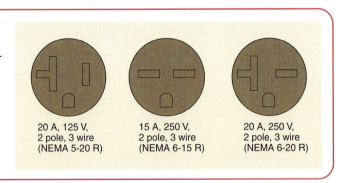

FIGURE 14.5 These receptacle types are used to supply cord-and-plug-connected window air conditioners.

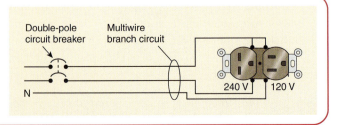

FIGURE 14.6 A combination 120/240-volt receptacle must be supplied by a multi-wire branch circuit.

rated 208Y/120 volts, 3-phase, 4-wire. Typically, baseboard heaters come from the factory wired for 208/240 volts and can be converted in the field for 120-volt operation by reconnecting a jumper.

Electric baseboard heaters put out different amounts of heat at different operating voltages. The manufacturer's literature will include wattage ratings at each voltage.

Determining Needed Wattage. Electric baseboard heaters come in different standard lengths (2, 2.5, 3, 4, 5, and 6 feet are common); and multiple units can be connected together. Heaters designed for mounting in continuous rows have an integral wiring channel, so that branch-circuit wiring can be run through one unit to supply another.

The number of baseboard heaters needed in a room depends on the size of the room. A rule of thumb is 6 watts per square foot of floor space. In colder climates or at higher altitudes, the allowance is 8 watts per square foot. Table 14.1 lists the total watts needed to heat a number of room sizes.

Thermostats

Many baseboard heaters have integral temperature controls built into the units. Typically, these controls are combined with ON/OFF switches (Figure 14.7). When baseboard heaters have built-in temperature controls, the supply conductors run directly from the panelboard to a junction box on the unit.

Baseboard heaters can also be controlled by separate wall-mounted thermostats. Both types of thermostats (unit-mounted and wall-mounted) come in line-voltage and low-voltage models.

Line-Voltage Thermostats. Using line-voltage thermostats is the most common approach for controlling electric baseboard heaters. Single-pole thermostats are used with 120-volt units (Figure 14.8) and double-pole thermostats with 208/240-volt units (Figure 14.9).

Table 14.1 **Electric Baseboard Heater Wattage**

Room Square Footage	Basic Wattage	Cold Climate Wattage
50	300	400
75	450	600
100	600	800
125	750	1000
150	900	1200
175	1050	1400
200	1200	1600
225	1350	1800
250	1500	2000
275	1650	2200
300	1800	2400
325	1950	2600
350	2100	2800
375	2250	3000
400	2400	3200

FIGURE 14.7 This electric baseboard heater has a built-in thermostat and integral duplex receptacle. The receptacle is supplied by a different branch circuit than the heating elements, to comply with *NEC* 210.52 and 424.9.

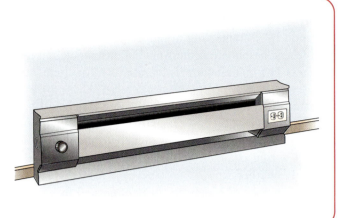

FIGURE 14.8 Single-pole thermostats are used to control 120-volt baseboard heaters.

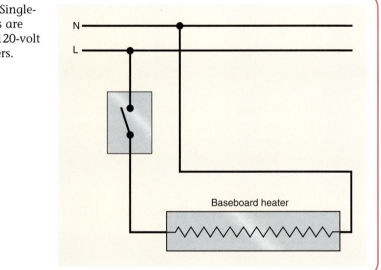

FIGURE 14.9 Double-pole thermostats are used to control 208- and 240-volt baseboard heaters.

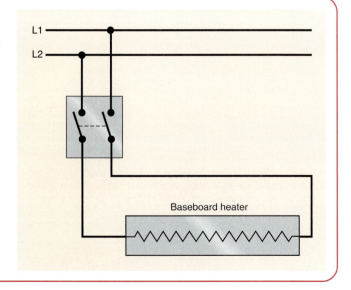

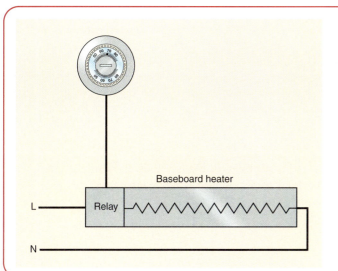

FIGURE 14.10 This low-voltage thermostat controls a 120-volt baseboard heater through a single-pole relay.

Low-Voltage Thermostats. When the total load of the baseboard heaters to be controlled exceeds that of a line-voltage thermostat, a low-voltage thermostat of the same type used to control central heating is installed. Low-voltage thermostats don't break the supply conductors. Instead, they send a low-voltage signal over Class 2 wiring that operates a relay to turn the heater(s) ON and OFF (Figure 14.10).

Branch Circuits for Baseboard Heaters

Branch circuits supplying two or more electric baseboard heaters must be rated 15, 20, 25, or 30 amperes [424.3(A)].

Baseboard heater nameplates usually specify that supply conductors be rated 90°C (194°F). These conductors can be Type THHN in conduit, or nonmetallic-sheathed (Type NM) or armored (Type AC) cable with conductors rated 90°C (194°F). One reason for the high temperature rating is that baseboard heaters often have a wiring channel running along the bottom. Branch-circuit conductors are run in this channel when two or more units are joined together end-to-end.

Reidentifying Grounded (White) Conductors. When a 240-volt baseboard heater is supplied by 2-wire cable such as Type NM or Type AC, the white wire must be reidentified as an ungrounded conductor at terminations and other points where it is visible and accessible [200.7(C)(1)]. This reidentification can be done using paint or electrician's tape.

Baseboard Heaters and Receptacle Outlets

Listed electric baseboard heaters may include instructions that prohibit installing them below receptacle outlets [210.52, FPN]. Heaters of this type come furnished with blank sections that can be installed below wall-mounted receptacle outlets.

Other electric baseboard heaters include factory-installed receptacles that can be used to meet the outlet spacing requirements of 210.52(A). (See Figure 14.7.)

However, these receptacles aren't permitted to be supplied by the heater circuit [210.52, 424.9]. Instead, they should be connected to the branch circuit supplying other receptacle outlets in the room where they're located.

WATER HEATERS

Electric Water Heaters

Water heaters are considered to be continuous loads, since they often operate for three hours or more. Thus, the load of an electric water heater must not exceed 80 percent of the rating of its branch circuit [210.19(A)].

Calculation Example

The nameplate rating for an electric water heater is 4500 watts, 240 volts:

$$\frac{4500 \text{ W}}{240 \text{ V}} = 18.75 \text{ A}$$

Dividing by 0.80 gives the continuous load amperes:

$$\frac{18.75}{0.80} = 23.44 \text{ A}$$

The result, 23.44 amperes, is 94 percent of 25 amperes, and the next higher standard overcurrent device rating is 30 amperes:

$$23.44\text{A} = 78 \text{ percent of } 25 \text{ A}$$

Therefore, a 30-ampere circuit breaker or fuse can be used to protect the electric water heater load in this example.

The additional 25 percent of the continuous load must also be added to feeder and service calculations [220.10, 230.42(A)]. (See Chapter 2, Planning the Installation—Required Branch Circuits and Load Calculations.)

Branch-Circuit Conductors

The water heater nameplate or terminal box may specify a temperature rating for the branch-circuit conductors. If no temperature rating is shown, then 60°C (140°F) conductors such as Type TW can be used. If the marking indicates 75°C (167°F), then wire with an "H" in its designation must be used, such as Type THW or Type THWN. If the nameplate or terminal box specifies 90°C (194°F) conductors, wire such as Type THHN must be used.

When higher temperature conductors are required for a water heater, one common practice is to mount a junction box on a wall near the unit. The ordinary branch-circuit conductors, such as 75°C (167°F) Type NM cable, are then spliced to shorter lengths of high-temperature conductors that connect to the water heater (Figure 14.11).

FIGURE 14.11 For permanently connected appliances rated over 300 volt-amperes, the branch circuit fuse or circuit breaker can serve as the disconnecting means where it is within sight from the appliance.

Disconnecting means: branch-circuit fuse or circuit breaker

Type NMB cable with conductors rated 167°F

Junction box

THHN conductors rated 194°F in flexible metal conduit

Flexible Connection. When branch-circuit conductors are installed in raceway, it is common practice to use a flexible raceway such as flexible metal conduit (FMC) to connect from the last junction box to the knockout on the water heater.

Cord-and-Plug Connection Not Permitted. Flexible cords are permitted for appliances only "where the fastening means and mechanical connections are specifically designed to permit ready removal for maintenance and repair, and the appliance is intended or identified for flexible cord connection" [400.7, 422.16(A)]. Since water heaters connected to rigid water piping aren't designed for ready removal, they can't have flexible cords with attachment plugs. Permanent wiring methods (cables, or conductors in raceways) must be used to connect water heaters.

Disconnecting Means

Typically, the circuit breaker or fuse in the service panelboard functions as the disconnecting means required by 422.31(B), as shown in Figure 14.11. If the panelboard is not within sight from the water heater location (for example, a furnace may be in the way), a separate non-fused disconnect switch must be provided.

Gas Water Heaters

Gas water heaters don't normally have an electrical connection.

SPECIAL RULES FOR TWO-FAMILY DWELLINGS

In two-family dwellings, the required second disconnecting means for fixed electric space-heating equipment isn't required to be located in the same dwelling unit as the equipment itself. The main circuit breaker or fused switch protecting each dwelling unit's panelboard can serve this function [424.19(C)(2)].

NOTE: In one-family dwellings, the service disconnecting means is permitted to serve as the required second disconnect for fixed electric space heating equipment [424.19(C)(3)].

CHAPTER REVIEW 14

MULTIPLE CHOICE

1. Fixed electric space-heating equipment is required to
   A. Be served by an individual branch circuit
   B. Have a disconnecting means
   C. Have the maximum overcurrent protection specified on the manufacturer's label or nameplate
   D. All of the above

2. Thermostat wiring is
   A. Class 2
   B. Class 3
   C. Power limited
   D. Both B and C

3. The following condition(s) applies to installing Class 2 wiring:
   A. It cannot be installed in the same raceway with power wiring.
   B. It can enter the same enclosure as power wiring only when the low-voltage circuit is derived from the power circuit.
   C. It shares a common ground with the power wiring.
   D. None of the above

4. The overcurrent protection and branch-circuit conductors for a continuous load must be rated at
   A. Not over 80 percent of the load
   B. Not less than 125 percent of the load
   C. One hundred percent of the load when copper conductors are used
   D. None of the above

5. When a 2-wire cable is used to supply 240-volt electric baseboard heaters
   A. The sheath must be bonded at all terminations and junction boxes.
   B. The grounded conductor must be reidentified as an ungrounded conductor.
   C. The 2-phase (hot) conductors must be identified using different colors of paint or tape.
   D. All of the above

6. A thermostatic control intended to serve as a required disconnecting means for fixed electric space-heating equipment must
   A. Directly interrupt all ungrounded conductors

B. Have a marked OFF position
   C. Not be capable of being overridden automatically
   D. All of the above

7. A disconnect switch installed for an outdoor air-conditioning unit must be
   A. Weatherproof
   B. Readily accessible
   C. Fused
   D. Both A and B

8. When a receptacle is installed as part of a baseboard heater assembly, it
   A. Is required to be sub-fed from the 240-volt heater circuit via an isolation transformer with a grounded secondary
   B. May not be mounted where subject to moisture damage
   C. Must be supplied by a branch circuit other than the heater circuit
   D. None of the above

9. The manufacturer's label or nameplate on central air-conditioning equipment typically specifies
   A. NEMA/UL enclosure type for the unit
   B. Minimum lead length for making connections in the terminal box
   C. Maximum overcurrent protection required
   D. All of the above

10. Electric baseboard heaters are prohibited from being located beneath wall-mounted receptacle outlets by
    A. *NEC* 110.14(A)
    B. NFPA *101®, Life Safety Code®*
    C. Baseboard heater listing instructions
    D. Both A and B

FILL IN THE BLANKS

1. A continuous load is a load for which the maximum current is expected to continue for _____ hours or more.

2. When the nameplate or terminal box of a heating appliance does not specify a temperature rating for the branch-circuit conductors, then conductors rated _____ are permitted to be used.

3. A _____ is required to have factory-installed LCDI or AFCI protection built into the power supply cord or attachment plug.

4. When it is unlikely that two loads will operate at the same time (such as heating and air-conditioning), they are called _____ loads.

5. Providing a device in a power supply cord or cordset that senses leakage current flowing between or from the cord conductors and interrupts the current at a predetermined level of leakage is known as a _____.

6. In residences, the _____ is permitted to serve as the required second disconnecting means for fixed electric space-heating equipment.

7. Low-voltage thermostats are typically used to control electric baseboard heating units when _____.

8. When a piece of utilization equipment isn't marked to indicate a temperature rating for the supply conductors, then conductors rated _____ can normally be used.

9. The _____ normally serves as the required disconnect for room air conditioners.

10. Fixed electric space-heating equipment in dwellings is required to be supplied by an individual branch circuit rated _____ or less.

TRUE OR FALSE

1. Thermostats can always serve as the required disconnecting means for electric space-heating equipment.
   ____ True   ____ False

2. Fixed gas space-heating equipment is required to have integral electrical backup protection, such as supplemental electric heating coils.
   ____ True   ____ False

3. A single receptacle installed on an individual branch circuit must have a rating not exceeding 80 percent of the branch circuit on which it is installed.
   ____ True   ____ False

4. A unit switch built into fixed electric space-heating equipment that has a marked OFF position is permitted to serve as the required disconnecting means.
   ____ True   ____ False

5. Line-voltage thermostats are required to have integral overcurrent protection unless they are inherently self-limiting.
   ____ True   ____ False

6. If a manufacturer's label or nameplate on air-conditioning equipment specifies maximum fuse size, then either fuses or HACR circuit breakers may be permitted to provide the required overcurrent protection.
   ____ True   ____ False

7. A cord-connected electric water heater must be supplied through a single receptacle installed on an individual branch circuit.
   ____ True   ____ False

8. Some baseboard heaters are suitable for use at more than one voltage.
   ____ True   ____ False

9. A 120- or 240-volt, single-phase room air conditioner equipped with an attachment cord and plug is required to have either LCDI or AFCI protection, but not both.
   ____ True   ____ False

10. Class 2 wiring can be installed in the same raceway with power wiring only when the two wiring systems are connected to the same piece of equipment.
   ____ True   ____ False

CHALLENGE QUESTIONS

1. Why shouldn't conventional circuit breakers that are not rated HACR be used to supply air-conditioning equipment that employs hermetic refrigerant motor-compressors?

2. Using a *Code* reference, explain why heating or air conditioning that is labeled only with a maximum fuse size must use fuses rather than a circuit breaker to provide overcurrent protection.

3. Why does the *Code* permit a kitchen waste disposer to be cord-and-plug connected but not permit this wiring method for a water heater

4. What is the reason that nameplates on some types of heating equipment such as baseboard heaters and water heaters specify the use of branch-circuit conductors with high temperature ratings?

5. Why are the disconnecting means for appliances required to be lockable in the open position if they are not within sight of the appliance?

Special Systems

15

OBJECTIVES

After completing this chapter, the student will be able to understand the following:

- What is meant by a *primary protector*
- Requirements for locating smoke detectors in dwellings
- Prohibitions on placing low-voltage and power conductors in the same enclosures
- The special dimensions of communications ground rods
- Situations where telephone wiring is required to be installed using the wiring methods outlined in Chapter 3 of the *National Electrical Code*®
- The difference between star and loop telephone wiring

INTRODUCTION

This chapter covers *NEC*® safety rules governing the installation of low-voltage systems, whether legally required, such as hard-wired smoke detectors in new homes, or optional, such as cable TV and door chimes. The following major topics are included:

- Telephones
- Cable television
- Smoke detectors and alarm systems
- Door chimes
- Residential structured cabling systems
- Special rules for two-family dwellings

IMPORTANT *NEC* TERMS

Branch Circuit

Branch Circuit, Multiwire

Cable

Circuit Breaker

Grounded Conductor

Outlet

Receptacle

Receptacle Outlet

Smoke Alarm

Smoke Detector

Wire

TELEPHONES

The telephone company installs its wiring to a terminal box called the *network interface unit (NIU),* or *network interface device (NID),* shown in Figure 15.1. In new residential construction, this terminal box is often located outside the dwelling and has up to four female jacks for plugging in phone wires. The electrical contractor is responsible for installing all interior telephone wiring beyond this service entrance.

The NIU/NID is also known as the "demarcation point," which is a concept similar to the "service point" for electric light and power wiring (Figure 15.2). Typically, it is installed by the telephone company. In some parts of the country, the electrical contractor installs an NIU/NID supplied by the telephone utility.

Primary Protector

The incoming telephone service cable is required to have a listed primary protector [800.90], which guards against high-voltage spikes caused by lightning. Sometimes this primary protector is installed by the telephone utility, but when installation is the responsibility of the electrical contractor, the steps that follow should be taken.

Step 1. *Install the type of listed primary protector supplied or specified by the local telephone utility. In many cases, this component comes built into the NIU/NID.*

Step 2. *Connect a 14 AWG insulated solid or stranded grounding conductor (copper or other corrosion-resistant material) from the primary protector to the dwelling's grounding electrode system [800.100(A)].* As shown in Figure

FIGURE 15.1 A telephone network interface unit (NIU) connects interior phone lines to the service from the telephone company's system.

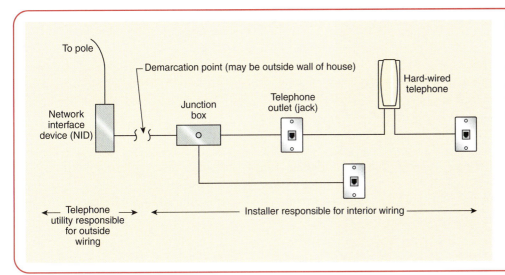

FIGURE 15.2 The "demarcation point" establishes responsibility for installing telephone wiring and equipment. It is similar to the "service point" for electric light and power.

15.3, a telephone network interface installed indoors is often connected with a clamp to the metal water piping within 5 ft of the point where this piping enters the building [800.100(B) and (C)].

Step 3. *Keep the grounding conductor under 20 ft in length and run it to the water piping in as straight a line as possible [800.100(A)(4) and (5)].*

Step 4. *Follow the usual practice of running the grounding conductor exposed. If it is run in a metal raceway to protect it from damage, both ends of the raceway must be bonded to the conductor or to the same terminals/electrodes to which the grounding conductor is connected.*

Communications (Telephone) Cables

Traditional telephone wiring has four conductors enclosed in a round or flat jacket. The conductors are color-coded red, green, yellow, and black. The red/green pair was originally intended for carrying telephone signals, and the black/yellow pair was intended for powering phone accessories such as lighted dials or pushbuttons. However, the black/yellow pair could also carry signals and was often used to add a second telephone line in an existing dwelling without rewiring. The second line could be made available at any outlet simply by connecting the black/yellow wires that were already present.

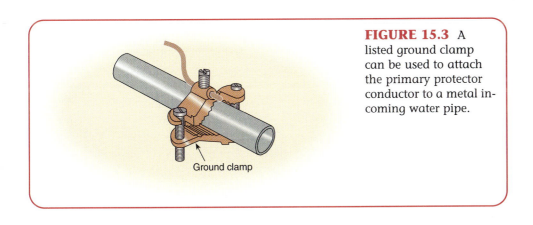

FIGURE 15.3 A listed ground clamp can be used to attach the primary protector conductor to a metal incoming water pipe.

Category 3. Since 2001, the U.S. Federal Communications Commission (FCC) has required that all new residential communications wiring be a minimum of two-pair Category 3 cable (Figure 15.4). Category 3 cable has two twisted pairs that are color-coded blue/blue-white and orange/orange-white. Both pairs are intended for carrying telephone signals, since many modern residences have more than one telephone line. Some residences use three- or four-pair Category 3 cable to provide either three or four telephone lines for specific use by parents, by children, for a home office, and for Internet access.

Listed Cables Required. All communications cables installed inside buildings are required to be listed [800.113, 800.179]. Type CMX limited-use cable can be used for all applications in one- and two-family dwellings [800.179(E)]. It can be installed exposed or in raceways [800.179(E)].

Installing Communications Wires and Cables

The *Code* doesn't require that communications wires and cables be installed in raceways. However, some jurisdictions may have their own local requirements. In doubtful situations, the authority having jurisdiction (AHJ) should be consulted.

When communications wires and cables are installed in a raceway, the raceway must be of a type permitted by Chapter 3 of the *NEC* and installed according to Chapter 3 rules, or it must be a listed nonmetallic communications raceway installed according to the rules of Article 362, Electrical Nonmetallic Tubing: Type ENT. However, conduit fill restrictions don't apply [800.110].

> **Loop and Radial Wiring:** Traditionally, residential telephone wiring was installed in serial or daisy-chain fashion, with multiple outlets (jacks) on each phone cable. This wiring method was known as "loop" wiring.

Most new communications wiring is installed in radial or star fashion, with each outlet connected by a homerun cable to the NIU/NID (Figure 15.5). This configuration agrees with the industry standard, EIA/TIA 570-A-1999, *Residential Telecommunications Cabling,* and is intended to work better for new applications such as home computer networks. However, *NEC* Article 800 has no rules for wiring communications (telephone) outlets. Installers should check with their local telephone company to determine whether it has installation guidelines for residential wiring.

Support of Conductors. Communications cables are required to be installed in a neat and workmanlike manner and must be attached to the building structure [800.24].

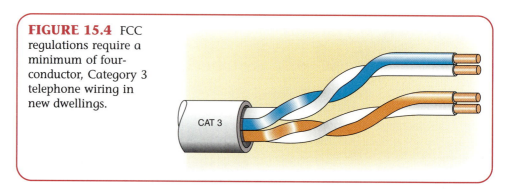

FIGURE 15.4 FCC regulations require a minimum of four-conductor, Category 3 telephone wiring in new dwellings.

CAT 3

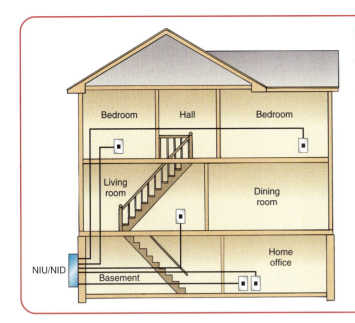

FIGURE 15.5 New residential telephone wiring should be installed in a radial or star configuration.

Raceways cannot be used to support telephone cables [800.133(C)]. Instead, they should be secured using plastic cable ties (not too tight) or stand-off staples.

It's very important not to crush communications cable or bend it too sharply. Four-conductor Category 3 cable has a diameter of about ⅛ in., and bends should have a radius of not less than five times the cable diameter—or ⅝ in.

Required Separations. Communications cables cannot be installed in the same raceways or enclosures as power wiring [800.133(A)(1)(c)].

The *Code* doesn't require a minimum separation between communications cables and power conductors run in cables or raceways [800.133(A)(2), Exception No. 1]. Many installers feel it isn't good practice to run low-voltage cables and nonmetallic-sheathed (Type NM) or armored (Type AC) power cables through the same holes in wood framing members, but there's no *NEC* rule that prohibits this practice.

When communications and power cables are run through the same bored holes, the larger power cables should be pulled in first and the smaller telephone wires should be pulled in later. This practice helps avoid damage to the Category 3 cables due to crushing or damage caused by snagging the insulation against the rough edges of bored holes.

Communications cables should also be kept away from hot-water pipes, hot-air ducts, and other sources of heat that might damage the insulation.

Communications (Telephone) Outlets

Telephone outlets are required to be listed [800.170], but need not be installed in outlet boxes. Nonmetallic brackets are normally used in residential construction to support telephone outlets (Figure 15.6). When communications cables are installed in raceways, then all the standard wiring rules of *NEC* Chapter 3 apply, including the rule that boxes must enclose telephone outlets. However, conduit fill restrictions don't apply [800.110].

Outlet Locations. The *NEC* doesn't specify required locations for telephone outlets. In new residential construction, it's common to provide a telephone outlet in most

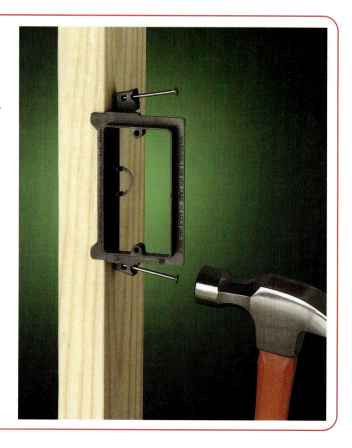

FIGURE 15.6 Communications (telephone) outlets are typically installed in nonmetallic brackets nailed to wooden studs. (Courtesy of Arlington Industries, Inc.)

habitable rooms: living room, bedrooms, den/study, family/recreation room, and kitchen (dining rooms don't usually have telephone outlets installed in them).

Some homes also have telephone outlets in other locations: bathrooms, workshop, laundry room, attached garage, covered porches, and patios. In doubtful situations, the owner or builder should be consulted.

Required Separations. Communications cables cannot be installed in the same boxes as power conductors [800.133(A)(1)(c), Exception No. 1]. When a telephone outlet and duplex receptacle are installed on the same faceplate, the receptacle must be installed in a device box. The telephone outlet can be installed on a bracket. Using a double-gang box with a barrier between the power conductors and telephone conductors is also permitted.

CABLE TELEVISION

Cable companies typically install a coaxial cable to some point inside the house near the electrical service equipment. In many parts of the country, a coordinated "single-trench" installation of electric-service lateral, telephone, and TV cables is used. No separation is required between the low-voltage and service-lateral conductors, as long as the power conductors are run in raceway, or are underground feeder and branch-circuit (Type UF) cable, or service-entrance (Type USE) cable [820.47(B), Exception No. 2].

Grounding

The outer metallic shield of the incoming coaxial cable is required to be grounded [820.93]. The following steps meet the grounding requirements in *NEC* 820.100.

Step 1. Connect an insulated solid or stranded grounding conductor to the cable shield [820.100(A)]. A special ground wire fitting that is made for this purpose looks something like a coaxial splitter. The incoming coaxial cable connects to one side, an interior coaxial cable connects to the other, and there is a screw terminal for a ground wire (Figure 15.7).

Step 2. Use a coaxial shield grounding wire that is a minimum of 14 AWG, but not larger than 6 AWG (copper or other corrosion-resistant material), and that has a current-carrying capacity approximately equal to that of the outer conductor (metallic shield) of the coaxial cable [820.100(A)]. The cable company's installation guidelines will specify the proper size grounding conductor to be used.

Step 3. Keep the grounding conductor specified in Step 2 under 20 ft in length, and run it in as straight a line as possible [800.100(A)(4) and (5)].

Step 4. Connect the coaxial cable shield grounding conductor to the metal water piping within 5 ft of the point where the piping enters the building, to a metal power service raceway, or to the service-equipment enclosure [820.100(B)(1)]. It can also be connected to a minimum 8-ft pipe electrode of the materials and diameters specified in 250.52(A)(5). The telephone service primary protector can also be grounded to the CATV ground rod, if desired.

CAUTION: When the coaxial system is connected to a separate grounding electrode, it is very important that a minimum 6 AWG copper bonding jumper be connected from the coaxial grounding system to the power grounding electrode system. Otherwise, potential differences may exist that can create a shock hazard and damage sensitive consumer electronic appliances (Figure 15.8) [820.100(D)].

Step 5. Follow the usual practice of running the coaxial cable shield grounding conductor exposed. If it is run in a metal raceway to protect it from damage, both ends of the raceway must be bonded to the conductor or to the same terminals/electrodes to which the grounding conductor is connected.

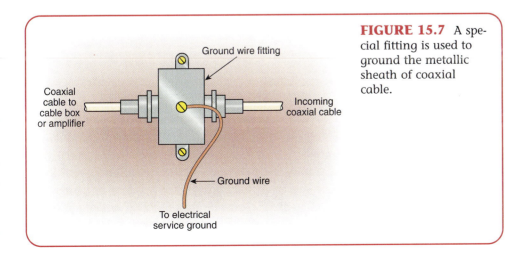

FIGURE 15.7 A special fitting is used to ground the metallic sheath of coaxial cable.

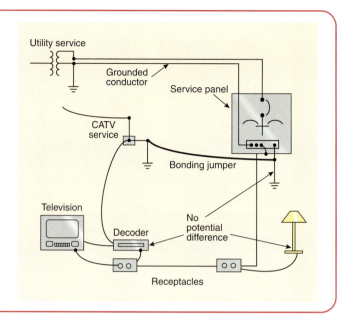

FIGURE 15.8 Dangerous potential differences can exist if a CATV system is grounded separately from the building electrical system. *NEC* 820.100(D) requires a bonding jumper between the TV system grounding electrode and the power grounding electrode system.

Listed Cables Required

All coaxial cables installed inside buildings are required to be listed [820.113, 820.179]. Type CATV and Type CATVX coaxial cables can be used for all applications in one- and two-family dwellings. Type CATVX cable can be installed exposed or in raceways [820.179(D)].

Installing Coaxial Cables

Coaxial cables are typically installed exposed. They are installed in radial or star fashion, with each outlet connected by a homerun cable to a splitter (Figure 15.9).

Support of Conductors. Coaxial cables are required to be installed in a neat and workmanlike manner and must be attached to the building structure [820.24]. Raceways cannot be used to support coaxial cables [820.133(C)]. Instead, they should be secured using plastic cable ties (not too tight) or depth-stop staples.

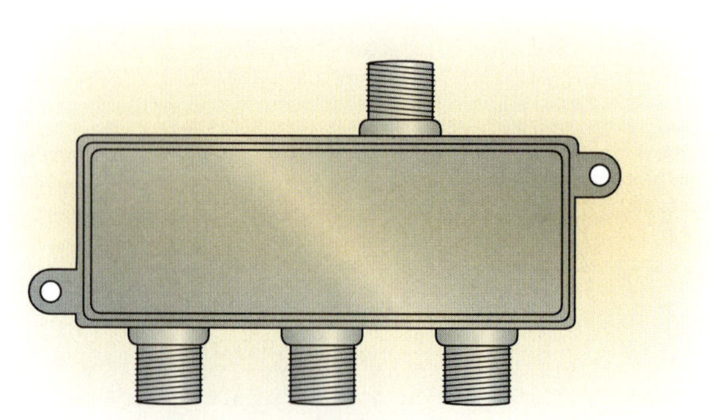

FIGURE 15.9
Splitters are used to connect coaxial cables. This unit has one incoming and three outgoing outlets.

It is very important not to crush coaxial cable or bend it too sharply. Coaxial cables used in residential TV systems have a diameter of about 5⁄16 in., and bends should have a radius of not less than five times the cable diameter—or 1 9⁄16 in.

Required Separations. Coaxial cables can't be installed in the same enclosures as power wiring [820.133(A)(2)]. The *Code* doesn't require a minimum separation between coaxial cables and power conductors run in cables or raceways [820.133(A)(2), Exception No. 1]. Many installers feel it isn't good practice to run low-voltage cables and Type NM or Type AC power cables through the same holes in wood framing members, but there's no *NEC* rule that prohibits this practice.

When coaxial and power cables are run through the same bored holes, the larger power cables should be pulled in first and the smaller coaxial cables should be pulled in later. This practice helps avoid damage to the coaxial cables due to crushing or damage caused by snagging the sheath against the rough edges of bored holes.

Coaxial cables should also be kept away from hot-water pipes, hot-air ducts, and other sources of heat that might damage the sheath.

Television Outlets

TV outlets (often called *F-connectors*) aren't required to be installed in outlet boxes. As with telephone outlets, they are usually installed in nonmetallic brackets nailed to wood studs.

Outlet Locations. The *NEC* doesn't specify locations in homes where TV outlets are required. In new construction, it is now common to provide a TV outlet in most habitable rooms: living room, bedrooms, den/study, family/recreation room, and kitchen (dining rooms don't usually have TV outlets installed in them).

Some homes also have TV outlets in other locations: bathrooms, workshop, laundry room, attached garage, covered porches, and patios. If in doubt, the installer should check with the owner or builder.

Required Separations. Coaxial cables cannot be installed in the same boxes as power conductors [820.133(A)(2)]. To create a neat appearance and minimize the number of separate "openings," a duplex receptacle is sometimes installed in a single faceplate with a telephone outlet and TV outlet. When this is done, the receptacle must be installed in a device box, while the low-voltage outlets can be installed on a bracket. Using a double-gang box with a barrier between the power conductors and low-voltage conductors is also permitted.

SMOKE DETECTORS AND ALARM SYSTEMS

New homes are required to have hard-wired smoke detectors supplied with 120-volt power. In one- and two-family dwellings, smoke detectors or smoke alarms are required in all sleeping rooms, in the immediate vicinity of the sleeping rooms, and on each additional story of the residence including basements [*NFPA 72*-2002, 11.5.1.1(1) through (3)]. Smoke detection can be provided either by approved single-station units (Figure 15.10) or as part of a complete household alarm system that includes separate smoke detectors along with other devices, such as heat detectors and security sensors.

Most smoke alarms produce an audible sound. Smoke alarms with flashing lights

FIGURE 15.10
Smoke detectors used in dwellings typically have integral alarms.

are also available (Figure 15.11). If the installer knows that a newly constructed dwelling will be occupied by people with hearing deficiencies, *NFPA 72* requires that visible alarms (either single-station smoke alarms or a fire alarm system with visible alarm appliances) be installed [*NFPA 72-2002*, 11.3.6].

Required Smoke Detector Locations

Smoke detectors or smoke alarms are required at the following dwelling locations.

• *Sleeping Areas.* Each bedroom is required to have a smoke detector. In homes where bedrooms are grouped along a single hall, one smoke detector is required in the hall. When bedrooms are located in different areas of the house, a smoke detector must be located outside each bedroom or cluster of bedrooms. Figure 15.12 shows smoke detector locations in a two-story house with bedrooms on the upper floor.

• *Each Story.* NFPA 72®, *National Fire Alarm Code®*, recommends that the living area smoke detector be installed in the living room or near the stairway to the upper level [*NFPA 72-2002*, A.11.8.3]. If the home has a basement, another

FIGURE 15.11
Smoke detectors with flashing lights are available for the hearing-impaired.

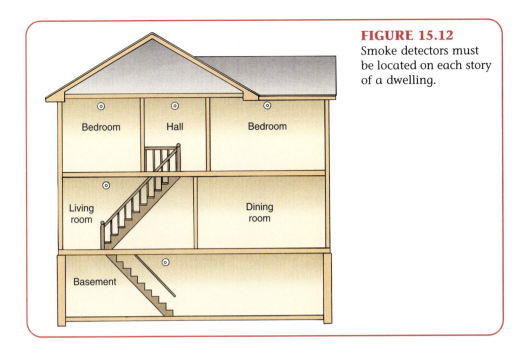

FIGURE 15.12
Smoke detectors must be located on each story of a dwelling.

smoke detector is required at that location, preferably near the stairway leading to the floor above. Smoke detectors aren't required in attics or crawl spaces, unless the attic is habitable and considered a "story." When it's doubtful where smoke detectors are required in a particular dwelling, the authority having jurisdiction should be consulted.

- *Bedroom in Basement or Attic.* If a bedroom is located in a basement or in a habitable attic, the smoke detector required to be located "in the immediate vicinity of the sleeping rooms" can also serve as the required unit for the "additional story of the residence," as noted in the opening paragraph to this section on smoke detectors.

- *Optional Detectors.* Additional smoke detectors can be provided at other locations, as desired. Figure 15.13 shows required smoke detector locations in a split-level house. The unit in the recreation room is optional, because there is no door between the living room and recreation room [*NFPA 72*-2002, A.11.5.1]. That means these rooms aren't considered to be on two different levels or stories.

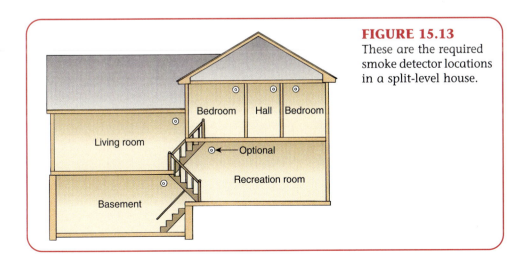

FIGURE 15.13
These are the required smoke detector locations in a split-level house.

- *Other Rooms.* Rooms identified on construction plans as other than bedrooms (such as dens, home offices, sewing rooms, and storage spaces) don't require smoke detectors. However, a den that could also serve as a bedroom may need smoke detectors. If in doubt, the installer should check with the authority having jurisdiction. *NFPA 72* doesn't prohibit installing additional smoke detectors to provide extra protection against fire.

General Guidelines for Locating Smoke Detectors

- *Avoid Dead-Air Spaces.* Smoke detectors should be placed to avoid "dead-air spaces," which are not easily penetrated by smoke and heat. Units in finished rooms shouldn't be placed close to locations where a wall meets the ceiling, because air doesn't circulate well in such locations (Figure 15.14). For the same reason, smoke detectors in unfinished basements should be mounted *on,* rather than *between,* joists.
- *Minimize Nuisance Alarms.* Cooking vapors or bathroom moisture can cause unwanted operation of smoke detectors. To minimize these so-called "nuisance alarms," *NFPA 72* requires that smoke detectors be located not closer than 3 ft from the door to a bathroom or kitchen. Smoke alarms and smoke detectors located within 20 ft of a cooking appliance should be of the photoelectric type, which are less likely to be sensitive to alarms caused by cooking vapors, or they should have a temporary silencing means, often called a "hush button" [*NFPA 72*-2002, 11.8.3.5(4)].

Power Supply

Smoke detectors and smoke alarms in new construction are required to be hard-wired [NFPA *101*-2003, 9.6.2.10] and to receive their power from the building

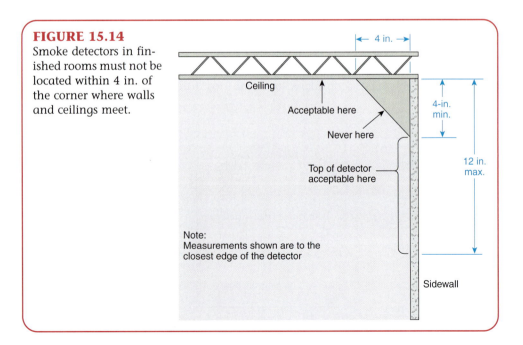

FIGURE 15.14
Smoke detectors in finished rooms must not be located within 4 in. of the corner where walls and ceilings meet.

electrical system [NFPA *101*-2003, 9.6.2.10.2]. Units also have backup batteries in case of utility power failure, but batteries cannot be the primary power source.

NOTE: Battery-powered smoke detectors are permitted only in existing homes.

Wiring Methods

No special wiring methods are required for smoke detectors. Type NM cable, Type AC cable, or individual conductors run in raceways can be used for the direct-connected, 120-volt smoke detectors required in new construction.

Ceiling- and wall-mounted smoke detectors are typically installed on single-gang outlet boxes (device boxes). These can be either metal or nonmetallic.

GFCI Protection Not Required for Alarm Systems. A receptacle in an unfinished basement supplying only a permanently installed fire alarm or burglar alarm system isn't required to have ground-fault circuit-interrupter (GFCI) protection. The reason is that the *Code* considers it more important to keep the fire or burglar alarm system energized than to protect that particular receptacle outlet against ground faults.

Exception No. 3 to 210.8(A)(5) permits a single receptacle for a fire alarm system, or a duplex receptacle supplying both a fire alarm and burglar alarm system, to be installed in an unfinished basement without GFCI protection.

However, a duplex receptacle can't be installed in an unfinished basement without GFCI protection if it supplies *only* the fire alarm system. The reason is that the other receptacle would still be available for plugging in utilization equipment such as lamps, electric heaters, or power tools. This situation would be a violation of the general requirement that receptacle outlets in unfinished basements (which are often damp) be provided with GFCI protection for personnel [210.8(A)(5)].

DOOR CHIMES

Many door chime systems have two different tones, one for the front door and another for the back door. Figure 15.15 shows a typical wiring diagram for a two-tone door chime installation. Door chime equipment and wiring are Class 2.

Transformers

Class 2 chime transformers are typically rated 120 volts primary, 16 volts secondary. Some transformers for residential use have multiple secondary taps for supplying low-voltage equipment operating at a range of 12 to 24 volts. Class 2 transformers must be supplied by a branch circuit rated 15 or 20 amperes [725.51].

Chime transformers are often constructed so that they can be installed at a knockout of a standard metal outlet box, with the primary 120-volt terminals located inside the box. These transformers can also be installed at other locations, such as near a panelboard or on a joist in an unfinished basement, as shown in Figure 15.16.

Wiring

Class 2 wiring used for door chimes is typically called *bell wire* or *thermostat wire* in the field. It comes in cables with two, three, four, or five conductors. The size

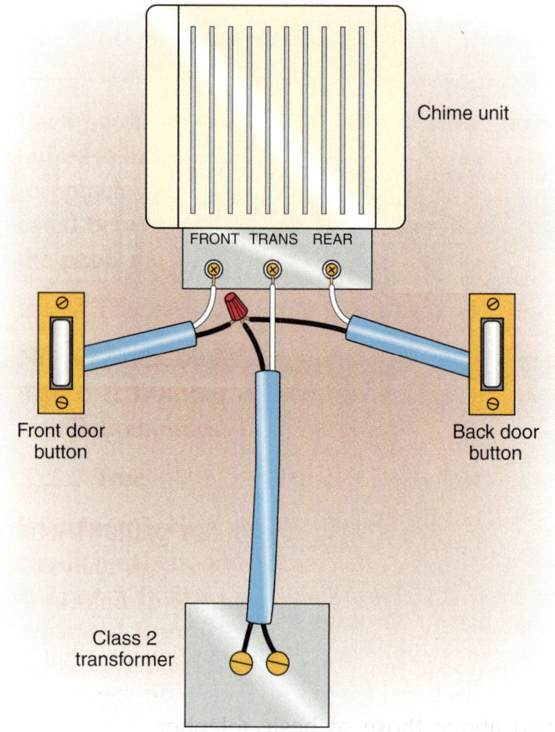

FIGURE 15.15
Many door chimes have operating buttons at more than one door.

Chime unit

FRONT TRANS REAR

Front door
button

Back door
button

Class 2
transformer

FIGURE 15.16 A single transformer can provide various voltages for low-voltage appliances.

most commonly installed for door chimes is 18 AWG, which is larger than the 22 and 24 AWG Class 2 wiring commonly used with thermostats for two reasons:

1. Most thermostats operate at 24 volts, which is higher than door chime voltage. Because door chime voltage is lower, the current is higher.

2. Mechanical door chimes require more power to operate than thermostats, which send a signal but don't have powered components.

Support of Conductors. Class 2 wiring must be installed in a neat and workmanlike manner and must be attached to the building structure [725.8]. Raceways cannot be used to support Class 2 conductors [725.58]. Instead, they should be secured using plastic cable ties or depth-stop staples. Using metal staples may damage Class 2 conductors.

Required Separations. The *Code* doesn't require a minimum separation between Class 2 wiring and power conductors run in cables or raceways [725.55(J)(1)]. Many installers feel it isn't good practice to run door chime wiring and Type NM or Type AC power cables through the same holes in wood framing members, but there's no *NEC* rule prohibiting this practice.

When Class 2 wiring and power cables are run through the same bored holes, the larger power cables should be pulled in first and the smaller low-voltage wiring pulled in later. This practice helps avoid damage to the Class 2 cables due to crushing or damage caused by snagging the insulation against the rough edges of bored holes.

Class 2 wiring should also be kept away from hot-water pipes, hot-air ducts, and other sources of heat that might damage the PVC insulation.

RESIDENTIAL STRUCTURED CABLING SYSTEMS

An increasing number of new homes have additional low-voltage wiring installed over and above those of basic telephone and cable TV. Usually called *structured cabling* or *structured wiring,* such systems are also advertised as *home automation, smart house,* or *intelligent home* wiring. Typically, this additional low-voltage wiring is used to create a home network for managing electrical, electronic, and HVAC functions, and/or for home theater applications. This textbook calls such additional low-voltage wiring *residential structured cabling systems.*

NOTE: This chapter doesn't cover Network-Powered Broadband Communications Systems described in NEC Article 830. Although this article has been in the National Electrical Code since 2002, cables, outlets, and other equipment for these systems are not yet commercially available.

Industry Standards

Residential structured cabling systems are available from a number of manufacturers. There is no *Code* article that applies specifically to them. However, this type of system generally uses Category-rated telephone cables (Article 800) and coaxial cables (Article 820).

Residential structured cabling systems aren't completely standardized, and features differ from one brand to another. However, most are based to some degree on EIA/TIA 570A, *Residential Telecommunications and Light Commercial Cabling Standard,* which classifies systems as either Grade 1 or Grade 2. This section describes the general characteristics of residential structured cabling systems based on the EIA/TIA 570A standard.

Description

Residential structured cabling systems consist of a central *distribution device* (sometimes called a *hub*) with cabling to outlets located throughout the home. Incoming services such as telephone, cable TV, direct broadcast satellite TV, or broadband are connected to the distribution device.

Signals are distributed from this central point over cables to wall-mounted information outlets. Each outlet has an individual homerun cable running back to the distribution device. Outlets are never "daisy chained" or wired in series.

Some residential structured cabling systems have one or more touch screen control panels for managing home function (Figure 15.17). Others are controlled through PCs equipped with proprietary software.

Cables

Telephone cables used in residential structured cabling systems are typically four-pair unshielded twisted-pair (UTP) cables rated Category 5 or 5E. This is the same type of cable installed for commercial telecommunications and computer networking applications.

Coaxial cables are typically quad-shielded RG-6 coaxial cable with a solid copper center conductor. This type of coax has higher bandwidth than the RG-59 cable with steel center conductor normally installed in dwellings for cable/satellite TV distribution. It provides better picture quality and reduces electromagnetic interference (EMI) from other signal sources.

Distribution Device (Hub)

The distribution device serves as a combination service entrance and distribution point for residential communications. It is an enclosure that contains splitters, terminal blocks, and other devices (Figure 15.18). It is typically classified as either *Grade 1* or *Grade 2.*

Grade 1. Distributes some or all of the following signals: telephone, data, and TV. Grade 1 distribution devices (hubs) may be either passive or active.

- *Passive* distribution devices handle only telecommunications signals, contain no powered components, and don't require a power supply.

FIGURE 15.17
Some residential structured cabling systems have touch screen control panels. (Courtesy of On-Q/Legrand)

FIGURE 15.18 The On-Q Service Center is a Grade 2 distribution device with many types of modules for managing telecom, video, security, audio, lighting, and heating/cooling systems. (Courtesy of On-Q/Legrand)

- *Active* distribution devices handle telecommunications plus other signals, such as TV and high-speed data. They may also have other functions such as low-voltage lighting control, security systems, surround sound for a home theater, and whole-house intercom with music. Active distribution devices contain powered components such as amplifiers, modulators, and controls that require a 120-volt power supply.

Grade 2. Distributes some or all of the following signals: TV, telephone, high-speed data, broadband multimedia, and computer network signaling. Grade 2 distribution devices are *active;* they contain powered components such as amplifiers, modulators, and Ethernet hubs that require 120-volt ac power (Figure 15.19).

Roughing-in a Distribution Device. Follow the manufacturer's instructions when installing the distribution device for residential home communications system. If roughing-in before the exact device has been selected or the manufacturer's

FIGURE 15.19 Outlets in residential structured cabling systems frequently have pairs of modular telephone jacks and F-type coaxial connectors. (Courtesy of On-Q/Legrand)

instructions aren't available, follow the space allocation and power guidelines in Table 15.1. They are based on EIA/TIA standard 570A.

Power Source

A source of power is recommended for Grade 1 distribution devices (some models may require it). All Grade 2 devices require power. Provide a 15-ampere, 120-volt duplex receptacle within 5 ft of the hub location. Active distribution devices typically contain a computer-type power strip with a plug and attachment cord; the internal components are plugged into this power strip.

TABLE 15.1 Distribution Device Space Allocation

Total Number of Outlets	Grade 1	Grade 2
1 to 8	16 in. wide × 24 in. high	32 in. wide × 36 in. high
9 to 16	16 in. wide × 36 in. high	32 in. wide × 36 in. high
17 to 24	16 in. wide × 60 in. high	32 in. wide × 48 in. high
More than 24	16 in. wide × 60 in. high	32 in. wide × 60 in. high

Source Cables to the Distribution Device (Hub)

Follow the manufacturer's instructions when connecting incoming communications cables to a distribution device. If the brand hasn't yet been selected or the manufacturer's instructions aren't available, follow these general guidelines when installing source cables:

Telephone. Run a 4-pair CAT 5 or 5E cable to the telephone company's *network interface device* (NIU) for standard telephone service. Place a label with the word "PHONE" on the end of this cable.

Broadband Data. Run a 4-pair CAT 5 or 5E cable to the NIU for DSL or other telephone-based digital data service. Place a label with the word "DATA" on the end of this cable.

Cable TV. Run an RG-6 coaxial cable from the distribution device to the cable TV service entrance area. Run two cables if the house is served by a CATV system having dual (A and B) cables. Place a label with the word "CATV" on the end of the cable(s).

TV Antenna Source. Run an RG-6 coaxial cable from the distribution device to the attic. Hang a coil of 25 ft of extra cable on an accessible roof truss for connecting to the off-air or satellite antenna. Place a label with the word "TV" on the end of this cable.

Outlets

Telecommunications Outlets. Residential structured cabling systems often use 8-conductor telecommunications outlets (also known as *modular jacks* or *RJ45 jacks*). These are the same type of outlets used for commercial telecommunications and computer networking applications. RJ45 8-conductor outlets are physically larger than the RJ11 two-position telephone outlets used in most residential construction (Figure 15.19).

However, the RJ11 plugs found on most telephone equipment, such as phones and answering machines, will fit into the larger RJ45 jacks because the two terminal conductors on an RJ11 plug are the same as the center pair of conductors on an RJ45 jack.

Installation of telephone outlets, along with *Code* rules applying to them, were covered earlier in this chapter. CAT 5 and 5E cables should always be installed using open brackets, not outlet boxes. This is because the 4-pair UTP cables are larger and stiffer than typical residential telephone wiring. When a CAT 5 or 5E cable is installed in an outlet box, it's almost impossible to not make tight bends or kinks that can degrade cable performance.

Coaxial Outlets. Residential structured cabling systems use the same F-type coaxial connectors and outlets normally used for TV systems of all types. Installation of coaxial outlets, along with *Code* rules applying to them, were covered earlier in this chapter.

National Electrical Code Rules

Residential structured cabling systems must be installed following the same *Code* rules as other telephone and TV systems. These were discussed earlier in this chapter.

Some systems also include Class 2 conductors for purposes such as low-voltage lighting control or home intercoms. Installation rules for all these types of conductors were discussed earlier in this chapter.

Underground Wiring

Underground communications and signaling conductors are installed at dwellings for purposes such as providing phone, computer network, or cable TV service at accessory buildings, such as a detached garage converted to a home office.

The minimum burial depths for underground wiring shown in Table 300.5 apply only to power conductors. The *Code* doesn't specify burial depths for communications and signaling conductors [90.3]. However Table 300.5 provides useful guidance; following its guidelines when installing direct-buried telephone cables or other communications and signal conductors will help prevent damage to these circuits from people digging in their yards or similar causes.

SPECIAL RULES FOR TWO-FAMILY DWELLINGS

There are no special *Code* rules that apply to communications (telephone), cable TV, fire alarm, or door chime wiring in two-family dwellings. However, residential branch circuits are only permitted to supply loads located within, or associated with, each individual dwelling unit [210.25].

This rule means that, if a common fire alarm system is installed to serve both units of a two-family dwelling, the control panel must be installed in a common space and powered from the "house loads panel." The bedrooms in a two-flat dwelling cannot have all of their smoke detectors wired together on a single branch circuit, or be controlled from a panelboard located inside one of the dwelling units.

15 CHAPTER REVIEW

MULTIPLE CHOICE

1. In dwelling units, smoke detectors must be located
   - **A.** In each bedroom
   - **B.** Outside of each sleeping area
   - **C.** On each additional story
   - **D.** All of the above

2. Residential telephone outlets are required to be
   - **A.** Installed in outlet boxes
   - **B.** Listed
   - **C.** Category 3
   - **D.** Both A and B

3. Coaxial cables are not permitted to be installed in the same raceways as
   - **A.** Class 2 and Class 3 conductors
   - **B.** Electric light and power wiring
   - **C.** Power-limited fire alarm wiring
   - **D.** None of the above

4. The communications (telephone) primary protector grounding conductor must be
   - **A.** Copper, aluminum, or copper-clad aluminum
   - **B.** 14 AWG
   - **C.** Installed in a raceway to provide physical protection
   - **D.** All of the above

5. A two-story house with three bedrooms requires how many smoke detectors?
   - **A.** Three
   - **B.** Four
   - **C.** Six
   - **D.** Cannot be determined from the available information

6. All communications (telephone) cables installed in new dwellings are required to be at least
   - **A.** Type III
   - **B.** Class 3
   - **C.** Category 3
   - **D.** Either A or B

7. The *NEC* requires cable television outlets in the following rooms:
   - **A.** Living room, bedrooms, and family room
   - **B.** All habitable rooms other than kitchens
   - **C.** Finished basements and den/study
   - **D.** None of the above

8. A single receptacle is defined as
   - **A.** A single wiring device
   - **B.** A single yoke (or strap)
   - **C.** A single contact device
   - **D.** All of the above

9. Smoke detectors are permitted to be wired using
   - **A.** Type AC cable
   - **B.** Individual conductors installed in raceways
   - **C.** Type NM cable
   - **D.** All of the above

10. The telephone company's wiring typically terminates at a component called the
    - **A.** Primary protector
    - **B.** Splitter
    - **C.** NIU/NID
    - **D.** None of the above

FILL IN THE BLANKS

1. Smoke alarms and smoke detectors located within _____ ft of a cooking appliance should be of the photoelectric type, which are less likely to be sensitive to alarms caused by cooking vapors, or they should have a temporary silencing means ("hush button").

2. Occupants in two-family dwellings are required to have _____ to the overcurrent devices protecting conductors in their occupancy.

3. All communications (telephone) and coaxial cables installed in dwelling units are required to be _____.

4. *NEC* 820.24 requires that coaxial cable be installed in a "neat and _____ manner."

5. *NFPA 72* requires that smoke detectors be located not closer than _____ ft from the door to a bathroom or kitchen.

6. The device used in place of a junction box to join coaxial cables is called a _____.

7. When the metal sheath of incoming coaxial cable is connected to a separate grounding electrode, a bond-

ing jumper of minimum size _____ or equivalent is required to be connected to the power grounding electrode system.

8. Smoke detectors must be placed to avoid _____, which smoke and heat have difficulty penetrating.

9. Battery-powered smoke detectors are permitted only in _____ dwelling units.

10. Low-voltage wiring and power wiring cannot be installed in the same outlet box or other enclosure unless there is a _____ separating the two types of wiring.

TRUE OR FALSE

1. When the length of the grounding conductor for a telephone primary protector would otherwise exceed 15 ft, a separate ground rod must be driven.

   _____ True   _____ False

2. Smoke detectors in new construction are permitted to be either direct-connected or stand-alone units with battery power.

   _____ True   _____ False

3. If a bedroom is located in a basement or in a habitable attic, the smoke detector located "in the immediate vicinity" can also serve as the required unit for the "additional story."

   _____ True   _____ False

4. Coaxial cables and power cables, such as Type AC or Type NM, are not permitted to occupy the same bored hole through a wood framing member.

   _____ True   _____ False

5. Raceways cannot be used to support communications (telephone) wiring.

   _____ True   _____ False

6. In one- and two-family dwellings, smoke detectors or smoke alarms are required in all habitable rooms.

   _____ True   _____ False

7. Smoke detectors in unfinished basements should be mounted between joists rather than on them, because smoke collects in these "dead-air spaces."

   _____ True   _____ False

8. Cooking vapors or bathroom moisture can cause unwanted operation of smoke detectors, often called "nuisance alarms."

   _____ True   _____ False

9. The *Code* classifies door chime wiring as Class 1, power-limited.

   _____ True   _____ False

10. Smoke detectors aren't required in attics, unless the attic is habitable and considered a "story."

   _____ True   _____ False

CHALLENGE QUESTIONS

1. In what way is the telecommunications "demarcation point" similar to the electrical "service point"?

2. Why doesn't the *National Electrical Code* specify locations for residential telephone and cable television outlets?

3. What is the rationale for not permitting telephone and cable TV conductors to be installed in the same raceways or enclosures as power wiring?

4. Why doesn't the *National Electrical Code* have an article that covers locations and installation requirements for residential smoke detectors?

5. Is Class 2 wiring for door chimes, thermostats, and other purposes required to comply with the general rules of *NEC* Chapters 1–4? Explain why or why not.

6. Can all smoke detectors protecting the sleeping areas of a two-family dwelling be powered by a single branch circuit?

Residential Generators

16

OBJECTIVES

After completing this chapter, the student will be able to understand the following:

- Sizing standby generators
- Separately derived systems
- Essential loads
- Transfer panels
- Grounding and bonding requirements for generators

INTRODUCTION

This chapter covers electrical generators installed at one- and two-family dwellings, to provide standby power during utility outages. It also covers installation of related items such as transfer switches (transfer panels), disconnect switches, control panels, and generator foundations. The following major topics are included:

- Separately derived systems
- Planning the installation
- Generator sizing
- Installing generator sets and associated components
- Generator grounding

IMPORTANT NEC TERMS

Ampacity

Automatic

Bonding Jumper, System

Branch Circuit

Disconnecting Means

Enclosure

Rainproof

Separately Derived System

Switch, Transfer

SEPARATELY DERIVED SYSTEMS

A residential generator is a separately derived system, when the grounded conductor is switched by transfer equipment, as defined in Article 100 of the *National Electrical Code®*.

> **Separately Derived System:** A premises wiring system whose power is derived from a source of electric energy or equipment other than a service. Such systems have no direct electrical connection, including a solidly connected grounded circuit conductor, to supply conductors originating in another system.

In this case, "supply conductors originating in another system" are the utility service that supplies the residence under normal circumstances. The generator operates only during a utility outage, when it supplies the premises wiring system through a transfer switch or transfer panel that interrupts both the ungrounded (phase) conductors and the grounded (neutral) conductor.

Solar photovoltaic systems (Article 690), fuel cells (Article 692), and wind-generated power are other types of separately derived systems sometimes found at dwellings.

Generator Voltage Ratings

Generator sets permanently installed at one-family dwellings to provide standby power are typically rated 120/240 volts, single-phase, 3-wire. Some have two separate 120-volt windings. Connecting the two windings in series results in 240-volt output. Residential generator sets are normally fueled by gasoline, natural gas, or liquefied petroleum (LP) gas.

NOTE: Some larger dwelling units have 3-phase electrical systems and use backup generators rated 120/208 volts, 3-phase, 4-wire. These larger, 3-phase generators are actually commercial units and normally use diesel fuel. This chapter doesn't discuss installation of 3-phase generators.

Regulatory Requirements

Residential generator installations must be coordinated with the local electric utility, including obtaining any necessary approvals and taking all safety precautions to prevent backfeeds onto the utility system.

Local building codes and land-use regulations often govern the placement of generators in proximity to property lines and rights-of-way, as well as on-site fuel storage.

Residential areas often have strict sound level requirements at the property line. It's best to check local codes for maximum sound level permitted. Meeting the required sound level may require the generator to have an optional sound-suppressing enclosure and residential muffler.

PLANNING THE INSTALLATION

Design Selection

There are two basic design approaches to providing residential generators at one- and two-family dwellings:

- Whole-house design (one panelboard)
- Essential loads design (two panelboards)

Whole-House Design. The generator powers all dwelling unit loads during a utility power outage. It supplies the service panelboard through a transfer switch, either manual or automatic, which selects between the two power sources: the utility and the generator.

The whole-house design approach requires a larger generator with a higher kVA rating to supply the total dwelling unit load. It also requires use of larger generator output conductors, transfer switch, and utility disconnect switch. For these reasons, whole-house generators are rarely installed.

Essential Loads Design. A more common approach is to divide the dwelling's circuits into essential and non-essential loads. Essential circuits have their own separate panelboard, and non-essential circuits are supplied from the service panelboard. This service panelboard subfeeds the smaller essential loads panel through a transfer switch (Figure 16.1).

If a utility power outage occurs, the transfer switch (either manual or automatic) changes over from utility power to generator power. The non-essential loads in the service equipment are "shed," and the generator supplies electricity only to the subset of essential loads connected to the essential loads panelboard.

The essential loads design approach requires a smaller generator with a lower kVA rating. It also allows use of smaller generator output conductors, transfer switch, and utility disconnect switch.

Other factors to be considered when planning an essential loads generator installation include the following:

- *Transfer Panels.* Many residential standby generators come with transfer panels that combine a transfer switch and circuit breakers for the essential loads within a single enclosure.
- *Essential Circuits.* Installers should determine which circuits are considered "essential" by consulting the builder or homeowner. The following are often considered essential:

  Refrigerator and freezer

  Sump pump and/or well pump

  Gas furnace blower

  Air-conditioning (in hot climates)

  Security system

  Selected lighting outlets

  Selected convenience receptacles, such as those supplying computers and a cable router or DSL hub in a home office

  Television(s)

Generator Sizing

The generator must be capable of supplying the maximum connected load, along with the required starting current of the largest motor. The minimum generator ratings to serve the connected load are determined as follows. (Also see "Generator Sizing Example.")

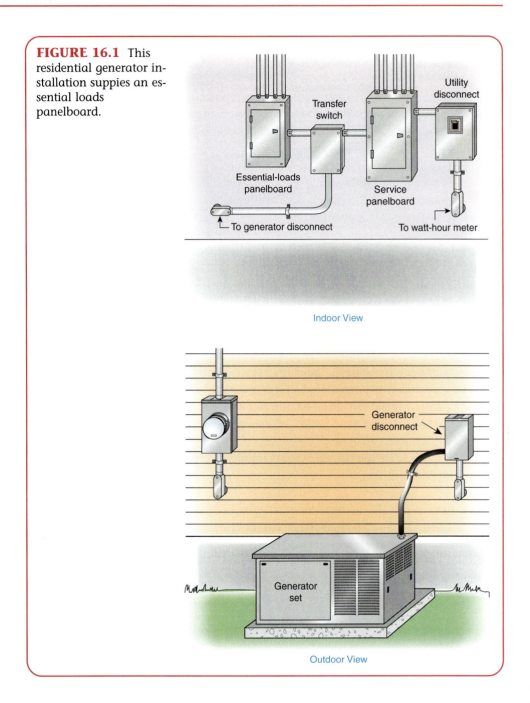

FIGURE 16.1 This residential generator installation suppies an essential loads panelboard.

Indoor View

Outdoor View

Step 1. *Determine the minimum generator wattage and ampere rating for the required loads.* The actual wattage of lighting and other loads are added, using nameplate information for appliances. When appliance nameplate information isn't available, current and power requirements of the actual loads are measured or taken from Table 16.1.

NOTE: *The starting wattage required by the* largest motor *only (not all motors in the dwelling) is included in this calculation.*

Table 16.1 Typical Appliance Power Requirements

Load	Running Watts	Starting Watts	Load	Running Watts	Starting Watts
Lights	15–150	0	Washing machine	1200	2300
Security system	250	0	Fan, attic	350	400
Refrigerator	900	2300	Fan, bathroom	100	
Freezer	900	2300	Fan, ceiling	300	200
Microwave	1450	0	Fan, fireplace	300	200
Furnace blower, 1/8 hp	300	500	Fan, kitchen exhaust	300	200
Furnace blower, 1/6 hp	500	900	Garage door opener, 1/4 hp	550	1100
Furnace blower, 1/4 hp	600	1100	Garage door opener, 1/3 hp	800	1700
Furnace blower, 1/3 hp	800	1400	Garage door opener, 1/2 hp	1000	2300
Furnace blower, 1/2 hp	1000	2350	Electric blanket	400	0
Pump, 1/3 hp	800	1400	De-humidifier, portable	650	800
Pump, 1/2 hp	1050	2150	Vacuum cleaner	800–1100	0
Pump, 3/4 hp	1400	2200	Coffeemaker	1750	0
Pump, 1 hp	2000	3000	Toaster	1050–1650	0
Electric range, 6 in. element	1500	0	Iron	1200	0
Electric range, 8 in. element	2100	0	Water heater	4500	0
Oven	6000	0	Hair dryer	300–1200	0
Air conditioner (window unit)	1000	2800	Television	100–500	0
Dishwasher (air dry)	700	1400	Radio	50–200	0
Dishwasher (heat dry)	1450	1400	Computer	600	0
Clothes dryer, gas	800	1800	Fax machine	220	0
Clothes dryer, electric	5750	1800	Copy machine	1500	0

Step 2. Determine the minimum current rating of the generator based on the power requirements of the connected load as determined in Step 1.

Step 3. First list the loads on a panelboard schedule and then balance them as closely as possible between phases. This step is necessary because the generator must be sized to supply the largest required phase current of the connected load.

Step 4. Select a generator with power and ampere ratings greater than the maximum connected load and with the maximum unbalanced load current.

Generator Sizing Example

This example determines the size of a generator needed to supply essential circuits only (not the entire dwelling unit load).

Step 1. Determine the minimum output power rating needed:

Essential Load	Power Demand
Lighting (4 × 60 watts, 6 × 75 watts, and 1 × 135 watts)	825 watts
Freezer (typical load from Table 16.1)	900 watts
Refrigerator (nameplate information)	675 watts
Sump pump (½ hp typical load from Table 16.1)	1050 watts
Furnace blower, ¼ hp (nameplate information)	600 watts
Microwave (nameplate information)	1200 watts
Attic fan (nameplate information)	350 watts
Television	500 watts
Computer	400 watts
Subtotal, connected load:	6500 watts
Add starting wattage for largest motor (sump pump, ½ hp):	+ 1100 watts
Total minimum generator output rating:	7600 watts

Step 2. Select the generator. A 10 kW (10,000 watts) model would be sufficient to support the essential loads in this dwelling.

Other System Components

Residential backup power systems have the following major components, which must be considered when planning the installation.

Generator Output Conductors. The ampacity of the field-installed conductors from the generator output terminals to the first overcurrent device must be at least 115 percent of the nameplate current rating of the generator [445.13].

Transfer Panel/Switch. Some packaged residential generator sets come equipped with a transfer panel. When a transfer switch is not furnished by the generator manufacturer, the installer must provide a UL 1008 transfer switch of the type (manual or automatic) and ampere rating specified in the generator instructions. Transfer switches mounted indoors normally have NEMA/UL Type 1 general-purpose enclosures. Switches mounted outdoors normally have Type 3R rainproof enclosures.

Generator Disconnecting Means. The generator disconnecting means is installed between the generator set and the transfer switch or transfer panel and is used to disconnect the generator set for servicing [445.18]. (See Figure 16.1.)

Some packaged residential generator sets come equipped with a disconnecting means, which is usually a circuit breaker. When a disconnecting means is not furnished by the generator manufacturer, a switch of the ampere rating specified in the generator instructions must be provided. Generator disconnect switches normally have a Type 3R enclosure for outdoor installation near the generator set.

Utility Disconnect Switch. The utility disconnecting means is installed between the utility watt-hour meter and the service equipment. It is used to disconnect the electrical power distribution system from the meter. (See Figure 16.1.)

When a utility disconnecting means isn't furnished by the generator manufacturer, the installer must provide a listed transfer switch of the type (manual or automatic) and ampere rating specified in the generator instructions. Transfer switches mounted indoors normally have Type 1 general-purpose enclosures. Switches mounted outdoors normally have Type 3R rainproof enclosures.

INSTALLING GENERATOR SETS AND ASSOCIATED COMPONENTS

Generator Set Installation

Residential generator sets are normally self-contained and intended for outdoor installation (Figure 16.2). The unit's location must comply with all applicable building codes and, at a minimum, the working space requirements of 110.26.

NOTE: Manufacturers' instructions may recommend larger working clearances.

The following are other factors to consider when locating generator sets.

* The unit should be located as close as possible to the transfer panel or transfer switch.
* Air inlet and exhaust openings shouldn't be obstructed by debris such as leaves, grass, and snow. If prevailing winds may cause blowing or drifting, a fence or windbreak can be used to protect the generator set.
* The unit must be at least 5 ft from all building openings (doors, windows, and vents) to ensure that exhaust fumes won't accumulate inside the dwelling unit.
* Residential areas often have strict noise level requirements. Meeting the required maximum sound level may require the generator set to have an optional sound-suppressing enclosure, exhaust muffler, or enclosing fence. In doubtful situations the authority having jurisdiction should be consulted.

Foundation. Some residential generators come supplied with prefabricated foundations or with shipping skids that can be used as mounting pads. These are installed in accordance with the manufacturer's instructions.

FIGURE 16.2 A residential generator set is intended for outdoor installation. (Courtesy of Kohler Power Systems)

If a mounting pad isn't provided, the generator installer should provide a level concrete slab at least 3 in. thick and extending at least 3 in. beyond the generator set enclosure on all sides. This pad should include J bolts (sometimes called L bolts) to anchor the unit enclosure on all sides.

Be careful when lifting heavy residential generators into place on foundations. Some manufacturers supply temporary lifting lugs that can be used to lift the generator (Figure 16.3).

Vibration Isolation. Bolting a residential generator set directly to its foundation or mounting pad will result in excessive noise and vibration and possible damage.

Transfer Switch and Essential Loads Panelboard. If the generator doesn't come with a transfer panel, the installer should locate and install the transfer switch and essential loads panel close together. (See Figure 16.1.)

Wiring Methods. Any indoor wiring method permitted by the *NEC* can be used for indoor feeders and circuits associated with residential generator installations. Individual conductors in electrical metallic tubing (EMT) are commonly used.

Liquidtight flexible metal conduit (LFMC) or liquidtight flexible nonmetallic conduit (LFNC) are often used to protect output power conductors from the generator set to the transfer switch or transfer panel, in order to provide vibration isolation.

Where LFMC or LFNC are installed to provide this flexibility, a separate equipment grounding conductor must be run inside the raceway with the circuit conductors [350.60, 356.60].

Generator Grounding

In most installations, a separate ground rod is not provided for the generator set. However, some local codes may require a separate generator ground rod.

No Separate Ground Rod. When no separate generator ground rod is installed, a system bonding jumper is run from the grounding lug on the generator frame to the

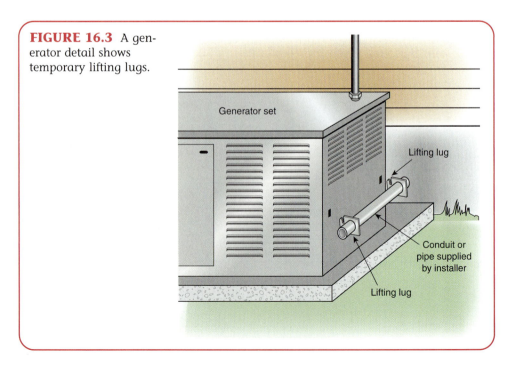

FIGURE 16.3 A generator detail shows temporary lifting lugs.

Generator set

Lifting lug

Conduit or pipe supplied by installer

Lifting lug

grounding terminal in the transfer switch or transfer panel [250.30(A)(1)]. The system bonding jumper is a copper conductor [250.28(A)] sized according to *NEC* Table 250.66, shown as Exhibit 16.1 [250.28(D)], and run in the same raceway with the generator output conductors.

NOTE: The grounding of residential generator sets should be completed before making the power circuit connections.

Non-Electrical Components

Residential generators are typically fueled by gasoline, natural gas, or LP-gas. Generator installers are normally responsible for supplying the engine fuel, coolant, lubricant, and battery. These subjects aren't governed by *NEC* rules and are outside the scope of this book. The installer should follow the manufacturer's instructions to complete the installation. Additional detailed information can be found in NECA 406-2003, *Recommended Practice for Installing Residential Generator Sets* (ANSI).

EXHIBIT 16.1

NEC Table 250.66 Grounding Electrode Conductor for Alternating-Current Systems

Size of Largest Ungrounded Service-Entrance Conductor or Equivalent Area for Parallel Conductors[a] (AWG/kcmil)		Size of Grounding Electrode Conductor (AWG/kcmil)	
Copper	Aluminum or Copper-Clad Aluminum	Copper	Aluminum or Copper-Clad Aluminum[b]
2 or smaller	1/0 or smaller	8	6
1 or 1/0	2/0 or 3/0	6	4
2/0 or 3/0	4/0 or 250	4	2
Over 3/0 through 350	Over 250 through 500	2	1/0
Over 350 through 600	Over 500 through 900	1/0	3/0
Over 600 through 1100	Over 900 through 1750	2/0	4/0
Over 1100	Over 1750	3/0	250

Notes:
1. Where multiple sets of service-entrance conductors are used as permitted in 230.40, Exception No. 2, the equivalent size of the largest service-entrance conductor shall be determined by the largest sum of the areas of the corresponding conductors of each set.
2. Where there are no service-entrance conductors, the grounding electrode conductor size shall be determined by the equivalent size of the largest service-entrance conductor required for the load to be served.
[a]This table also applies to the derived conductors of separately derived ac systems.
[b]See installation restrictions in 250.64(A).

16 CHAPTER REVIEW

MULTIPLE CHOICE

1. Generator sets installed at residences are normally rated
   - **A.** 120/208-volt, 3-phase, 4-wire
   - **B.** 120/240-volt, single-phase, 3-wire
   - **C.** 277/480-volt, 3-phase, 4-wire
   - **D.** Any of these

2. When sizing a generator to supply essential loads, the following must be included:
   - **A.** Appliance loads
   - **B.** General lighting load from Table 220.12
   - **C.** Starting load of largest motor
   - **D.** Both A and C

3. Residential generator installers are typically responsible for providing the following non-electrical items:
   - **A.** Fuel
   - **B.** Building permit
   - **C.** Lock or other security measure
   - **D.** None of these

4. Sizing a standby generator to supply only "essential circuits" has the following advantages:
   - **A.** Smaller generator with lower kVA rating
   - **B.** Output conductors of smaller AWG size
   - **C.** Transfer switch of smaller ampere rating
   - **D.** All of these

5. Residential generators may need optional enclosures, mufflers, or fences in order to meet local codes governing
   - **A.** Generator exhaust
   - **B.** Generator sound level
   - **C.** Generator appearance
   - **D.** Both A and B

FILL IN THE BLANKS

1. Loads should be balanced to the extent possible on generator phases. This is because the generator must be sized to supply the _____ of the connected load.

2. The _____ is installed between the generator set and the transfer switch or transfer panel, and is used to disconnect the generator set for servicing.

3. Transfer switches mounted indoors normally have NEMA/UL Type _____ enclosures.

4. *NEC* section _____ specifies minimum working space requirements around residential generator sets.

5. The ampacity of field-installed conductors from the generator output terminals to the first overcurrent device must be at least _____ percent of the nameplate current rating of the generator.

TRUE OR FALSE

1. "Essential loads" are required by 760.13 to include fire alarm and burglar alarm systems.
   _____ True   _____ False

2. Generators with dual 120-volt windings are used only to supply extremely small dwellings with two 120-volt branch circuits.
   _____ True   _____ False

3. A transfer panel combines a transfer switch with a circuit breaker panelboard for supplying essential loads.
   _____ True   _____ False

4. A generator is one example of a separately derived system.
   _____ True   _____ False

5. Vibration isolation is used to prevent generator damage due to unbalanced loads on generator phases.
   _____ True   _____ False

CHALLENGE QUESTIONS

1. Discuss the pluses and minuses of installing a generator that can supply the total house loads versus one that can supply only selected loads.

2. Explain the difference between a *transfer switch* and a *transfer panel*.

3. Can a single emergency generator be used to supply two dwelling units?

4. What are the two distinguishing characteristics of a "separately derived system"?

5. Who is generally responsible for installing non-electrical components of emergency generators foundations and fuel supply systems?

Old Work

17

INTRODUCTION

This chapter covers a fairly small number of specialized *NEC*® rules that apply to electrical work performed in existing residences. The following major topics are included:

- Securing and supporting requirements for fished cables
- Old work boxes
- Replacing existing receptacles of the nongrounding type
- Replacing non-GFCI receptacles in locations where the 2005 *NEC* requires them
- Determining when a grounding means exists in a outlet box
- *Code* rules for old work low-voltage wiring

BASIC CONCEPTS

This textbook primarily describes new residential construction in accordance with the 2005 *National Electrical Code*. However, many one- and two-family dwellings are remodeled and/or expanded at some point in their long useful lifetimes. And almost all residences have electrical equipment replaced or repaired from time to time. These kinds of electrical construction and maintenance are called *old work*—as opposed to the new work of building houses, condominiums, and apartment buildings from the ground up.

Old work electrical construction is a big subject. The most difficult aspect of old work is analyzing and making sense of existing electrical systems: locating cables and raceways hidden inside walls and ceilings, figuring out which outlet is protected by which fuse or circuit breaker, determining whether existing branch circuits have spare capacity to support new outlets, and so on. These topics are outside the scope of this book.

NEC **Rules for Old Work.** Instead, this chapter concentrates on explaining the wiring rules that apply to old work in existing homes. As a general rule, existing electrical systems don't have to be updated each time a new edition of the *National Electrical Code* is published.

Once a home or other building has been wired in accordance with the *Code* in effect at the time of construction, it remains "legal" indefinitely. Thus, a house built 30 years ago (or even much longer) is still considered *Code*-compliant today although it certainly won't be wired in accordance with all the rules of the 2005 *NEC*.

However, when changes are made to an existing electrical system, there are a few 2005 *National Electrical Code* rules that apply to the changes. The reason for this is to ensure that the existing home meets a certain minimum, modern, level of safety.

INSTALLING NEW ELECTRICAL BOXES IN OLD WORK

NEC 314.23 has a number of rules for securely supporting outlet boxes, device boxes, and other enclosures. In new construction, boxes are typically secured to structural members such as wall studs, floor joists, and ceiling rafters before the wall materials are applied. Once the wall materials such as sheetrock or plaster are installed, they appear to be recessed. Electrical boxes are also mounted exposed on the surface of rigid surfaces such as concrete and brick walls.

Additions and Renovations.

In the case of additions and major renovations, electrical construction techniques are essentially the same as in new construction. Outlet and device boxes are attached to the exposed building structure before the wall materials are applied.

What most people consider "old work" consists of more minor renovations. In these cases, homeowners want to have new wiring and outlets installed with minimum damage to existing walls, floors, and ceilings. Or the existing structural members may not be spaced conveniently for mounting outlet and device boxes in the locations they're wanted.

Old Work Boxes.

In these cases, so-called "old work" boxes are often installed. Various devices are used to secure old work boxes to flat surfaces without being nailed or screwed to framing members such as studs and joists [314.23(C)]. Some old work boxes have springy brackets so that, when the box is inserted into an opening cut in an existing wall, it is rigidly supported and secured. Others use ears that swing up behind the wallboard and hold the box in place by friction when the ear screws are tightened (Figure 17.1). Old work boxes are manufactured in various sizes and designs (device box, multigang, or round) for both wall and ceiling installation.

INSTALLING NEW CABLES IN OLD WORK

Almost all residential wiring is nonmetallic-sheathed cable (Type NM) and armored cable (Type AC), except in a few areas such as Chicago and New York City where individual conductors run in electrical metallic tubing (EMT) are widely used.

In new construction, AC and NM cables are typically stapled to structural members such as wall studs, floor joists, and ceiling rafters before the wall materials are applied. Once the wall materials such as sheetrock or plaster are installed, they are concealed. In unfinished spaces such as attics and basements, they are frequently stapled and are exposed to the building structure.

Chapter 4, Wiring Methods, describes cable installation techniques for new construction in greater detail. Generally speaking, Type NM and AC cables are required to be supported and secured at intervals not exceeding 4½ ft and within 12 in. of every outlet box, junction box, cabinet, or fitting [320.30, 334.30].

Relaxed Securing and Supporting Requirements for Fished Cables.

However, when doing old work in existing dwellings, pushing and pulling the flexible cables through concealed wall and ceiling spaces is a common technique

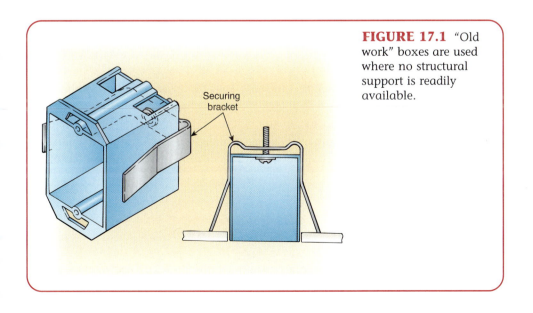

FIGURE 17.1 "Old work" boxes are used where no structural support is readily available.

Securing bracket

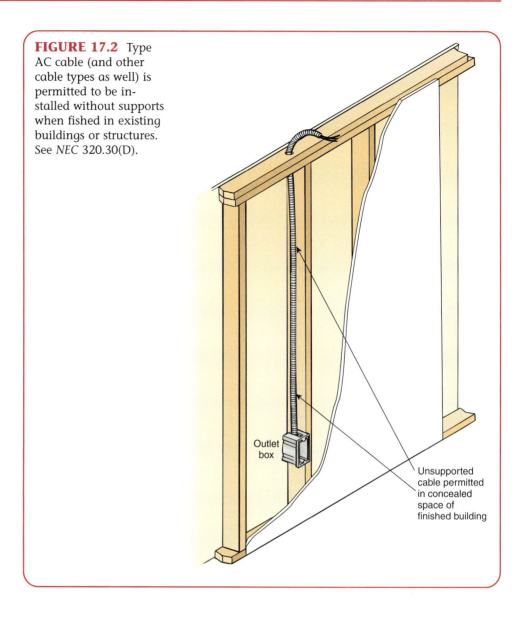

FIGURE 17.2 Type AC cable (and other cable types as well) is permitted to be installed without supports when fished in existing buildings or structures. See *NEC* 320.30(D).

Outlet box

Unsupported cable permitted in concealed space of finished building

for minimizing damage to existing walls and floors. When Type AC and NM cables are "fished" in this way, the *NEC* permits them to be unsupported (Figure 17.2) [320.30(D)(1), 334.30(B)(1)].

Since fished cables aren't secured to studs, they are also exempt from the *Code* requirement that they not be installed closer than 1¼ in. from the edge of framing members where screws are likely to penetrate [300.4(D)]. This wiring rule is intended to prevent cables from being damaged during installation of drywall, when nails are driven to hang pictures, and so on. However, fished cables can "wiggle" out of the way of nails, drywall screws, and similar fasteners that penetrate walls.

SURFACE RACEWAYS

Surface raceways come in both metallic and nonmetallic versions. They are frequently known in the field by the trade name Wiremold®. Surface raceways are commonly used in old work to extend power from an existing receptacle or lighting outlet using new Type AC or NM cables to feed a new outlet(s) without breaking holes in the walls or ceilings (Figures 17.3 and 17.4).

Installation Procedure. Instead, a new extension ring is mounted over an existing recessed box [314.22], and a complete system consisting of surface-mounted raceways and boxes is attached to the walls and ceilings (Figures 17.3 and 17.4). After this system is complete, individual conductors are installed [300.18(A)] and spliced to the existing branch-circuit conductors. Snap-on covers are fitted in place to finish the installation.

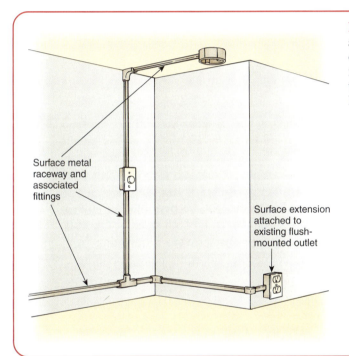

Surface metal raceway and associated fittings

Surface extension attached to existing flush-mounted outlet

FIGURE 17.3 This surface metal raceway extends from an existing receptacle outlet to a new dimmer and lighting outlet.

FIGURE 17.4 This surface metal raceway supplies a speed control switch and a paddle fan outlet. (Courtesy of The Wiremold® Co.)

FIGURE 17.5 Surface metal raceways contain a means for terminating the equipment-grounding conductor from another wiring method.

Flexible metal conduit

Equipment-grounding conductor

Grounding terminal screw, lug, or other approved device

Surface metal raceway

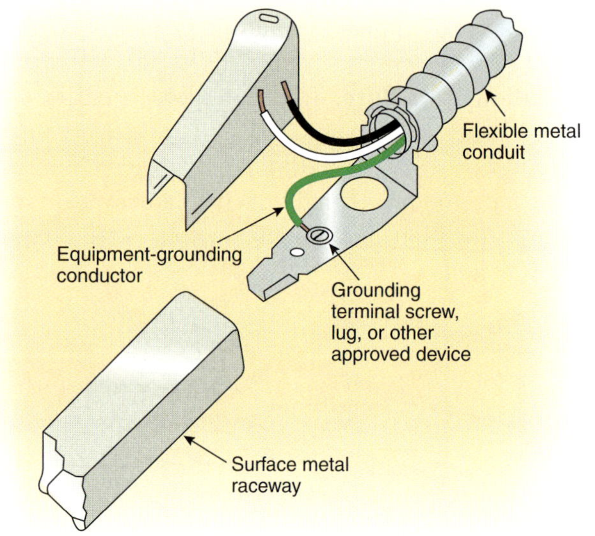

Both metal and nonmetallic surface raceways are permitted in dry locations. They can't be installed in concealed locations but are permitted through walls and ceilings if access to the conductors is maintained on both sides of the wall [386.10, 388.10]. Surface metal raceway enclosures that provide a transition from another wiring method have a means for connecting an equipment-grounding conductor [386.60] (Figure 17.5).

SURFACE EXTENSIONS

Power can also be extended from a flush box using other wiring methods. Figure 17.6 shows a surface extension using electrical metallic tubing (EMT) connected to an extension ring over an existing flush box. Whatever type of cable or raceway is used must be installed and supported according to the *Code* rules for that wiring method.

Flush-Mounted Box Cover. Surface extensions can also be made from the cover of a flush-mounted box [314.22, Exception]. In this type of surface extension, the wiring method must be flexible for a length sufficient to permit removal of the cover to gain access to the wiring in the box. Figure 17.7 shows a flexible extension from a flush box using conductors installed in flexible metal conduit (FMC).

When making a flexible extension from the cover of a flush-mounted box, the *Code* requires that grounding or bonding continuity must be independent of the connection between box and cover. This means Type AC cable cannot be used for this type of flexible extensions because it doesn't contain a separate equipment-grounding conductor. The metal armor of Type AC cable serves as the equipment-grounding path [250.118(8), 320.108], so if the metal box cover were removed, this would interrupt the bonding continuity.

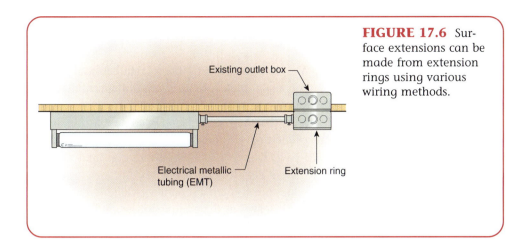

FIGURE 17.6 Surface extensions can be made from extension rings using various wiring methods.

Existing outlet box

Electrical metallic tubing (EMT)

Extension ring

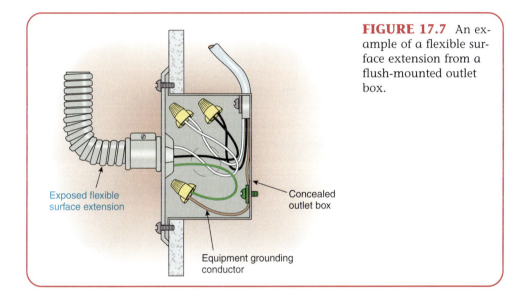

FIGURE 17.7 An example of a flexible surface extension from a flush-mounted outlet box.

Exposed flexible surface extension

Concealed outlet box

Equipment grounding conductor

REPLACING WIRING DEVICES IN OLD WORK

Replacing Existing Receptacles

NEC 406.3(A) requires that receptacles installed on 15- and 20-ampere branch circuits must be of the grounding type. These three-prong receptacles have a green screw for attaching an equipment-grounding conductor and have been required by the *Code* for many years. However, many older residences have nongrounding receptacles of the two-blade type with no terminal for connecting a ground wire.

When the existing receptacles in a home are this older nongrounding type, they can be replaced with new nongrounding receptacles under certain circumstances—or sometimes with ground-fault circuit-interrupter (GFCI) receptacles. Let's take a close look at the *Code* rules governing receptacle replacement when doing old work:

Existing Grounding-Type Receptacles. These must be replaced with new grounding-type receptacles when a grounding means exists in the receptacle enclosure [406.3(D)]. But if a grounding means does *not* exist in the outlet box, then a new grounding-type receptacle cannot be installed. See the section below on *Nongrounding-Type Receptacle*.

Determining When a Grounding Means Exists. *NEC* 406.3(D) talks about ''where a grounding means exists in the receptacle enclosure or a grounding conductor is installed.'' The reason for this either/or language is as follows:

Type NM cable—Older nonmetallic-sheathed cable found in some existing dwellings does not contain a separate grounding equipment conductor. When existing outlet boxes are supplied by an old-style Type NM cable without a ground wire, *no grounding means exists in the receptacle enclosure.*

Type AC cable—Armored cable does not contain a separate ground wire. Instead, the interlocked steel or aluminum armor serves as the equipment-grounding conductor. There are two possibilities:

- Existing metal outlet boxes supplied by Type AC cable are grounded by metal-to-metal contact between the box and the cable armor. In this case, *a grounding means exists in the receptacle enclosure.*

- Existing nonmetallic outlet boxes supplied by Type AC cable are not grounded. In this case, *no grounding means exists in the receptacle enclosure.*

CAUTION: People living in older homes sometimes replace nongrounding receptacles (two-prong) with grounding-type receptacles (three-prong) for convenience just to avoid using use two-to-three prong adapters when plugging in modern appliances.

But this unsafe practice can give homeowners a false sense of security by making it seem as if they have the safety benefits of grounding when they really don't. If you are doing old work and find that a grounding-type receptacle has been installed with no connection to an equipment-grounding means, then proceed as described in the section entitled Nongrounding-Type Receptacle.

Nongrounding-Type Receptacle. When a grounding means doesn't exist in the outlet box, the 2005 *National Electrical Code* provides several different replacement options:

- An existing nongrounding-type receptacle can be replaced with a new nongrounding-type receptacle [406.3(D)(3)(a)]. These are still sold in stores with the wiring device itself or its packaging marked ''For Replacement Purposes Only'' (Figure 17.8). *NOTE:* Nongrounding-type receptacles *are not permitted* to be installed in new dwellings or in locations of existing dwellings where there is a grounding means in the outlet box.

- An existing nongrounding-type receptacle can be replaced with a new GFCI receptacle [406.3(D)(3)(b)]. The new GFCI receptacle must be marked ''No Equipment Ground.'' GFCI receptacles normally come supplied with small adhesive labels for use in this type of installation situation (Figure 17.9).

- An existing nongrounding-type receptacle can be replaced with a new grounded-type receptacle that is supplied from a GFCI receptacle or circuit breaker [406.3(D)(3)(c)]. The new GFCI receptacle must be marked ''GFCI Protected'' and ''No Equipment Ground.'' GFCI receptacles normally come supplied with small adhesive labels for use in this type of installation situation.

No Equipment-Grounding Conductor. However, when a GFCI receptacle is used to protect new grounding-type receptacles installed without a ground conductor in old work, the *NEC* doesn't permit an equipment-grounding conductor to be connected from the GFCI receptacle downstream to the grounding-type receptacle outlets [406.3(D)(3)(c)]. Instead, only two wires are connected between the GFCI and non-GFCI receptacles: the ungrounded (hot) conductor and grounded (neutral) conductor.

Replacing Existing Wall Switches

NEC 404.9(B) requires that general-use snap switches, including dimmers and similar controls (such as occupancy sensors that turn lights on and off), must be effectively grounded and must provide a means to ground metal faceplates. This is a relatively recent *Code* requirement.

Thus, many older homes have existing snap switches, dimmers, and similar controls without terminals for connecting a ground wire. Some older branch-circuit wiring methods didn't include a separate equipment-grounding conductor.

However, when the existing snap switches in a home are of the older nongrounding type, they can be replaced with new nongrounding switches under certain circumstances.

Existing General-Use Snap Switch. Existing switches must be replaced by new grounding-type snap switches when a grounding means exists in the snap-switch enclosure. But if a grounding means does *not* exist in the device box or the wiring method does not provide an equipment ground, then an existing snap switch can be replaced by a new snap switch without a grounding means [404(9)(B)(2) Exception]. These are still sold in stores (Figure 17.10) with the wiring device itself or packaging marked "For Replacement Purposes Only."

When a nongrounding snap switch is installed for replacement purposes in old work that is located within reach of earth, grade conducting floors (such as a concrete slab on earth), or other conducting surfaces (such as metal plumbing fixtures and pipes), the faceplate can't be metal. Instead, it must be made of nonconducting,

FIGURE 17.8 Nongrounding-type receptacles with two slots are permitted by the *NEC* for replacement purposes only. (Courtesy of Pass & Seymour/Legrand)

FIGURE 17.9 This GFCI receptacle is labeled "No Equipment Ground" to indicate that no equipment-grounding conductor is present in the wiring method. (Courtesy of Pass & Seymour/ Legrand)

noncombustible material [404(9)(B)(2) Exception]. Listed plastic faceplates for snap switches meet these *National Electrical Code* requirements.

Installing Low-Voltage Outlets in Old Work

Installing TV, telephone, and other low-voltage outlets when doing old work is similar to installing power cables and outlets, except that boxes are not required [300.15]. Old work brackets can be installed in finished wall surfaces without nailing or screwing them to studs, in a manner similar to old work boxes (Figure 17.11). Low-voltage cables are fished through concealed wall spaces to the outlets mounted in these brackets.

Relaxed Securing and Supporting Requirements for Fished Cables. In new construction, the *NEC* does not specify the intervals at which communications (telephone-type) cables, coaxial cables, and broadband cables must be secured to framing members [800.24, 820.24, and 830.24]. However all three types of low-voltage cables must be installed according to 300.11, which requires that they be adequately secured and supported, and 300.4(D), which requires that they be kept back 1¼ in. from the edge of framing members.

Low-voltage cables fished through walls and ceilings when doing old work are

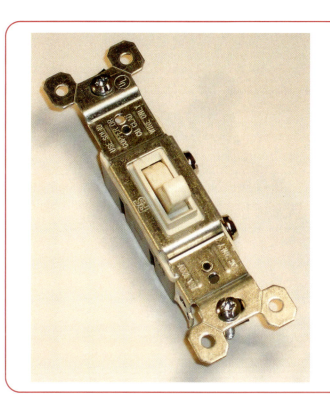

FIGURE 17.10 Non-grounding-type general-use snap switches are permitted by the *NEC* for replacement purposes only. (Courtesy of Pass & Seymour/ Legrand)

FIGURE 17.11 Old work brackets for telephone and TV outlets can be installed in a finished wall surface without being secured to framing members. (Courtesy of Carlon/ Lamson & Sessions)

exempt from these installation rules, just as described previously for Type AC and NM cables used for branch-circuit wiring.

SPECIAL RULES FOR TWO-FAMILY DWELLINGS

There are no special *Code* rules for old work that apply only in two-family dwellings. If there is a fire-rated wall, floor, or ceiling between two units in the same structure, any electrical box or cable penetrating that barrier is required to be fire stopped [300.21]. However, two-family dwellings aren't normally separated by fire walls. Residential branch circuits are only permitted to supply loads located within, or associated with, each individual dwelling unit [210.25].

CHAPTER REVIEW   17

MULTIPLE CHOICE

1. A dwelling built to *NEC* rules is considered to be *Code*-compliant for how long?

   A. 10 years
   B. 30 years
   C. Indefinitely
   D. None of these

2. Fished cables are required to be secured and supported

   A. At intervals not exceeding 6 ft
   B. Within 12 in. of outlet boxes and terminations
   C. As installation conditions permit
   D. None of these

3. Boxes that can be installed in finished surfaces without being attached to structural members are known as

   A. Jiffy boxes
   B. Nonmetallic boxes
   C. Old work boxes
   D. Both A and B

4. In old work, an existing nongrounding-type receptacle is permitted to be replaced by

   A. A nongrounding-type receptacle
   B. A GFCI receptacle
   C. A grounded-type receptacle that is supplied from a GFCI receptacle or circuit breaker
   D. Any of these

5. A surface extension from an existing outlet box can be installed using

   A. Electrical metallic tubing
   B. Surface nonmetallic raceway
   C. Armored cable, Type AC
   D. Both A and B

6. The *National Electrical Code* specifies intervals (distances) at which the following conductors must be secured and supported

   A. Class 2 and 3 cables (Article 725)
   B. Communications cables (Article 800)
   C. Coaxial cables (Article 820)

   D. None of these

7. In old work, when a grounding means exists in the box, a general-use snap switch must be replaced by

   A. A GFCI-protected snap switch
   B. A grounding-type snap switch
   C. A nongrounding-type snap switch
   D. Both A and B

8. Surface metal raceways and surface nonmetallic raceways are permitted to be installed

   A. In damp locations
   B. Concealed in old work
   C. As surface extensions from existing outlet boxes
   D. Any of these

9. The commonest wiring method(s) used in residential old work is (are)

   A. Nonmetallic-sheathed cable, Type NM
   B. Armored cable, Type AC
   C. Rigid metal conduit, Type RMC
   D. Both A and B

10. The following technique(s) is (are) permitted in old work but not in new construction

    A. Fishing cables between access points
    B. Replacing existing nongrounding receptacles with the same type
    C. Replacing existing nongrounding snap switches with the same type
    D. Any of these

FILL IN THE BLANKS

1. Old work boxes are designed to be installed in flat surfaces without being attached to _____.

2. Pushing and pulling cables through concealed spaces in existing walls and ceilings is called _____.

3. Type NM and AC cables must normally be supported and secured at intervals not exceeding _____ and within _____ of every termination.

4. When new cables are fished in the course of doing old work, they are not required to be _____.

5. Surface raceways are manufactured in both _____ _____ and _____ versions.

6. Surface metal and nonmetallic raceways are permitted to extend through _____ so long as the conductors are accessible.

7. Typically, surface extensions are used to _____ _____ power from an existing outlet.

8. Wiring can be installed in surface metal and nonmetallic raceway only after the raceway installation is _____.

9. The *Code* requires that general-use snap switches and dimmers be effectively _____.

10. If a grounding means doesn't exist in a switch box or the wiring method doesn't provide an equipment ground, then an existing snap switch can be replaced by a new snap switch without a _____. This type of switch is sold for replacement purposes only.

TRUE OR FALSE

1. When existing outlet boxes are supplied by an old-style Type NM cable that doesn't have a separate equipment-grounding conductor, no wire nut exists in the receptacle enclosure.

____ True   ____ False

2. When an existing nongrounding-type receptacle is replaced with a new GFCI receptacle, the new GFCI receptacle is required to be marked "Caution – No GFCI."

____ True   ____ False

3. The steel or aluminum armor of Type AC cable is permitted to serve as an equipment-grounding conductor in accordance with 250.118(6).

____ True   ____ False

4. Nongrounding-type receptacles have two slots that accept receptacle blades.

____ True   ____ False

5. When a nongrounding snap switch is installed for replacement purposes and is located within reach of earth or conducting surfaces, the faceplate isn't permitted to be made of metal.

____ True   ____ False

6. The *Code* requires that low-voltage cables be secured at intervals not exceeding 4½ ft and within 12 in. of every bracket or other termination.

____ True   ____ False

7. A raceway with conductors to the cover of an existing flush box is called a *fixture whip*.

____ True   ____ False

8. When a GFCI receptacle is used to protect new grounding-type receptacles installed without a ground conductor in old work, the *Code* doesn't permit an equipment-grounding conductor to be connected between the GFCI and non-GFCI receptacles.

____ True   ____ False

9. Type AC cable can't be used to make surface extensions because the wiring method doesn't contain an anti-short bushing ("redhead").

____ True   ____ False

10. When existing outlet boxes are supplied by old-style Type NM cable that doesn't have a separate equipment-grounding conductor, no grounding means exists in the receptacle enclosure.

____ True   ____ False

CHALLENGE QUESTIONS

1. When a GFCI receptacle is used to protect new grounding-type receptacles installed without a ground conductor in old work, why doesn't the *Code* permit an equipment-grounding conductor to be connected between the GFCI and the downstream non-GFCI receptacles?

2. Why does the *Code* permit cables and flexible raceways to be fished between access points through concealed spaces in old work, rather than requiring that these wiring methods be secured and supported in accordance with normal *NEC* rules?

3. Why does the *NEC* require that grounding or bonding continuity be independent of the connection between box and cover when making a flexible extension from the cover of a flush-mounted box?

4. Why aren't boxes required to enclose for low-voltage outlets and terminations involving Class 2 and 3 cables, communications conductors, and coaxial cables?

5. Explain why electrical work performed in exposed areas of existing structures must comply with all normal *NEC* rules.

NEC Definitions

Article 100 of the *National Electrical Code*® contains definitions. Many apply broadly to all types of electrical installations. Others are primarily related to commercial, industrial, and occupational construction and have little to do with residential wiring. Official *NEC*® definitions that apply to electrical systems in dwellings are listed here and are extracted from Article 100, unless otherwise noted in brackets. For convenience in referring to this list, important *NEC* terms related to individual chapters are noted at the beginning of each chapter.

Accessible (as applied to equipment). Admitting close approach; not guarded by locked doors, elevation, or other effective means.

Accessible (as applied to wiring methods). Capable of being removed or exposed without damaging the building structure or finish or not permanently closed in by the structure or finish of the building.

Accessible, Readily (Readily Accessible). Capable of being reached quickly for operation, renewal, or inspections without requiring those to whom ready access is requisite to climb over or remove obstacles or to resort to portable ladders, and so forth.

Ampacity. The current, in amperes, that a conductor can carry continuously under the conditions of use without exceeding its temperature rating.

Appliance. Utilization equipment, generally other than industrial, that is normally built in standardized sizes or types and is installed or connected as a unit to perform one or more functions such as clothes washing, air conditioning, food mixing, deep frying, and so forth.

Approved. Acceptable to the authority having jurisdiction.

Arc-Fault Circuit Interrupter. A device intended to provide protection from the effects of arc fault by recognizing characteristics unique to arcing and by functioning to de-energize the circuit when an arc fault is detected. *[210.12(A)]*

Armored Cable, Type AC. A fabricated assembly of insulated conductors in a flexible metallic enclosure. See 320.100. *[320.2]*

Attachment Plug (Plug Cap) (Plug). A device that, by insertion in a receptacle, establishes a connection between the conductors of the attached flexible cord and the conductors connected permanently to the receptacle.

Authority Having Jurisdiction. The organization, office, or individual responsible for approving equipment, materials, an installation, or a procedure.

Automatic. Self-acting, operating by its own mechanism when actuated by some impersonal influence, as, for example, a change in current, pressure, temperature, or mechanical configuration.

Bathroom. An area including a basin with one or more of the following: a toilet, a tub, or a shower.

Bonding (Bonded). The permanent joining of metallic parts to form an electrically conductive path that ensures electrical continuity and the capacity to conduct safely any current likely to be imposed.

Bonding Jumper. A reliable conductor to ensure the required electrical conductivity between metal parts required to be electrically connected.

Bonding Jumper, Equipment. The connection between two or more portions of the equipment grounding conductor.

Bonding Jumper, Main. The connection between the grounded circuit conductor and the equipment grounding conductor at the service.

Bonding Jumper, System. The connection between the grounded circuit conductor and the equipment grounding conductor at a separately derived system.

Branch Circuit. The circuit conductors between the final overcurrent device protecting the circuit and the outlet(s).

Branch Circuit, Appliance. A branch circuit that supplies energy to one or more outlets to which appliances are to be connected and that has no permanently connected luminaires (lighting fixtures) that are not a part of an appliance.

Branch Circuit, General-Purpose. A branch circuit that supplies two or more receptacles or outlets for lighting and appliances.

Branch Circuit, Individual. A branch circuit that supplies only one utilization equipment.

Branch Circuit, Multiwire. A branch circuit that consists of two or more ungrounded conductors that have a voltage between them, and a grounded conductor that has equal voltage between it and each ungrounded conductor of the circuit and that is connected to the neutral or grounded conductor of the system.

Building. A structure that stands alone or that is cut off from adjoining structures by fire walls with all openings therein protected by approved fire doors.

Cabinet. An enclosure that is designed for either surface mounting or flush mounting and is provided with a frame, mat, or trim in which a swinging door or doors are or can be hung.

Cable. A factory assembly of two or more conductors having an overall covering. *[800.2]*

Circuit Breaker. A device designed to open and close a circuit by nonautomatic means and to open the circuit automatically on a predetermined overcurrent without damage to itself when properly applied within its rating.

Concealed. Rendered inaccessible by the structure or finish of the building. Wires in concealed raceways are considered concealed, even though they may become accessible by withdrawing them.

Conductor, Bare. A conductor having no covering or electrical insulation whatsoever.

Conductor, Covered. A conductor encased within material of composition or thickness that is not recognized by this *Code* as electrical insulation.

Conductor, Insulated. A conductor encased within material of composition and thickness that is recognized by this *Code* as electrical insulation.

Conduit Body. A separate portion of a conduit or tubing system that provides access through a removable cover(s) to the interior of the system at a junction of two or more sections of the system or at a terminal point of the system. Boxes such as FS and FD or larger cast or sheet metal boxes are not classified as conduit bodies.

Connector, Pressure (Solderless). A device that establishes a connection between two or more conductors or between one or more conductors and a terminal by means of mechanical pressure and without the use of solder.

Continuous Load. A load where the maximum current is expected to continue for 3 hours or more.

Controller. A device or group of devices that serves to govern, in some predetermined manner, the electric power delivered to the apparatus to which it is connected.

Cooking Unit, Counter-Mounted. A cooking appliance designed for mounting in or on a counter and consisting of one or more heating elements, internal wiring, and built-in or mountable controls.

Copper-Clad Aluminum Conductors. Conductors drawn from a copper-clad aluminum rod with the copper metallurgically bonded to an aluminum core. The copper forms a minimum of 10 percent of the cross-sectional area of a solid conductor or each strand of a stranded conductor.

Cord-and-Plug-Connected Lighting Assembly. A lighting assembly consisting of a luminaire (lighting fixture) intended for installation in the wall of a spa, hot tub, or storable pool, and a cord-and-plug-connected transformer. *[680.2]*

Cutout Box. An enclosure designed for surface mounting that has swinging doors or covers secured directly to and telescoping with the walls of the box proper.

Dead Front. Without live parts exposed to a person on the operating side of the equipment.

Demand Factor. The ratio of the maximum demand of a system, or part of a system, to the total connected load of a system or the part of the system under consideration.

Device. A unit of an electrical system that is intended to carry or control but not utilize electric energy.

Disconnecting Means. A device, or group of devices, or other means by which the conductors of a circuit can be disconnected from their source of supply.

Dry-Niche Luminaire (Lighting Fixture). A luminaire (lighting fixture) intended for installation in the wall of a pool or fountain in a niche that is sealed against the entry of pool water. *[680.2]*

Dwelling Unit. A single unit, providing complete and independent living facilities for one or more persons, including permanent provisions for living, sleeping, cooking, and sanitation.

Dwelling, One-Family. A building that consists solely of one dwelling unit.

Dwelling, Two-Family. A building that consists solely of two dwelling units.

Dwelling, Multifamily. A building that contains three or more dwelling units.

Enclosed. Surrounded by a case, housing, fence, or wall(s) that prevents persons from accidentally contacting energized parts.

Electrical Metallic Tubing (EMT). An unthreaded thinwall raceway of circular cross section designed for the physical protection and routing of conductors and cables and for use as an equipment grounding conductor when installed utilizing appropriate fittings. EMT is generally made of steel (ferrous) with protective coatings or aluminum (nonferrous). *[358.2]*

Electrical Nonmetallic Tubing (ENT). A nonmetallic pliable corrugated raceway of circular cross section with integral or associated couplings, connectors, and fittings for the installation of electric conductors. ENT is composed of a material that is resistant to moisture and chemical atmospheres and is flame retardant A pliable raceway is a raceway that can be bent by hand with a reasonable force but without other assistance. *[362.2]*

Enclosure. The case or housing of apparatus, or the fence or walls surrounding an installation to prevent personnel from accidentally contacting energized parts or to protect the equipment from physical damage.

Energized. Electrically connected to, or is, a source of voltage.

Equipment. A general term including material, fittings, devices, appliances, luminaires (fixtures), apparatus, and the like used as a part of, or in connection with, an electrical installation.

Equipment, Fixed. Equipment that is fastened or otherwise secured at a specific location. *[680.2]*

Equipment, Portable. Equipment that is actually moved or can easily be moved from one place to another in normal use. *[680.2]*

Equipment, Stationary. Equipment that is not easily moved from one place to another in normal use. *[680.2]*

Exposed (as applied to live parts). Capable of being inadvertently touched or approached nearer than a safe distance by a person. It is applied to parts that are not suitably guarded, isolated, or insulated.

Exposed (as applied to wiring methods). On or attached to the surface or behind panels designed to allow access.

Externally Operable. Capable of being operated without exposing the operator to contact with live parts.

Feeder. All circuit conductors between the service equipment, the source of a separately derived system, or other power supply source and the final branch-circuit overcurrent device.

Fitting. An accessory such as a locknut, bushing, or other part of a wiring system that is intended primarily to perform a mechanical rather than an electrical function.

Flexible Metal Conduit (FMC). A raceway of circular cross section made of helically wound, formed, interlocked metal strip. *[348.2]*

Forming Shell. A structure designed to support a wet-niche luminaire (lighting fixture) assembly and intended for mounting in a pool or fountain structure. *[680.2]*

Fountain. Fountains, ornamental pools, display pools, and reflection pools. The definition does not include drinking fountains. *[680.2]*

Garage. A building or portion of a building in which one or more self-propelled vehicles can be kept for use, sale, storage, rental, repair, exhibition, or demonstration purposes.

Ground. A conducting connection, whether intentional or accidental, between an electrical circuit or equipment and the earth or to some conducting body that serves in place of the earth.

Grounded. Connected to earth or to some conducting body that serves in place of the earth.

Grounded, Effectively. Intentionally connected to earth through a ground connection or connections of sufficiently low impedance and having sufficient current-carrying capacity to prevent the buildup of voltages that may result in undue hazards to connected equipment or to persons.

Grounded Conductor. A system or circuit conductor that is intentionally grounded.

Ground-Fault Circuit Interrupter (GFCI). A device intended for the protection of personnel that functions to de-energize a circuit or portion thereof within an established period of time when a current to ground exceeds the values established for a Class A device.

Grounding Conductor. A conductor used to connect equipment or the grounded circuit of a wiring system to a grounding electrode or electrodes.

Grounding Conductor, Equipment. The conductor used to connect the non–current-carrying metal parts of equipment, raceways, and other enclosures to the system grounded conductor, the grounding electrode conductor, or both, at the service equipment or at the source of a separately derived system.

Grounding Electrode Conductor. The conductor used to connect the grounding electrode(s) to the equipment grounding conductor, to the grounded conductor, or to both, at the service, at each building or structure where supplied by a feeder(s) or branch circuit(s), or at the source of a separately derived system.

Hermetic Refrigerant Motor-Compressor. A combination consisting of a compressor and motor, both of which are enclosed in the same housing, with no external shaft or shaft seals, the motor operating in the refrigerant. *[440.2]*

Hydromassage Bathtub. A permanently installed bathtub equipped with a recirculating piping system, pump, and associated equipment. It is designed so it can accept, circulate, and discharge water upon each use. *[680.2]*

Identified (as applied to equipment). Recognizable as suitable for the specific purpose, function, use, environment, application, and so forth, where described in a particular *Code* requirement.

In Sight From (Within Sight From, Within Sight). Where this *Code* specifies that one equipment shall be "in sight from," "within sight from," or "within sight," and so forth, of another equipment, the specified equipment is to be visible and not more than 15 m (50 ft) distant from the other.

Interrupting Rating. The highest current at rated voltage that a device is intended to interrupt under standard test conditions.

Labeled. Equipment or materials to which has been attached a label, symbol, or other identifying mark of an organization that is acceptable to the authority having jurisdiction and concerned with product evaluation, that maintains periodic inspection of production of labeled equipment or materials, and by whose labeling the manufacturer indicates compliance with appropriate standards or performance in a specified manner.

Leakage Current Detection and Interruption (LCDI) Protection. A device provided in a power supply cord or cord set that senses leakage current flowing between or from the cord conductors and interrupts the circuit at a predetermined level of leakage current. *[440.2]*

Lighting Outlet. An outlet intended for the direct connection of a lampholder, a luminaire (lighting fixture), or a pendant cord terminating in a lampholder.

Lighting Track. A manufactured assembly designed to support and energize luminaires (lighting fixtures) that are capable of being readily repositioned on the track. Its length can be altered by the addition or subtraction of sections of track. *[410.100]*

Liquidtight Flexible Nonmetallic Conduit (LFNC). A raceway of circular cross section of various types as follows:

- A smooth seamless inner core and cover bonded together and having one or more reinforcement layers between the core and covers, designated as Type LFNC-A

- A smooth inner surface with integral reinforcement within the conduit wall, designated as Type LFNC-B

- A corrugated internal and external surface without integral reinforcement within the conduit wall, designated as Type LFNC-C *[356.2]*

Listed. Equipment, materials, or services included in a list published by an organization that is acceptable to the authority having jurisdiction and concerned with evaluation of products or services, that maintains periodic inspection of production of listed equipment or materials or periodic evaluation of services, and whose listing states that the equipment, material, or services either meets appropriate designated standards or has been tested and found suitable for a specified purpose.

Live Parts. Energized conductive components.

Location, Damp. Locations protected from weather and not subject to saturation with water or other liquids but subject to moderate degrees of moisture. Examples of such locations include partially protected locations under canopies, marquees, roofed open porches, and like locations, and interior locations subject to moderate degrees of moisture, such as some basements, some barns, and some cold-storage warehouses.

Location, Dry. A location not normally subject to dampness or wetness. A location classified as dry may be temporarily subject to dampness or wetness, as in the case of a building under construction.

Location, Wet. Installations under ground or in concrete slabs or masonry in direct contact with the earth; in locations subject to saturation with water or other liquids, such as vehicle washing areas; and in unprotected locations exposed to weather.

Luminaire. A complete lighting unit consisting of a lamp or lamps together with the parts designed to distribute the light, to position and protect the lamps and ballast (where applicable), and to connect the lamps to the power supply.

Maximum Water Level. The highest level that water can reach before it spills out. *[680.2]*

Multioutlet Assembly. A type of surface, flush, or freestanding raceway designed to hold conductors and receptacles, assembled in the field or at the factory.

Nonautomatic. Action requiring personal intervention for its control. As applied to an electric controller, nonautomatic control does not necessarily imply a manual controller, but only that personal intervention is necessary.

No-Niche Luminaire (Lighting Fixture). A luminaire (lighting fixture) intended for installation above or below the water without a niche. *[680.2]*

Nonmetallic-Sheathed Cable. A factory assembly of two or more insulated conductors enclosed within an overall nonmetallic jacket. *[334.2]*

Outlet. A point on the wiring system at which current is taken to supply utilization equipment.

Overcurrent. Any current in excess of the rated current of equipment or the ampacity of a conductor. It may result from overload, short circuit, or ground fault.

Overload. Operation of equipment in excess of normal, full-load rating, or of a conductor in excess of rated ampacity that, when it persists for a sufficient length of time, would cause damage or dangerous overheating. A fault, such as a short circuit or ground fault, is not an overload.

Packaged Spa or Hot Tub Equipment Assembly. A factory-fabricated unit consisting of water-circulating, heating, and control equipment mounted on a common base, intended to operate a spa or hot tub. Equipment can include pumps, air blowers, heaters, lights, controls, sanitizer generators, and so forth. *[680.2]*

Panelboard. A single panel or group of panel units designed for assembly in the form of a single panel, including buses and automatic overcurrent devices, and equipped with or without switches for the control of light, heat, or power circuits; designed to be placed in a cabinet or cutout box placed in or against a wall, partition, or other support; and accessible only from the front.

Permanently Installed Decorative Fountains and Reflection Pools. Those that are constructed in the ground, on the ground, or in a building in such a manner that the fountain cannot be readily disassembled for storage, whether or not served by electrical circuits of any nature. These units are primarily constructed for their aesthetic value and are not intended for swimming or wading. *[680.2]*

Permanently Installed Swimming, Wading, and Therapeutic Pools. Those that are constructed in the ground or partially in the ground, and all others capable of holding water in a depth greater than 1.0 m (42 in.), and all pools installed inside of a building, regardless of water depth, whether or not served by electrical circuits of any nature. *[680.2]*

Plenum. A compartment or chamber to which one or more air ducts are connected and that forms part of the air distribution system.

Pool. Manufactured or field-constructed equipment designed to contain water on a permanent or semipermanent basis and used for swimming, wading, or other purposes. *[680.2]*

Pool Cover, Electrically Operated. Motor-driven equipment designed to cover and uncover the water surface of a pool by means of a flexible sheet or rigid frame. *[680.2]*

Premises Wiring (System). That interior and exterior wiring, including power, lighting, control, and signal circuit wiring together with all their associated hardware, fittings, and wiring devices, both permanently and temporarily installed, that extends from the service point or source of power, such as a battery, a solar photovoltaic system, or a generator, transformer, or converter windings, to the outlet(s). Such wiring does not include wiring internal to appliances, luminaires (fixtures), motors, controllers, motor control centers, and similar equipment.

Qualified Person. One who has skills and knowledge related to the construction and operation of the electrical equipment and installations and has received safety training on the hazards involved.

Raceway. An enclosed channel of metal or nonmetallic materials designed expressly for holding wires, cables, or bus-bars, with additional functions as permitted in this *Code*. Raceways include, but are not limited to, rigid metal conduit, rigid nonmetallic conduit, intermediate metal conduit, liquidtight flexible conduit, flexible metallic tubing, flexible metal conduit, electrical nonmetallic tubing, electrical metallic tubing, underfloor raceways, cellular concrete floor raceways, cellular metal floor raceways, surface raceways, wireways, and busways.

Rainproof. Constructed, protected, or treated so as to prevent rain from interfering with the successful operation of the apparatus under specified test conditions.

Raintight. Constructed or protected so that exposure to a beating rain will not result in the entrance of water under specified test conditions.

Receptacle. A receptacle is a contact device installed at the outlet for the connection of an attachment plug. A single receptacle is a single contact device with no other contact device on the same yoke. A multiple receptacle is two or more contact devices on the same yoke.

Receptacle Outlet. An outlet where one or more receptacles are installed.

Remote-Control Circuit. Any electric circuit that controls any other circuit through a relay or an equivalent device.

Rigid Nonmetallic Conduit (RNC). A nonmetallic raceway of circular cross section, with integral or associated couplings, connectors, and fittings for the installation of electrical conductors and cables. *[352.2]*

Self-Contained Spa or Hot Tub. Factory-fabricated unit consisting of a spa or hot tub vessel with all water-circulating, heating, and control equipment integral to the unit. Equipment can include pumps, air blowers, heaters, lights, controls, sanitizer generators, and so forth. *[680.2]*

Self-Contained Therapeutic Tubs or Hydrotherapeutic Tanks. A factory-fabricated unit consisting of a therapeutic tub or hydrotherapeutic tank with all water-circulating, heating, and control equipment integral to the unit. Equipment may include pumps, air blowers, heaters, light controls, sanitizer generators, and so forth. *[680.2]*

Separately Derived System. A premises wiring system whose power is derived from a source of electric energy or equipemnt other than a service. Such systems have no direct electrical connection, including a solidly connected grounded circuit conductor, to supply conductors originating in another system.

Service. The conductors and equipment for delivering electric energy from the serving utility to the wiring system of the premises served.

Service Cable. Service conductors made up in the form of a cable.

Service Conductors. The conductors from the service point to the service disconnecting means.

Service Drop. The overhead service conductors from the last pole or other aerial support to and including the splices, if any, connecting to the service-entrance conductors at the building or other structure.

Service-Entrance Conductors, Overhead System. The service conductors between the terminals of the service equipment and a point usually outside the building, clear of building walls, where joined by tap or splice to the service drop.

Service-Entrance Conductors, Underground System. The service conductors between the terminals of the service equipment and the point of connection to the service lateral.

Service Equipment. The necessary equipment, usually consisting of a circuit breaker(s) or switch(es) and fuse(s) and their accessories, connected to the load end of service conductors to a building or other structure, or an otherwise designated area, and intended to constitute the main control and cutoff of the supply.

Service Lateral. The underground service conductors between the street main, including any risers at a pole or other structure or from transformers, and the first point of connection to the service-entrance conductors in a terminal box or meter or other enclosure, inside or outside the building wall. Where there is no terminal box, meter, or other enclosure, the point of connection is considered to be the point of entrance of the service conductors into the building.

Service Point. The point of connection between the facilities of the serving utility and the premises wiring.

Signaling Circuit. Any electric circuit that energizes signaling equipment.

Smoke Alarm. A single or multiple station alarm responsive to smoke. *[NFPA 72, 3.3.179]*

Smoke Detector. A device that detects visible or invisible particles of combustion. *[NFPA 72, 3.3.43.17]*

Solar Photovoltaic System. The total components and subsystems that, in combination, convert solar energy into electrical energy suitable for connection to a utilization load.

Spa or Hot Tub. A hydromassage pool, or tub for recreational or therapeutic use, not located in health care facilities, designed for immersion of users, and usually having a filter, heater, and motor-driven blower. It may be installed indoors or outdoors, on the ground or supporting structure, or in the ground or supporting structure. Generally, a spa

or hot tub is not designed or intended to have its contents drained or discharged after each use. *[680.2]*

Special Permission. The written consent of the authority having jurisdiction.

Storable Swimming or Wading Pool. Those that are constructed on or above the ground and are capable of holding water to a maximum depth of 1.0 m (42 in.), or a pool with nonmetallic, molded polymeric walls or inflatable fabric walls regardless of dimension. *[680.2]*

Structure. That which is built or constructed.

Switch, General-Use. A switch intended for use in general distribution and branch circuits. It is rated in amperes, and it is capable of interrupting its rated current at its rated voltage.

Switch, General-Use Snap. A form of general-use switch constructed so that it can be installed in device boxes or on box covers, or otherwise used in conjunction with wiring systems recognized by this *Code*.

Switch, Motor-Circuit. A switch rated in horsepower that is capable of interrupting the maximum operating overload current of a motor of the same horsepower rating as the switch at the rated voltage.

Switch, Transfer. An automatic or nonautomatic device for transferring one or more load conductor connections from one power source to another.

Through-Wall Lighting Assembly. A lighting assembly intended for installation above grade, on or through the wall of a pool, consisting of two interconnected groups of components separated by the pool wall. *[680.2]*

Utilization Equipment. Equipment that utilizes electric energy for electronic, electromechanical, chemical, heating, lighting, or similar purposes.

Ventilated. Provided with a means to permit circulation of air sufficient to remove an excess of heat, fumes, or vapors.

Voltage (of a circuit). The greatest root-mean-square (rms) (effective) difference of potential between any two conductors of the circuit concerned.

Voltage, Nominal. A nominal value assigned to a circuit or system for the purpose of conveniently designating its voltage class (e.g., 120/240 volts, 480Y/277 volts, 600 volts). The actual voltage at which a circuit operates can vary from the nominal within a range that permits satisfactory operation of equipment.

Voltage to Ground. For grounded circuits, the voltage between the given conductor and that point or conductor of the circuit that is grounded; for ungrounded circuits, the greatest voltage between the given conductor and any other conductor of the circuit.

Watertight. Constructed so that moisture will not enter the enclosure under specified test conditions.

Weatherproof. Constructed or protected so that exposure to the weather will not interfere with successful operation.

Wet-Niche Luminaire (Lighting Fixture). A luminaire (lighting fixture) intended for installation in a forming shell mounted in a pool or fountain structure where the luminaire (fixture) will be completely surrounded by water. *[680.2]*

Wire. A factory assembly of one or more insulated conductors without an overall covering. *[800.2]*

Answers

CHAPTER 1

Multiple Choice

1. The *National Electrical Code* chapter that covers special occupancies is

   C. Chapter 5

2. The first four chapters of the *NEC*

   A. Apply generally to all electrical installations

3. Which of the following meets the *Code* definition of *dwelling unit*?

   D. All of the above

4. The following *NEC* article defines wiring practices for temporary power on construction sites:

   C. Article 590

5. The National Electrical Installation Standards (NEIS) are

   C. Quality standards for electrical construction

6. *BX, greenfield, guts,* and *toggle switch* are all examples of

   C. Field names

7. The Comprehensive Consensus Codes (C3) are

   D. A series of building codes that includes the *NEC*

8. When an AHJ conducts more than one inspection of a dwelling under construction, these are typically called

   D. Both A and B

9. Listed electrical products are

   D. All of the above

10. A *qualified person,* as defined in Article 100, is required to have

    C. Skills, knowledge, and safety training

Fill in the Blanks

1. The title of NFPA 70 is *National Electrical Code.*

2. A housekeeping unit with space for eating, living, and sleeping, and permanent provisions for cooking and sanitation is known as a dwelling unit.

3. Products can only be approved by the Authority Having Jurisdiction. *(AHJ is also an acceptable answer.)*

4. Grounded conductor is the official *Code* name for what is commonly called a "neutral."

5. A qualified person is one who has skills and knowledge related to the construction and operation of the electrical equipment and has received safety training on the hazards involved.

6. The quality and workmanship standards published by the National Electrical Contractors Association (NECA) are known as National Electrical Installation Standards. *(NEIS is also an acceptable answer.)*

7. Subjects covered by Chapter 2 of the *National Electrical Code* are wiring and protection.

8. Romex® is a common brand name for a wiring method whose proper *Code* name is nonmetallic-sheathed cable. *(NM cable, Type NM, or Type NM cable are also acceptable answers.)*

9. The purpose of the *National Electrical Code,* as stated in 90.1(A) *or* Section 90.1(A), is "the practical safeguarding of persons and property from hazards arising from the use of electricity."

10. Article 100 defines *dwelling unit* as "a single unit, providing complete and independent living facilities for one or more persons, including permanent provisions for living, sleeping, cooking, and sanitation."

True or False

1. The *NEC* requires a minimum of 12 AWG conductors for branch-circuit wiring.

   False

2. The *NEC* is a building code.

   True

3. The *NEC* refers to product testing and listing agencies as Nationally Recognized Testing Laboratories (NRTLs).

   False

4. "Listed" means the same as "labeled."

   False

5. All electricians are required to be licensed.

   False

6. The *National Electrical Code* automatically supersedes (is enforced in place of) local electrical codes.

> False

7. Other NFPA codes besides the *NEC* have electrical requirements that apply to dwelling units.

> True

8. Organizations such as Underwriters Laboratories Inc. approve electrical products for use according to applicable *Code* rules.

> False

9. The AHJ is responsible for approving equipment, materials, installations, and procedures.

> True

10. Article 527 requires that all 125-volt, single-phase, 15-, 20-, and 30-ampere receptacles used for construction purposes have GFCI protection for personnel.

> True

Challenge Question Answers

1. The *NEC* isn't an inspection agency and doesn't have an approval mechanism. Only the authority having jurisdiction (AHJ) can approve electrical installations and equipment. However, the AHJ will normally approve only electrical installations and equipment that comply with *Code* rules.

2. An Article is a major division of the *Code* that covers a particular subject such as lighting or transformers. A Section is a *Code* rule contained within an Article. Articles are made up of Sections.

3. The *NEC* specifies safety rules for electrical construction. It describes what must be done rather than who must do it. Not all states and localities require electrician licensing, and other technically knowledgeable people sometimes do work within the scope of the NEC. The two key concepts behind "qualified person" are that a person must have skills and knowledge related to electrical installations, and that the person must have received safety training.

4. Properly speaking, a neutral conductor is one that doesn't carry current under normal operating conditions, such as the neutral of a three-phase, four-wire circuit. Since the grounded conductor of a two-wire circuit does carry current in normal operation, it isn't a neutral. An ungrounded conductor is what electricians typically call a "hot" or "phase" conductor.

5. *NEC* Article 100 defines a dwelling unit as "including permanent provisions for living, sleeping, cooking, and sanitation." The typical hotel/motel room lacks permanent cooking provisions, so it isn't a dwelling unit. But a hotel room with a kitchenette would be considered a dwelling

unit since it satisfies all four conditions of the Article 100 definition.

6. *Listed,* as defined in Article 100, means that the equipment or materials are included in a list (usually called a *directory*) by an organization that evaluates products to ensure that they meet appropriate test methods and standards such as the *National Electrical Code. Labeled* means that a product bears an identifying mark from such an organization.

CHAPTER 2

Multiple Choice

1. The minimum wire size permitted by the *NEC* for branch circuits is

> **D.** Both A and C

2. A dwelling unit, as defined in the *NEC*, must have

> **A.** Permanent provisions for living, sleeping, cooking and sanitation

3. Branch-circuit ratings depend on

> **C.** The rating or setting of the overcurrent protective device

4. The most common service voltage for one- and two-family dwellings is

> **C.** 120/240-volt, single-phase, 3-wire service

5. The maximum number of receptacle outlets permitted to be served by a 15-ampere branch circuit in a dwelling unit is

> **D.** None of the above

6. The *Code* specifies a general lighting load for dwellings of

> **A.** 3 volt-amperes per square foot

7. The following branch circuit(s) is (are) required in every dwelling unit:

> **D.** Both A and B

8. As a general rule of thumb for planning residential services, it is a good idea to allow this (these) minimum(s):

> **D.** All of the above

9. General-purpose branch circuits of 120 volts are rated

> **D.** Both A and B

10. A 20-ampere branch circuit can supply a maximum continuous load of

> **C.** 1920 volt-amperes

Fill in the Blanks

1. A *continuous load* is a load in which the maximum current is expected to continue for <u>three</u> *or* <u>3</u> hours or more.

2. The minimum 3-wire size service permitted by the *NEC* for a one-family dwelling is 100 amperes.

3. Branch circuits supplying multiple outlets are permitted to have the following ampere ratings: 15, 20, 30, 40, 50.

4. When computing usable floor area for branch circuit, feeder, and service calculations according to Article 220, name at least one type of area that is permitted to be excluded if "not adaptable for future use": open porches, garages, other unused or unfinished spaces.

5. The demand loads for electric ranges and cooking appliances in dwellings are determined according to Table 220.19.

6. Each 240-volt branch circuit requires 2 pole space(s).

7. Small-appliance branch circuits and laundry branch circuits are required to be rated 20 amperes.

8. Section 220.53 permits applying a demand factor of 75 percent to the loads of four or more appliances fastened in place. (This doesn't apply to cooking, laundry, space heating, or air-conditioning equipment, which have their own *Code* rules governing how loads are computed.)

9. Large custom homes sometimes require services rated 208Y/120-volt, 3-phase, 4-wire *(120/208-volt and 120/208-volt, 3-phase, 4-wire are also acceptable answers)*, particularly if they are all-electric or have large loads such as guest houses and snow-melting equipment.

10. The *NEC* recommends a maximum total voltage drop of 5 percent *or* 5% on both feeders and branch circuits.

True or False

1. A branch circuit that supplies energy only for outlets to which appliances and permanently installed luminaires (lighting fixtures) are connected is called an *appliance branch circuit*.

    False

2. A building that consists of two apartments, one above the other, is a two-family dwelling.

    True

3. A one-family dwelling with two laundry rooms or areas is required to have two laundry branch circuits.

    False

4. Voltage drop causes electrical equipment to run cooler and last longer.

    False

5. When calculating connected load, outdoor lighting is required to be added to the interior lighting load of 3 volt-amperes per square foot before the demand factor is applied.

    False

6. In a two-family dwelling, the common area loads are divided between the two dwelling unit panelboards.

    False

7. Townhouses divided from the units on either side by fire-walls are considered to be individual buildings (i.e., one-family dwellings).

    True

8. The general lighting load for dwellings includes loading for receptacle outlets.

    True

9. When gas cooking appliances will be installed in a residence, a load of 180 volt-amperes must be included in the service calculation to cover such items as an integral clock, convenience receptacle, and control circuitry.

    False

10. The *National Electrical Code* requires a minimum of three branch circuits to supply appliances.

    False

Challenge Question Answers

1. Circuit breakers are listed for use at a maximum 80 percent of their rating. Thus, they must be sized at 125 percent of continuous loads, defined as those that continue for three hours or more. For noncontinuous loads, which are on for shorter periods of times, circuit breakers can be sized to handle the complete load.

2. Bathroom, small-appliance, and laundry branch circuits are rated at 20 amperes to carry high-load appliances.

3. Voltage drop causes incandescent lamps to operate less brightly. And because abnormally low voltage causes abnormally high current, voltage drop can cause motors and resistance heaters to overheat.

4. Compared to receptacle outlets in commercial buildings, those in dwellings are often lightly loaded. For this reason, it makes sense to use a per-square-foot electrical load in dwellings for purposes of calculating service and feeder size. Assigning a load of 180 volt-amperes per receptacle outlet would result in service and feeder load that were unrealistically high.

5. When multiple cooking or laundry appliances are installed in a dwelling, they may not all be in use simultaneously. Demand factors allow panelboards, feeders, and services to be sized to handle the actual demand load, rather than the higher connected load.

6. Electric heat and air conditioning do not normally operate at the same time. For this reason, they are considered "noncoincident loads," and only the larger of the two loads needs to be considered when calculating service and feeder size.

7. *NEC* 220.12 permits open porches, garages, and unfinished spaces to be omitted from floor area computations to determine general lighting load because there is very little electric service in these areas of dwellings.

8. Article 220 load calculations, like other *NEC* requirements, represent minimums required for safety. However, user convenience and satisfaction may be improved by providing additional capacity over the minimums required by *Code*.

CHAPTER 3

Multiple Choice

1. Lighting and appliance panelboards have
   D. Both A and B

2. Grounded and equipment grounding conductors are connected together at
   C. Service equipment only

3. Service equipment is required to withstand
   B. Available fault current

4. Split-bus panelboards are permitted to have
   A. A maximum of 42 branch-circuit pole spaces

5. Meter pedestals are used with
   B. Underground services

6. Where a single grounding electrode doesn't have a resistance to ground of 25 ohms or less, it must be augmented by the following number of additional electrodes, each located not closer than 6 ft away:
   A. One

7. Watt-hour meters are installed by the
   D. Both A and B

8. Pad-mounted transformers are used with the following type of residential service:
   B. Underground

9. Residential panelboards without main overcurrent devices are known as
   C. MLO

10. The conductor used to connect non—current-carrying metal parts of equipment, raceways, and other enclosures to the system grounded conductor is called the
   C. Equipment grounding conductor

Fill in the Blanks

1. The conductor that connects the service equipment to the grounding electrode is known as a grounding electrode conductor.

2. Grounded conductors in 120/240-volt, single-phase, 3-wire systems typically have insulation that is white in color.

3. The most common grounding electrode used in residential construction is metal underground water pipe.

4. Service-entrance conductors are those that connect the service-drop conductors or service-lateral conductors with the service equipment.

5. A bonding jumper is used to ensure the continuity of the grounding path around a water meter or water heater.

6. It isn't usually necessary to calculate the fault current available at a dwelling, because this information is normally provided by the serving utility *(utility, utility company, electric utility, power company, etc., are also acceptable answers)*.

7. Panelboards and disconnect switches listed for use as service equipment come equipped with a(n) main bonding jumper or MBJ, which connects the grounding busbar to the panelboard enclosure.

8. Wiring and equipment downstream of the service point are called premises wiring, which must be installed according to *NEC* rules and are subject to approval by the AHJ.

9. Grounding electrode conductors smaller than 6 AWG *(No. 6 is also an acceptable answer)* and run exposed are required to be protected by metallic conduit or raceway armor.

10. Service equipment must have an interrupting rating sufficient to interrupt the fault current *or* short-circuit current available at its terminals.

True or False

1. Equipment grounding conductors are required to be copper.

   False

2. The minimum size conductor permitted for connecting a supplemental grounding electrode is 6 AWG copper or 4 AWG aluminum or copper-clad aluminum.

   True

3. Some large all-electric houses use 480Y/277-volt, 3-phase, 4-wire services.

   False

4. The service grounding electrode conductor is required to be connected to the metal underground water piping system within the first 10 ft after it enters the house, unless a bonding jumper is installed around the water heater.

   False

5. Rigid metal conduit (RMC) and intermediate metal conduit (IMC) are commonly used for service masts.

   True

6. A 120/240-volt, single-phase, 3-wire residential service consists of two ungrounded (hot) conductors and one grounded conductor.

   True

7. The *NEC* has a special table for selecting conductor sizes for 120/240-volt, 3-wire, single-phase services for dwellings.

   True

8. The sizes of service conductors and raceways selected according to *National Electrical Safety Code* (NESC) rules sometimes may be smaller than those of the service entrance conductors (selected under *NEC* rules) to which they are spliced.

   True

9. Grounding electrode conductors and bonding jumpers are required to be copper or copper-clad aluminum conductors only.

   True

10. When a two-family dwelling has common area facilities such as stairway lighting, storage areas, and laundry rooms, branch circuits serving these areas must originate at service equipment protected by fuses.

    False

Challenge Question Answers

1. The equipment grounding conductor (EGC) is permitted to be connected to the grounded (neutral) conductor at only one place, at the service. This is the purpose of the MBJ. Panelboards other than those listed as service equipment don't have main bonding jumpers because the EGC isn't allowed to be connected to the neutral at those panelboards.

2. The reason for this rule is to ensure that operating handles of switches and circuit breakers are readily accessible so that they can be turned off quickly in an emergency.

3. Boxes, furniture, and other items must not be stored in such a way that they block panelboard doors or be placed on the floor in front of panelboards. Doing so violates the working space requirements of *NEC* 110.26.

4. *Service-drop conductors* are installed overhead from the transformer to the meter. *Service-lateral conductors* are installed underground from the transformer to the meter. *Service-entrance conductors* are those that run from the meter to the service panelboard.

5. Water heaters, meters, and other equipment may be removed from the piping for maintenance or replacement

purposes. For this reason, the *Code* requires continuity of the bonding path and requires the bonding connection to interior piping to be independent of such equipment.

6. Molded-case circuit breakers with ½-in. frames are rated 10,000 A.I.R. So when interrupting currents with higher ratings must be protected against, these 10,000 A.I.R branch circuit breakers are used in series with a main circuit breaker rated at either 22,000 A.I.R. or 42,000 A.I.R.

CHAPTER 4

Multiple Choice

1. Nonmetallic-sheathed cables are permitted to be installed above dropped or suspended ceilings only

   B. In dwellings

2. Electrical metallic tubing (EMT), flexible metal conduit (FMC), liquidtight flexible metal conduit (LFMC), and rigid nonmetallic conduit (RNC) all

   D. Both B and C

3. The *Code* has conductor fill rules intended to ensure that boxes don't become overcrowded. Trying to jam too many conductors into a too-small box creates which of the following problems?

   D. All of the above

4. When most types of cables and raceways used for dwellings are run parallel to framing members or through bored holes, how far must they be kept from the face of the stud when protection is not provided for the cables or raceways?

   C. 1¼ in.

5. Type AC cable armor can be removed by using

   D. Both A and B

6. The minimum radius of the curved inner edge of a Type NM or Type AC cable bend is not permitted to be

   A. Less than five times the diameter of the cable

7. When nonmetallic-sheathed cables pass through openings in steel studs and framing members, they must be protected from damage by sharp metal edges by installing

   D. Both A and B

8. The total volume of a 3 × 2 × 3½ in. device box is

   B. 18 in.^{3}

9. The following raceway type(s) is (are) rarely used in residential construction:

   D. Rigid metal conduit (RMC)

10. Type NM cable must be secured as follows:

    D. Both A and B

Fill in the Blanks

1. Type AC cable is permitted to be used unsupported where fished, and in whips up to 6 ft long for connecting to luminaires (lighting fixtures) or equipment installed within accessible ceilings.

2. Raceways are permitted to bend a maximum of 360 degrees between pull points such as boxes, conduit bodies, and panelboard cabinets. This total includes offsets at boxes and enclosures.

3. Type ACTHH armored cable has conductors rated 90°C with thermoplastic insulation.

4. When exposed Type NM cable passes through a floor, it must be enclosed in rigid metal conduit, intermediate metal conduit, electrical metallic tubing, Schedule 80 PVC conduit, listed metal or nonmetallic surface raceway, or other metal pipe extending at least 6 in. above the floor.

5. The tables in *NEC* Annex C can be used to determine the maximum conductor fill for raceways when all wires are the same size (the most common situation in house wiring).

6. Type NM cable is permitted to be used unsupported where fished, and in whips up to 4½ ft long for connecting to luminaires (lighting fixtures) or equipment installed within accessible ceilings.

7. FNMC is an alternate designation for the raceway type also known as liquidtight flexible nonmetallic conduit, Type LFNC, LFNC *(any of these answers are acceptable)*.

8. Volumes of common box sizes are listed in Table 314.16(A), along with the maximum number of conductors (all of the same size) permitted in each box.

9. Conductors with gray, white *(either or both of these answers are acceptable)* insulation cannot be used as switch legs, unless reidentified with tape or paint.

10. Type AC or Type NM cables run across the top of floor joists must be protected by guard strips when the attic is accessible by a permanent ladder or stairs, or within 6 ft of a scuttle hole.

True or False

1. Type AC cable consists of two or three insulated conductors, 12 AWG through 2000 kcmil, that are individually wrapped in waxed paper.

   False

2. Flexible metal conduit (FMC) is permitted to serve as an equipment grounding conductor in lengths up to 12 ft where the conductors are protected at 20 amperes or less.

   False

3. Increasing the size of conductors for long branch circuits and homeruns can help reduce voltage drop.

   True

4. At boxes and other terminations, nonmetallic-sheathed cables are required to be secured by cable clamps listed for the purpose.

   False

5. Electrical boxes are required to project a minimum of ¹⁄₁₆ in. beyond any combustible wall surface.

   False

6. At boxes and other terminations, armored cables are required to be secured by cable clamps listed for the purpose.

   True

7. *Total box volume* is determined by adding up the volumes of the box itself plus any attachment such as a plaster ring, extension ring, domed (as opposed to flat) cover, or luminaire canopy.

   True

8. Types NM, NMC, and NMS are all types of nonmetallic-sheathed cable defined in *NEC* Article 334.

   True

9. An insulating bushing (often called a "red head" in the field) must be installed at every cable termination to protect conductors from damage by sharp edges of cut metal, wood splinters, and similar items.

   False

10. Nonmetallic-sheathed cable contains conductors rated 90°C (194°F), but the ampacity of those conductors is based on the 60°C (140°F) column of *NEC* Table 310.16.

    True

Challenge Question Answers

1. Limiting the number of bends makes it easier to pull conductors into the raceways without damaging them.

2. Cable wiring methods are easier and less expensive to install. And compared to commercial installations, wires are replaced less often in dwelling occupancies, so there is less reason to install individual conductors in raceways.

3. This *Code* rule protects cables and raceways from being penetrated by nails and screws used to attached wall finishes such as gypsum board (drywall).

4. In one- and two-family dwellings, raceways are commonly used as service masts, where underground cables emerge from grade, and other places where physical protection is needed for cables.

5. The *NEC* specifies minimum bending radii for cables to avoid damaging conductor insulation and cable jackets.

6. All conductors in the same raceway segment should be pulled in the same operation to avoid damaging cable insulation by forcing fish tapes and additional conductors into raceways that already contain conductors.

7. A building with two grade-level dwelling units separated by fire walls is actually two one-family dwellings located side by side. A similar building with no fire wall between the dwelling units is a two-family dwelling.

CHAPTER 5 _____

Multiple Choice

1. The highest voltage permitted by the *NEC* for lighting in dwellings is
   A. 120 volts

2. The following stairways are required to have a wall switch at every floor or landing that includes an entry:
   C. Interior stairways with six or more risers

3. The following are not permitted to be located within a zone extending 3 ft horizontally and 8 ft vertically from the top of the bathtub rim or shower stall threshold:
   D. All of the above

4. Most listed luminaires are designed to limit the temperature to which supply conductors are exposed to a maximum of
   B. 75°C (167°F)

5. An incandescent lamp rated 130 volts will operate longest on a circuit operating at
   A. 115 volts

6. Electric-discharge lighting sources include the following lamp types:
   D. All of the above

7. Switch-controlled receptacle outlets are permitted to be used as the required lighting outlets in
   B. Living rooms

8. Many flush- and recessed-mounted incandescent luminaires use remote junction box construction to
   C. Prevent overheating of the supply conductors

9. When lighting track is installed in dwellings, the following load must be used for purposes of branch-circuit, feeder, and service calculations:
   D. None of the above

10. Switches with pilot lights
    D. Both A and B

Fill in the Blanks

1. Ceiling-suspended (paddle) fans weighing more than 35 lb must be independently supported. Lighter-weight pad-

dle fans must be supported by outlet boxes identified for the purpose.

2. Track lighting that operates at 30 volts or higher is not permitted to be located less than 5 ft above the finished floor.

3. Class P ballasts are associated with the following type of electric-discharge luminaires: fluorescent

4. Three-way switches control a lighting outlet from two different locations.

5. So-called "long life" or "extended service" incandescent lamps are rated at higher operating voltages than conventional incandescent lamps.

6. Tungsten-halogen, quartz-halogen *(either is a correct answer)* lamps use lamp-within-a-lamp construction to minimize temperatures on the exterior bulb or glass envelope.

7. General-use snap switches rated 15 and 20 amperes and designed for connection to either copper or aluminum conductors are listed and marked CO/ALR.

8. Equipment, materials, or services included in a list published by an organization that is acceptable to the authority having jurisdiction and concerned with evaluation of products or services is (are) considered to be listed.

9. The inside of a fluorescent lamp is coated with a powder called phosphor that glows when excited by ultraviolet light, thus producing visible light.

10. Recessed-mounted incandescent and HID luminaires installed in poured concrete are not required to have integral thermal protection.

True or False

1. Switch-controlled receptacle outlets are permitted to function as lighting outlets in all habitable rooms of a dwelling.
   False

2. Fluorescent luminaires mounted end-to-end must be marked "Identified for Through-Wiring" if the conductors of other branch circuits pass through them.
   False

3. Luminaires equipped with integral thermal protectors are marked Type IC.
   False

4. Luminaires in clothes closets must contain a switch, or be controlled by a switch, with the point of control located at the usual point of entry to the closet.
   False

5. Receptacles located within a zone extending 3 ft horizontally and 8 ft vertically from the top of the bathtub rim or shower stall threshold are required to have ground-fault circuit-interrupter protection for personnel.
   False

6. Single-pole wall switches have marked ON and OFF positions.

True

7. Lighting track is permitted to be cut to length in the field when listed for this application.

False

8. A *luminaire* is a complete lighting unit consisting of a lamp or lamps together with the parts designed to distribute the light, to position and protect the lamps and ballast (where applicable), and to connect the lamps to the power supply.

True

9. The *National Electrical Code* prohibits the use of pull-chain switches to control luminaires (lighting fixtures).

False

10. Section 600.32(I) prohibits neon lighting in dwellings.

False

Challenge Question Answers

1. To make it easier for people with disabilities, such as those in wheelchairs or those using crutches, to reach wall switches and receptacles more easily.

2. All are intended to prevent luminaires or their components from overheating and possibly causing a fire.

3. They all require ballasts.

4. Fluorescent lamps have longer life and are more energy-efficient than incandescent lamps, and are available in a variety of colors. Compact fluorescent lamps offer a range of new options and are available in a variety of fixture designs; some types can be used as a direct replacement for incandescent lamps.

5. (a) Running power conductors to the outlet first is generally the most logical approach, and this type of wiring is easier to figure out later for troubleshooting and renovations when the wiring is covered by wall finishes. (b) Running power conductors to switch boxes may in some cases result in shorter wiring runs. Also, some types of switching devices such as electronic dimmers, occupancy sensors, and pilot-light switches require both a hot and grounded conductor in the switch box to power the device.

6. The *National Electrical Code* is a minimum safety standard and not a design guide, as described in 90.1(A). So it requires that lighting outlets be capable of being de-energized, but their location is not discussed.

7. In general, the *NEC* doesn't have requirements for colors of conductor insulation, other than grounded conductors (white or gray) and grounding conductors (green), and this is for safety reasons. There is no safety reason for specifying required colors of travelers.

CHAPTER 6

Multiple Choice

1. The minimum mounting height permitted by the *Code* for wall-mounted receptacles is

D. None of the above

2. The maximum number of duplex receptacles permitted on a 15-ampere branch circuit in a dwelling is

D. None of the above

3. Listed or labeled equipment is required to be installed and used in accordance with

D. Both A and B

4. Having more than one branch circuit supply lighting outlets and receptacle outlets in a room offers the following advantages:

D. Both A and B

5. A habitable room with a lighting outlet and three entrances is required by the *Code* to have how many wall switches or equivalent controls such as dimmers and occupancy sensors?

A. One

6. A lighting outlet with a permanently installed luminaire (lighting fixture) is required in the following room(s) of a home:

B. Kitchen

7. A single receptacle can have the following number of contact devices on the same yoke:

A. 1

8. Permanently installed electric baseboard heaters are prohibited from being located beneath wall-mounted receptacle outlets by

B. Baseboard heater listing instructions

9. Receptacles located higher above the floor than the following are not permitted to be counted among the receptacle outlets required by 210.52(A)(1) and (2):

D. 66 in.

10. Floor receptacles used as required receptacle outlets along floor-to-ceiling windows or fixed panels of sliding glass doors are required to be installed within what distance of the wall line (fixed glass panel)?

D. 18 in.

Fill in the Blanks

1. The 125-volt, 15- and 20-ampere receptacle outlets in living rooms must be installed so that no point along the floor in any wall space is more than 6 ft, measured horizontally, from another receptacle outlet in that space.

2. The *Code* requires 20-ampere branch circuits in dwellings for small-appliance branch circuits and bathroom branch circuits.

3. Any wall space 2 ft or wider must have its own receptacle outlet.

4. Most 120-volt branch circuits in a typical dwelling unit are rated 15 amperes.

5. According to the definition in Article 100, a multiple (duplex) receptacle is considered to be the same as two receptacle(s).

6. Wall-mounted luminaires (lighting fixtures) weighing not more than six pounds *or* 6 lb are permitted to be supported by standard outlet boxes using No. 6 or larger screws.

7. Small-appliance branch circuits serving receptacles in kitchens, dining rooms, pantries, and breakfast nooks are required to be protected by overcurrent devices rated 20 amperes (*20A or 20 amps are also acceptable answers*) and have GFCI protection.

8. When firewalls are penetrated by electrical cables or conduits, the openings are required to be firestopped.

9. A 20-ampere branch circuit requires 12 AWG conductors.

10. In habitable rooms, 125-volt, 15- and 20-ampere receptacle outlets must be installed so that no point along the floor in any wall space is more than six *or* 6 ft, measured horizontally, from another receptacle outlet in that space.

True or False

1. Built-in receptacles in electric baseboard heater units are supplied from one leg of the 240-volt heater circuit and the grounded (neutral) conductor.

    False

2. Duplex receptacles mounted 5 ft above the floor can be counted among the receptacle outlets required by 210.52(A)(1) and (2).

    True

3. A 125-volt, 15-ampere single receptacle can be installed on either a 15-ampere or 20-ampere branch circuit.

    False

4. Section 210.25 permits certain common-area loads of two-family dwellings, such as outdoor receptacles, to be supplied from either unit.

    False

5. In other than kitchens and bathrooms, the lighting outlet required by 210.70(A) is permitted to be a receptacle outlet controlled by a wall switch rather than a lighting outlet.

    True

6. Small-appliance branch circuits, laundry branch circuits, and bathroom branch circuits are permitted to supply only receptacle outlets.

    True

7. Living room receptacle outlets are permitted to be connected to small-appliance branch circuits for the kitchen and dining room, as long as the receptacles are rated 125 volts, 20 amperes.

    False

8. The *National Electrical Code* permits lighting and other receptacle loads to be mixed together on the same branch circuits.

    True

9. Living rooms in dwelling units are required to have lighting outlets with permanently installed luminaires (lighting fixtures).

    False

10. Overcurrent protective devices (OCPD) for branch circuits may be either circuit breakers or fuses.

    True

Challenge Question Answers

1. The *NEC* is a safety standard rather than a design guide, as made clear in 90.1(B) and 90.1(C). This means that every rule in the *NEC* is supposed to have a safety rationale. Heights of switches and receptacles are seen as a design and convenience issue.

2. Tripping a circuit breaker doesn't result in losing all the power to a room. If one branch circuit is turned off to service or replace an item such as a luminaire or receptacle, energized receptacles are still available for portable lamps and power tools.

3. Extension cords indoors are seen as inherently hazardous since they can create a tripping hazard. Also, when receptacles aren't conveniently placed to serve lamps and appliances, homeowners sometimes run extension cords underneath rugs or carpets. In effect, the homeowners are using these flexible cords as a substitute for permanent wiring methods, which the *NEC* discourages for safety reasons [400.8].

4. GFCIs are generally required where the presence of moisture increases potential shock hazard, for example, in areas such as bathrooms, unfinished basements, outdoors, and serving kitchens and wet bar countertops. Living rooms generally aren't characterized by excessive moisture.

5. Receptacles in dwellings tend to be lightly loaded, so that it isn't necessary to allow 180 volt-amperes (VA) per receptacle. When calculating service size, residential

receptacle outlets are included in the 3 VA per square foot general lighting load.

6. Commercial installations typically have more branch circuits than dwelling installations and usually carry higher loads as well. Commercial branch circuits are also changed more frequently than residential circuits. Both reasons make it more logical and convenient to keep lighting and receptacles wired on separate circuits. Also, lighting in commercial occupancies frequently operates at 277 volts while general-purpose receptacle outlets operate at 120 volts.

CHAPTER 7

Multiple Choice

1. Receptacle outlets for refrigerators are allowed to be supplied by

   D. All of the above

2. A small-appliance branch circuit is permitted to supply receptacle outlets in how many kitchens?

   A. One

3. Receptacles serving kitchen countertop areas are

   D. All of the above

4. Small-appliance branch circuits are permitted to serve lighting outlets only in

   D. None of the above

5. An appliance branch circuit is a branch circuit that

   B. Supplies energy to one or more outlets to which appliances are to be connected

6. Which of the following statement(s) about receptacle ratings is false?

   D. Receptacle rating is not dependent upon branch-circuit rating.

7. In addition to supplying receptacle outlets serving kitchen countertop areas, small-appliance branch circuits are permitted to supply the following:

   D. All of the above

8. Switched receptacles are not permitted to serve as the required lighting outlet in the following room(s):

   C. Kitchen

9. Kitchen waste disposers and other motor loads are permitted to be controlled by a general-use snap switch under the following conditions:

   B. The load doesn't exceed 80 percent of the switch's rating.

10. Which statement about built-in dishwashers and trash compactors with attachment cords and plugs is true?

   A. The plug and receptacle are permitted to serve as the required disconnecting means for servicing the appliance.

Fill in the Blanks

1. Kitchen receptacle outlets not serving countertop areas are required to be located according to the rules of 210.52(A)(1).

2. Multiwire branch circuits consisting of either two or three ungrounded conductors plus a grounded (neutral) conductor cannot be used to supply GFCI-protected receptacle outlets.

3. Each kitchen wall space measuring 12 *or* twelve in. or wider must have a receptacle outlet installed.

4. Receptacle outlets located more than 20 *or* twenty in. above a kitchen countertop are not considered to be receptacles serving countertop surfaces.

5. Most branch circuits serving electric cooking appliances are rated either 40 or 50 *or* forty or fifty amperes at 240 volts.

6. Built-in dishwashers and trash compactors aren't permitted to be supplied by small-appliance *or* 20-ampere small-appliance branch circuits.

7. Receptacle outlets serving kitchen counters *or* kitchen countertop areas are required to have GFCI protection.

8. All 25-volt, 15- and 20-ampere receptacle outlets in dining rooms must be installed so that no point along the floor in a wall space is more than 6 ft from another receptacle outlet in the same space.

9. Receptacles aren't permitted to be installed in a face-up *or* facing-upward position in a countertop or similar surface, in order to prevent them from collecting liquid, crumbs, and other debris.

10. A countertop area that stands in the middle of a kitchen and isn't connected to another kitchen countertop is known as an island.

True or False

1. Kitchen counter peninsulas 48 in. × 12 in. or larger are required to have receptacle outlets located so that no point along the edge of the peninsula is more than 24 in. measured horizontally from a receptacle outlet.

   False

2. All receptacle outlets located in kitchens are required to have GFCI protection for personnel.

   False

3. Switched receptacle outlets used in place of lighting outlets are permitted to be supplied by the small-appliance

branch circuit in all rooms of dwellings except kitchens, as permitted by 210.70(A)(1).

False

4. The neutral conductor of a 120/240-volt, single-phase, 3-wire branch circuit supplying an electric range, wall-mounted oven, or cooktop can be smaller than the un-grounded (phase) conductors when the maximum demand is 8¾ kW, as computed according to *NEC* Table 220.55, column 6.

True

5. An "appliance garage" is a dedicated space for storing non-fixed-in-place appliances when they are not in use.

False

6. A receptacle that supplies power for the operation of igniters, clocks, timers, or controls on a gas cooking appliance is permitted to be connected to a 20-ampere small-appliance branch circuit.

True

7. Range hoods and kitchen exhaust fans rated not more than $^1/_8$ horsepower are permitted to be supplied by kitchen small-appliance branch circuits.

False

8. Electric cooking appliances with a total demand exceeding 60 amperes are constructed so that they can be supplied by two branch circuits, each rated a maximum of 40 amperes.

False

9. Ranges, wall-mounted ovens, and cooktops are permitted to be either hard-wired (permanently connected to the branch-circuit conductors) or cord-and-plug connected.

True

10. Each wall along a counter space measuring 12 in. or wider must have a receptacle outlet.

True

Challenge Question Answers

1. Kitchen countertops have sinks, and food preparation also involves liquids. The *NEC* typically requires GFCI protection where the presence of moisture increases shock hazard. Since other areas of dwellings served by 20-ampere small-appliance branch circuits (dining rooms, pantries, breakfast nooks) typically don't have sinks and aren't used for food preparation, GFCI protection isn't required for receptacle outlets in those areas.

2. The reason for this requirement is to provide sufficient capacity to supply high-load, cord-and-plug connected appliances such as toasters, coffee makers, slow cookers, and electric frying pans. Not only do these kitchen appli-

ances have high loads, but two or more of them may be used simultaneously.

3. Clock hanger outlets have only a single receptacle to discourage plugging in other kitchen appliances. Doing so would circumvent the required GFCI protection for receptacle outlets serving kitchen countertop areas. It might also create safety problems with trailing appliance power cords and/or extension cords.

4. (a) *NEC* 210.21(B)(2) permits 15-ampere receptacles to be installed on 20-ampere branch circuits that supply two or more receptacles. (b) Kitchen appliances are manufactured with 15-ampere attachment plugs.

5. *Connected load* is an individual appliance's nameplate rating, the maximum load that it can draw when turned on. *Demand load* recognizes the fact that when two or more cooking appliances are installed in the same dwelling, they may not operate at full connected load simultaneously, and so the branch circuit rating and service/feeder load calculation don't have to be based on the total connected load of the multiple cooking appliances.

6. The *Code* requires that tap conductors to kitchen cooking appliances must not be longer than needed to service the appliances so that they cannot be spliced or tapped to serve other outlets or utilization equipment.

7. Kitchen waste disposers are installed within kitchen cabinets and are not provided with on/off switches on the appliance. Dishwashers and trash compactors have operating controls on the units themselves, so general-use snap switches aren't necessary.

8. The *Code* requires that a means be provided to disconnect appliances for servicing. When an appliance is cord-and-plug connected, that connection must be accessible so the appliance can be disconnected for servicing.

CHAPTER 8

Multiple Choice

1. The following method(s) can be used to provide GFCI protection for receptacle outlets installed in bathrooms:

   D. All of the above

2. Ground-fault circuit interrupters of the type used to protect bathroom receptacle outlets are known as

   A. Class A

3. Table 210.21(B)(2) specifies that when a branch circuit supplies two or more receptacle outlets, a 15-ampere receptacle shall not supply a total cord-and-plug-connected load exceeding

   C. 12 amperes

4. The minimum number of receptacle outlets required in a bathroom is

A. One

5. The easiest GFCI protection approach for homeowners to understand is

B. Installing a GFCI receptacle at each outlet required to have such protection

6. Pump motors, metal piping, and metal parts of electrical equipment associated with hydromassage tubs must be bonded together using

A. A minimum 8 AWG solid copper conductor

7. GFCI-protected receptacle outlets in the following locations *only* are permitted to be connected to the 20-ampere bathroom branch circuit

D. None of the above

8. *National Electrical Code* rules that apply to residential construction cover only

D. All of the above

9. Article 422 states that the following types of utilization equipment must be constructed to provide protection for personnel against electrocution when immersed:

D. Both A and B

10. Ground-fault circuit interrupters (GFCIs) are available in the following type(s) of construction:

D. Both A and B

Fill in the Blanks

1. The required branch circuit for bathroom receptacle outlets must be rated 20 amperes *or* 20 A.

2. Wiring requirements for hydromassage bathtubs are found in *NEC* Article 680.

3. A GFCI-protected receptacle is required to be located within 3 ft *or* 36 in. of the edge, on a wall or partition adjacent to the basin or countertop.

4. Installation of some types of luminaires is prohibited within a zone measured 8 ft vertically and 3 ft horizontally from the bathtub rim or shower stall threshold.

5. The official *Code* name for "whirlpool bath" is hydromassage tub.

6. The only types of luminaires permitted to be installed within a zone measured 8 ft vertically and 3 ft horizontally from the bathtub rim or shower stall threshold are recessed and surface mounted.

7. To protect construction workers from electric shock, all 125-volt, single-phase, 15-, 20-, and 30-ampere receptacles used for construction purposes such as supplying power tools and portable work lights are required to be protected by ground-fault circuit-interrupters *or* GFCIs.

8. Timer switches are sometimes used to control heat lamps installed in bathrooms, to insure that they aren't left ON accidentally.

9. According to the definition in Article 100, a tub or shower space in a bathroom must be considered a wet location.

10. Although a ground-fault circuit interrupter opens a circuit quickly enough to prevent electrocution of a healthy adult, a GFCI cannot prevent the sensation of receiving a shock, being shocked, getting shocked, feeling an electric shock *(any of these, or other similar language, is considered a correct answer)*.

True or False

1. GFCI-protected receptacle outlets are prohibited within a bathtub or shower space, unless equipped with enclosures that remain weatherproof when an attachment plug is inserted.

False

2. Bathtub spaces and shower enclosures are considered to be wet locations.

True

3. GFCI circuit breakers must be installed in accordance with manufacturer's special instruction to provide only protection against faults between phase and grounded (neutral) conductors, or between two different phases.

False

4. Lighting outlets are permitted to be supplied by the 20-ampere bathroom branch circuit, but must be connected ahead (upstream) of any ground-fault circuit-interrupter device.

False

5. Only bathroom receptacle outlets located within 3 ft of the outside edge of the basin are required to have GFCI protection for personnel.

False

6. Section 406.8(C) prohibits installation of receptacles within a zone measured 8 ft vertically and 3 ft horizontally from the bathtub rim or shower stall threshold.

False

7. It is common practice to circuit together all receptacle outlets in a dwelling that are required to have GFCI protection, and supply them from a single GFCI circuit breaker.

False

8. A bathroom is required to contain a basin, a toilet, and a bathtub or shower.

False

9. A receptacle built into a luminaire (light fixture) is permitted to serve as the bathroom receptacle required by 210.52(D) if it is located within 3 ft of the outside edge of the basin and furnished with GFCI protection.

   False

10. GFCIs don't prevent the sensation of receiving an electric shock.

    True

Challenge Question Answers

1. There is no restricted space for installing receptacles near bathtubs because the potential shock hazard from touching a receptacle is less than that from touching a luminaire or ceiling fan. The hazard of touching a receptacle is perhaps similar to the hazard presented by a surface-mounted or recessed-mounted luminaire, and the *NEC* permits both of these types of fixtures to be installed anywhere in bathrooms.

2. The average kitchen has only a sink, while bathrooms have a sink, toilet, and tub or shower. People use more water in bathrooms and are also more likely to be wet and barefoot simultaneously. For this reason, all receptacle outlets in bathrooms must have GFCI protection because the presence of moisture increases the risk of electric shock.

3. In a two-wire circuit, the currents in the line and grounded (neutral) conductors are equal under normal operating conditions. A GFCI works by measuring current in these two circuit conductors and trips when it senses a current imbalance of about 5 milliamperes. Proper operation of a GFCI doesn't depend on the presence of an equipment-grounding conductor.

4. If a branch circuit supplying all the outlets in a single bathroom is protected by a GFCI circuit breaker and it trips, all the outlets in the bathroom will be deactivated, including the lighting. This may create a safety hazard of tripping or falling, particularly in a bathroom without windows. For this reason, safety for occupants is improved by keeping bathroom lighting outlets and GFCI-protected receptacle outlets on separate branch circuits.

5. A GFCI circuit breaker de-energizes a complete branch circuit, a GFCI receptacle de-energizes just that receptacle, and an IDCI de-energizes the appliance of which it is a part.

CHAPTER 9

Multiple Choice

1. Arc-fault circuit-interrupter protection is required for
   B. 15- and 20-ampere branch circuits in bedrooms

2. All 125-volt, 15- and 20-ampere receptacle outlets must be installed so that no point along the floor in any wall space is more than how many feet, measured horizontally, from another receptacle outlet in that space?
   B. 6 ft

3. The following type of room often found in close proximity to bedrooms requires AFCI protection for all 120-volt, 15- and 20-ampere branch circuits:
   D. None of the above

4. Circuits supplying the following items in bedrooms are required to have AFCI protection:
   D. All of the above

5. A bedroom with two entrances is required to have how many wall switches (or equivalent controls such as dimmers and occupancy sensors) controlling the lighting outlet(s) installed in that room?
   A. One

6. When listed electric baseboard heaters are installed in bedrooms
   D. All of the above

7. AFCI protection is permitted for 120-volt, 15- and 20-ampere branch circuits supplying outlets in which of the following?
   B. Bedrooms

8. A duplex receptacle has the following number(s) of contact devices on the same yoke:
   B. 2

9. Receptacles located at heights above the floor greater than which of the following are not permitted to be counted among the receptacle outlets required by 210.52(A)(1) and (2)?
   C. 5½ ft

10. The following types of fluorescent luminaires (fixtures) are permitted to be installed inside closets:
    D. Both A and B

Fill in the Blanks

1. Accessories known as hickeys *or* fixture studs *or* both (*studs is also an acceptable answer*) are used to attach luminaires to outlet boxes and help support their weight, and often are supplied with the luminaire itself.

2. AFCIs will not perform properly if installed on multiwire branch circuits.

3. Standard ceiling outlet boxes can be used to support luminaires weighing up to 50 lb.

4. Any wall space 2 ft or wider must have its own receptacle outlet.

5. Occupants in a two-family dwelling are required to have <u>ready access</u> to the overcurrent devices protecting conductors in their occupancy.

6. A minimum of <u>one</u> lighting outlet(s), controlled by a wall switch, must be installed in every habitable room of a dwelling.

7. Most 120-volt branch circuits in a typical dwelling unit are rated <u>15</u> amperes.

8. According to the definition of the term *receptacle* in Article 100, a duplex receptacle is considered to be the same as <u>two</u> receptacle(s).

9. Wall-mounted luminaires (lighting fixtures) weighing not more than 6 lb are permitted to be supported by standard outlet boxes using No. <u>6</u> or larger screws.

10. On arcing faults to ground, listed AFCIs are required to trip at <u>5</u> amperes.

True or False

1. The *National Electrical Code* requires that closets have two independent sources of lighting, so that the failure of any single source does not leave the closet in darkness.

False

2. Duplex receptacles mounted up to 6 ft above the floor can be counted among the receptacle outlets required by 210.52(A)(1) and (2).

False

3. A 125-volt, 15-ampere duplex receptacle can be installed on either a 15-ampere or 20-ampere branch circuit.

True

4. Special outlet boxes are required for ceiling-suspended (paddle) fans only when the fan weighs 35 lb or more.

False

5. Bedroom lighting outlets are permitted to be of the type defined by 210.70(A).

True

6. AFCI protection is required for all branch circuits in bedrooms, associated closets, and associated bathrooms.

False

7. Receptacle outlets are permitted to be located above electric baseboard heaters in bedrooms only when supplied from a different branch circuit than the heater.

False

8. Bedroom circuits and other outlets/loads can be mixed together on the same AFCI-protected branch circuits.

True

9. Bedrooms in dwelling units are required to have lighting outlets with permanently installed luminaires (lighting fixtures).

False

10. All types of incandescent luminaires (lighting fixtures) are prohibited in closets unless protected by an AFCI.

False

Challenge Question Answers

1. The rule against installing receptacles above electric baseboard heaters is intended to prevent damage by overheating to power cords of appliances that might be draped over the electric baseboard heater. Hot-water baseboard heaters do not reach the same high temperatures as electric-resistance heaters.

2. Hard-wired smoke detectors are required in bedrooms of dwellings to protect sleeping occupants against the dangers of fire. However, if the branch-circuit wiring supplying a smoke detector is faulty, this may cause a fire. Simultaneously, it would also disable the smoke detector, thereby preventing an alarm from sounding. For this reason, 15- and 20-ampere branch circuits supplying power to hard-wired smoke detectors are not exempted from the general requirement for AFCI protection of bedroom branch circuits.

3. Exposed or partially exposed incandescent lamps are prohibited because the hot filaments create a potential fire hazard for flammable clothing and other items stored in closets if the outer glass bulb is broken. Pendant luminaires and lampholders are prohibited because they would tend to interfere with shelves and clothes hangers in closets. This in turn creates a potential shock hazard for people using the closets.

4. Article 100 defines *outlet* as "a point on the wiring system at which current is taken to supply utilization equipment." Thus, a lighting outlet and a receptacle outlet both clearly meet this definition. A typical general-use snap switch in a switch box doesn't; it contains only an ungrounded (hot) conductor and a switch leg. Also, a switch is considered a wiring device, and Article 100 defines *device* as "a unit of an electrical system that is intended to carry or control but not utilize electric energy." Since a device is not the same as utilization equipment, this appears to mean that a switch location does not qualify as an outlet. However, some devices such as pilot-light switches, electronic dimmers, and programmable lighting controllers require an ungrounded (neutral) conductor in the switch box to provide 120-volt power to operate their electronics or pilot lights. Products such as these seem to occupy a sort of gray area between traditional *devices* and *utilization equipment*, and make it more reasonable to regard switch locations as outlets.

5. Homeowners often replace bedroom luminaires with ceiling fans after the original construction. To avoid future safety problems that fan vibration can cause for conventional outlet boxes, boxes listed for use with fans should be installed during the original construction.

6. The purpose of a GFCI is to protect people from shock or electrocution. The purpose of an AFCI is to protect against fires caused by damaged branch-circuit wiring.

7. Receptacle outlet spacing rules for bedrooms are intended to ensure that utilization equipment (such as lamps and entertainment equipment) can be plugged into a receptacle safely without requiring the use of extension cords and without requiring attachment cords to cross doorways or be concealed under rugs.

CHAPTER 10 _____

Multiple Choice

1. An unfinished basement area used for storage or equipment is required to have (a) lighting outlet(s)

    A. At or near equipment requiring servicing

2. The following areas of a basement are required to have receptacle outlets:

    D. All of the above

3. GFCI protection is required for receptacles installed in the following area(s) of a basement:

    C. Within 6 ft of a wct bar sink

4. Wall switch-controlled lighting outlets are required in the following area(s) of a basement:

    D. None of the above

5. AFCI protection is required for basement receptacle outlets in the following area(s):

    C. Bedroom

6. Three-way switches are required to control lighting outlets when a basement

    D. None of the above

7. The following receptacle outlets in a basement do not require GFCI protection:

    B. Those supplying appliances occupying dedicated space that are not easily moved from one place to another

8. The following *NEC* rule(s) applies to laundry rooms located in basements:

    B. One or more laundry receptacles must be supplied by a 20-ampere branch circuit that has no other outlets.

9. The following statement about the electric supply to a gas clothes dryer located in an unfinished area of a basement is true:

    D. All of the above

10. The following wiring method(s) is permitted to be used in the basement of a one-family dwelling:

    D. All of the above

Fill in the Blanks

1. At least one control for basement lighting is required to be located at the usual point of entry.

2. The minimum circuit ampacity and maximum overcurrent protection for an electric clothes dryer is marked on the appliance nameplate.

3. The laundry branch circuit is required to supply at least 1, one receptacle outlet(s).

4. Receptacles installed more than 6½ ft above floor level are considered to be not readily accessible.

5. Receptacle outlets installed in basement bedrooms are required to have arc-fault circuit-interrupter *or* AFCI protection.

6. A wall switch-controlled lighting outlet is required on the exterior *or* outside *or* outdoor side of any outdoor entrance to a basement.

7. All clothes dryer wiring methods are required to include a(n) equipment grounding conductor.

8. A single receptacle that supplies a freezer in an unfinished basement area is not required to have ground-fault circuit-interrupter *or* GFCI protection.

9. Electric clothes dryer receptacles are typically rated 240 volts, 30 or 50 amperes.

10. Receptacles serving wet bar countertops located within 6 *or* six ft of the sink are required to have GFCI protection.

True or False

1. Receptacle outlets in finished areas of basements are required to have GFCI protection.

    False

2. Article 100 defines *bedroom* as an area intended for sleeping and provided with a window and closet.

    False

3. The *Code* requires a wall switch-controlled lighting outlet near equipment requiring servicing in a utility room.

    False

4. A duplex receptacle supplying only a refrigerator located in an unfinished part of a basement requires GFCI protection.

    True

5. All branch-circuit wiring installed in unfinished basement areas is required to be run in EMT or Schedule 40 PVC for physical protection.

    False

6. A bedroom located in a basement is required to include a smoke detector.

    True

7. Lampholders or luminaires (lighting fixtures) may be installed in all areas of basements.

    False

8. Each unfinished area of a basement is required to have a minimum of one receptacle outlet in addition to any required for laundry equipment.

    True

9. Receptacle outlets in unfinished areas of basements must be supplied by separate branch circuits from those in finished areas.

    False

10. A bathroom located in a basement area below grade is required to have AFCI protection on all receptacle outlets.

    False

Challenge Question Answers

1. Receptacle outlets located more than 6½ feet above grade or floor level are considered not readily accessible. It's assumed they won't be used to plug in portable utilization equipment and thus aren't required to have GFCI protection for personnel.

2. The reason for this rule is to prevent homeowners from plugging other items such as lamps or power tools into the other, unused receptacle without having the benefits of GFCI protection.

3. Accessible ceilings (often called *suspended* or *dropped ceilings*) have panels that fit into support grids. Luminaires such as lay-in fluorescent fixtures also fit into these grids. The *NEC* allows unsupported lengths of wiring methods (often called *fixture whips*) within accessible ceilings to provide the flexibility necessary to lift fixtures into and out of the ceiling grids.

4. Yes, although 15- and 20-ampere branch circuits supplying outlets in a bedroom are required to have AFCI protection, the *NEC* doesn't prohibit using AFCI-protected branch circuits to supply outlets located in other rooms.

5. GFCI circuit breakers and receptacles protect against the hazards of current leaking to ground (perhaps establishing a ground path through a human body). These hazards are more extreme in unfinished areas of basements, which typically have concrete slab floors in direct contact with, the earth, and exposed metal water piping.

6. Yes, *NEC* 210.70(A)(3) requires that basements and utility rooms have at least one lighting outlet containing a switch or controlled by a wall switch where these spaces are used for storage or contain equipment requiring servicing. "Lighting outlet containing a switch" permits the use of lampholders and luminaires with pull chains or other built-in switches.

7. The receptacle placement rules of 210.52(A) apply to habitable rooms in basements such as family rooms, bedrooms, and home offices. Separate receptacle outlet rules apply to unfinished basement areas [210.52(G)], laundry rooms [210.52(F)], and areas containing HVAC equipment (210.63).

CHAPTER 11

Multiple Choice

1. The following hallway(s) is (are) required to have a receptacle outlet:

    C. Hallway 10 ft or more in length

2. Attics are required to have more than one lighting outlet if they

    D. None of the above

3. Interior stairway lighting is permitted to be controlled by the following:

    D. All of the above

4. The following area(s) is (are) required to have a receptacle outlet:

    C. Hallway 12 ft in length

5. An attic luminaire (lighting fixture) mounted near a scuttle hole can be controlled by a

    D. All of the above

6. If an attic contains heating, air-conditioning, or refrigeration equipment, a 15- or 20-ampere, 125-volt receptacle must be installed

    B. On the same level and within 25 ft of the equipment

7. A hallway more than 10 ft long, with three points of entry, is required to have a lighting outlet controlled by the following number of switches:

    A. One

8. The following types of interior stairways are required to have wall switch-controlled lighting:

    D. All of the above

9. A lighting outlet located away from an attic entrance can be controlled by the following type(s) of switch:

    D. Both B and C

10. Receptacle outlets are required to be installed in attics when they

    D. Both A and B

Fill in the Blanks

1. An attic used for storage is required to have at least 1 *or* one lighting outlet(s).

2. A hallway with three entrances must have a minimum of 1 *or* one wall switch(es).

3. A wall switch to control interior stairway lighting is required at each floor or landing level with a point of entry, when these levels are separated by 6 *or* six or more risers.

4. The *Code* requires a minimum of 1 *or* one lighting outlet(s) for an interior stairway that changes direction at a landing.

5. Branch circuits serving common areas in a two-family dwelling must be supplied from a house loads panel *or* house panel.

6. The required receptacle outlet in a hallway 10 ft or more in length is also permitted to serve as the required lighting outlet.

7. Lighting outlets in hallways of dwellings are permitted to be controlled by wall switches, by remote, central, or automatic means, or by occupancy sensors.

8. When an attic contains heating, air-conditioning, or refrigeration equipment, a receptacle rated 15 or 20 amperes, 125 volts *(similar wording with the same meaning is also acceptable),* must be installed at an accessible location on the same level and within 25 ft of the equipment.

9. When a lighting outlet is installed in an interior stairway, a wall switch must be installed at each floor level, and at each landing that includes a point of entry where the stairway between the floor levels has six or more risers.

10. The receptacle outlet required in attics with mechanical equipment requiring servicing must be located on the same level and within 25 *or* twenty-five ft of the equipment.

True or False

1. Wall switch-controlled lighting outlets are required at the top and bottom of an interior stairway.

   False

2. A minimum of one receptacle outlet is required in an attic containing equipment that requires servicing.

   False

3. A split-wired receptacle controlled by three-way switches can serve as the lighting outlet in a hallway 14 ft long.

   True

4. A stairway between two floors with a single landing requires three wall switches (two three-way and one four-way) to control lighting.

   False

5. Occupancy sensors are permitted to be used to control hallway and interior stairway lighting.

   True

6. An attic that contains mechanical equipment must have a separate lighting outlet for each piece of equipment requiring maintenance.

   False

7. Lighting outlets in nonhabitable spaces such as attics used for storage purposes are required by *Code* to have switches with pilot lights.

   False

8. General-use snap switches are constructed so that they can be installed in device boxes or on box covers.

   True

9. *NEC* 210.70(A)(2)(c), Exception, permitting automatic control of lighting in hallways, in stairways, and at outdoor entrances, applies only to common areas of two-family dwellings.

   False

10. Attic lighting is permitted to be controlled by an occupancy sensor equipped with a manual override that allows the sensor to function as a wall switch.

   False

Challenge Question Answers

1. Automatic control of lighting may be safer than switch control in places like stairways and hallways where people's arms may be full of packages. It's also a good solution for two-family dwellings because it makes lighting instantly available without leaving lights burning, thereby wasting energy.

2. The *NEC* is a safety document, not a design guide. Thus, it requires switches to control stairway lighting but doesn't specify the switching arrangements because this is a design consideration. There is no place that the *NEC* requires the use of three- or four-way switches to control lighting.

3. Receptacle outlets are normally used to supply power to appliances, portable lighting, and other utilization equipment. Since such equipment isn't normally used in stairways, receptacle outlets aren't required.

4. Most HVAC equipment operates at 240 volts. A receptacle connected to the load side of an equipment disconnecting means would not have 120-volt overcurrent protection. Similar prohibitions exist elsewhere in the *Code,* such as in 210.52, which prohibits built-in receptacles of baseboard heaters from being connected to the heater circuit.

CHAPTER 12

Multiple Choice

1. Receptacle outlets in the following locations are required to have GFCI protection:

   D. All of the above

2. The following structures are required to have lighting outlets:

   A. All attached garages

3. The following cable types are permitted to be used for branch-circuit wiring inside accessory buildings:

   D. Both A and B

4. When free-standing boxes containing wiring devices are supported by two or more raceways, each raceway must be secured

   B. Within 18 in. of the enclosure

5. Direct-burial cables or raceways can be installed under dwelling driveways and parking areas at a minimum depth of

   C. 24 in.

6. Low-voltage luminaires generally cannot be located closer than which of the following to swimming pools, spas, fountains, or similar bodies of water?

   C. 10 ft

7. The following raceway(s) is (are) permitted to be installed below grade:

   D. Rigid metal conduit (Type RMC)

8. Receptacles in accessory buildings other than garages must have GFCI protection, except when they

   D. Both A and B

9. The following cable type(s) is (are) rarely used in residential construction:

   B. Mineral-insulated, metal-sheathed cable (Type MI)

10. Type UF cable used for interior wiring must be secured as follows:

    D. Both A and B

Fill in the Blanks

1. All 15- and 20-ampere receptacles installed outdoors in wet locations are required to have covers that are weatherproof, regardless of whether an attachment plug is <u>inserted</u> *(connected, attached, or plugged in are also acceptable answers)*.

2. Residential branch circuits rated 20 amperes or less with GFCI protection can be direct-buried at a minimum depth of <u>12 in.</u> below driveways and parking areas.

3. When Type SE cable emerges from the ground at an indoor location, it can be installed as interior wiring, following the *Code* rules for Type <u>NM</u> cable.

4. Garages attached to houses are required to contain at least <u>one</u> wall switch-controlled lighting outlet(s).

5. GFCI-protected outdoor receptacles cannot be supplied from <u>small-appliance</u> branch circuits serving dining rooms and kitchens.

6. Both RMC and IMC can function as equipment grounding conductors for the branch-circuit or feeder conductors installed in them. Some installers also choose to run a separate <u>equipment grounding conductor</u> along with the circuit conductors.

7. Most outdoor receptacle outlets installed at dwellings are required to have <u>GFCI</u> protection for personnel.

8. When freestanding weatherproof boxes are supported by conduits, at least two of them must be threaded into the enclosure and be secured within <u>18</u> in. of the enclosure.

9. When Type UF cable is installed indoors, it must be secured and supported in the same way as Type <u>NM</u> cable.

10. A <u>vehicle</u> door in a garage is not considered an outdoor entrance, and isn't required to have a wall switch-controlled lighting outlet installed on its exterior side.

True or False

1. A two-family dwelling is required to have two HACR maintenance receptacle outlets.

   False

2. Luminaires (lighting fixtures) are permitted to be mounted in trees.

   True

3. Receptacles of 15 and 20 amperes installed outdoors in damp locations are required to have a cover that is weatherproof only when an attachment plug is inserted.

   False

4. When a duplex receptacle is located within dedicated space for a cord-and-plug-connected appliance that can't be easily moved from one place to another, the outlet isn't required to have GFCI protection.

   False

5. Nonmetallic elbows are often used with Type RMC installed below grade, because they make it easier to pull conductors into the raceway.

   False

6. Lighting systems operating at 30 volts or less must be supplied from a branch circuit rated either 15 or 20 amperes.

   True

7. When receptacles or luminaires (lighting fixtures) are installed on device boxes recessed into exterior walls and under roofs, Types NM and AC cables are the most common wiring methods used in residential construction.

   True

8. Types NM, NMC, and NMS are all types of nonmetallic-sheathed cable defined in *NEC* Article 334.

   True

9. Pole-mounted luminaires (lighting fixtures), bollards, luminaires mounted in trees, and luminaires mounted on outside walls without a roof or overhang to protect them must be listed for use in wet locations.

   True

10. All 15- and 20-ampere, 125-volt receptacles installed outdoors in wet locations are required to have covers that are weatherproof even when an attachment plug is inserted.

    True

Challenge Question Answers

1. A minimum of four GFCI-protected receptacles is required, two per dwelling unit (one at the front and one at the back of each unit).

2. An attached garage is treated like a part of a one- or two-family dwelling. Thus, it must have lighting for safety because residents may enter the dwelling by way of the garage. It must have a minimum of one receptacle for plugging in utilization equipment so that residents won't attempt unsafe wiring practices such as using extension cords from the dwelling or screwing plug adapters into lamp sockets.

3. Service-lateral conductors are owned by the serving utility and thus are not subject to *NEC* rules. Utilities follow the *National Electrical Safety Code (NESC)*.

4. Burial depths for underground circuits are intended to prevent accidental damage from digging that might result in shock or electrocution. Breaking an underground cable with a shovel or other tool would normally cause a ground fault that would result in a GFCI device disconnecting the branch circuit. Because of this extra protection against digging hazards, GFCI-protected branch circuits are permitted to be buried at shallower depths.

5. Outlet boxes supported by conduits must have two conduits threaded into the enclosure for safety. An outlet box supported by a single conduit might tend to turn on the threads and come loose, exposing live wires or terminals inside.

CHAPTER 13 _____

Multiple Choice

1. The following types of underwater lighting are used in permanently installed swimming pools:

   D. All of the above

2. Wiring to a dry-niche luminaire (lighting fixture) is required to be installed in

   D. Both A and B

3. The power supply to pool cover motors is permitted to

   A. Use a cord-and-plug connection

4. When nonmetallic conduit is used to enclose branch-circuit wiring to a wet-niche luminaire, the equipment grounding conductor must be

   B. Minimum 8 AWG insulated copper

5. All underwater pool luminaires are required to be installed at least 18 in. below the normal water level, unless listed and identified for use at lesser depths, because

   A. They are cooled by water.

6. Pool pump motors installed outdoors are required to have

   C. A maintenance disconnecting means

7. Underground wiring that does not supply pool equipment is not permitted under swimming pools or within 5 ft horizontally from an inside wall of the pool, unless installed in

   A. RMC

8. Each permanently installed pool at a dwelling is required to have at least how many 125-volt, 15- or 20-ampere receptacles on a general-purpose branch circuit located between 10 and 20 ft from the inside walls of the pool, measured horizontally, and no more than 6½ ft above floor, platform, or grade level?

   A. One

9. Pool motor branch-circuit wiring is permitted to be installed in

   D. All of the above

10. Power conductors and cables for network-powered broadband communications systems are required to be installed at the following minimum distance above the water level or deck area immediately surrounding the swimming pool:

    D. 22½ ft

Fill in the Blanks

1. When the branch circuit supplying equipment is GFCI protected, then totally enclosed luminaires and ceiling-suspended (paddle) fans are permitted to have a minimum

vertical clearance of 7½ ft above the maximum pool water level.

2. Pool lighting that operates at 15 volts or less is required to be supplied through a(n) <u>isolation transformer *or* isolation transformers with an ungrounded secondary and a grounded metal barrier between the primary and secondary windings</u> to reduce the chance of an accidental short circuit that could energize the secondary wiring to the underwater luminaires (lighting fixtures) at higher voltage.

3. A junction box connected to a conduit from the forming shell of a wet-niche luminaire (lighting fixture) must be installed either <u>4 in.</u> above ground level or <u>8 in.</u> above the maximum pool water level, whichever provides the higher elevation.

4. Switches and receptacle outlets must be located at least 5 ft from the inside walls of a swimming pool, unless separated from them by a <u>solid wall, a fence, or other permanent barrier.</u>

5. Pool area junction boxes are frequently installed above grade using conduits for support. A junction box is not permitted to be supported by <u>a single *(one or only one are also acceptable answers)*</u> conduit.

6. Lighting systems operating at 15 volts or less are permitted to use <u>*flush-mounted deck boxes (deck boxes or flush-mounted junction boxes are also acceptable answers)*</u> rather than junction boxes raised above grade level.

7. Branch-circuit breakers supplying 120-volt pool-lighting luminaires (lighting fixtures) must have <u>ground-fault circuit-interrupter *or* GFCI</u> protection.

8. No receptacle is permitted to be installed closer than <u>5 ft</u> from an outdoor pool.

9. Underwater pool-lighting luminaires used in residential construction typically are of three types: *wet niche, dry niche,* and <u>no niche.</u>

10. Receptacles that supply pool pump motors and are rated <u>15- and 20-amperes, single-phase, 120 or 240 volts</u> are required to have GFCI protection.

True or False

1. Typically, pool area junction boxes are manufactured of aluminum, are equipped with threaded hubs or conduit entries, and have extra grounding terminals.

   False

2. An insulated copper equipment grounding conductor, 12 AWG or larger, must be run with the branch-circuit conductors for dry-niche underwater pool lighting.

   True

3. Each permanently installed pool at a dwelling is required to have at least one 125-volt, 15- or 20-ampere receptacle

on a general-purpose branch circuit located between 5 and 10 ft from the inside walls of the pool, measured horizontally.

   False

4. No receptacle outlet is permitted to be installed closer than 6 ft from an outdoor pool, fountain, or other installation, unless GFCI protected.

   True

5. Switching devices such as general-use snap switches, circuit breakers, and automatic timers must be located at least 5 ft from an inside wall of a pool, unless separated from it by a solid wall, a fence, or other permanent barrier.

   True

6. All branch circuits supplying underwater pool lighting operating at more than 15 volts are required to be GFCI protected.

   True

7. Luminaires (lighting fixtures) and ceiling-suspended (paddle) fans cannot be installed within 12 ft vertically above the maximum water level or within 5 ft horizontally from the pool's inside walls.

   True

8. Reinforcing steel used for a pool bonding grid is required to have a corrosion-resistant nonmetallic coating.

   False

9. Receptacles and switching devices located within buildings are considered to be separated from the swimming pool by a permanent barrier.

   True

10. The operating mechanism for an electrically operated pool cover is required to be located within sight of the machinery controlled.

   True

Challenge Question Answers

1. They do not because utility-owned conductors are outside the scope of the *National Electrical Code,* as described in 90.2(B)(4) and (5). However, the conductor clearance rules prohibit pools, fountains, and similar installations from being constructed beneath existing overhead conductor spans.

2. Five feet is greater than the reach of a person in a swimming pool, so this *Code* rule helps prevent shock and electrocution.

3. The smaller clearances for GFCI-protected lighting allow swimming pools, fountains, and similar installations to be constructed at existing dwellings where less space is available than the amount required by the 2005 *NEC.*

4. Raceways between junction boxes and wet-niche-forming shells aren't sealed to keep out moisture. For this reason, they must be located above the maximum water level to prevent entry of water into the junction box.

5. Bonding is particularly important around pools to reduce shock hazard. Also, the presence of water and pool chemicals creates a corrosion hazard, so larger solid conductors are used to ensure that bonding conductors remain intact.

6. The grounding terminal bar in the pool equipment panelboard provides a local ground reference connection for equipment-grounding conductors running with branch circuits that supply pool equipment. However, as with all installations, the grounding and grounded conductors are permitted to be connected only at one place, at the service equipment.

7. The purpose of equipotential bonding is to eliminate voltage gradients in the pool area, which people perceive as "touch voltage."

CHAPTER 14 _____

Multiple Choice

1. Fixed electric space-heating equipment is required to
   D. All of the above

2. Thermostat wiring is
   A. Class 2

3. The following condition(s) applies to installing Class 2 wiring:
   A. It cannot be installed in the same raceway with power wiring.

4. The overcurrent protection and branch-circuit conductors for a continuous load must be rated at
   B. Not less than 125 percent of the load

5. When a two-wire cable is used to supply 240-volt electric baseboard heaters
   B. The grounded conductor must be reidentified as an ungrounded conductor.

6. A thermostatic control intended to serve as a required disconnecting means for fixed electric space-heating equipment must
   D. All of the above

7. A disconnect switch installed for an outdoor air-conditioning unit must be
   D. Both A and B

8. When a receptacle is installed as part of a baseboard heater assembly, it
   C. Must be supplied by a branch circuit other than the heater circuit

9. The manufacturer's label or nameplate on central air-conditioning equipment typically specifies
   C. Maximum overcurrent protection required

10. Electric baseboard heaters are prohibited from being located beneath wall-mounted receptacle outlets by
    C. Baseboard heater listing instructions

Fill in the Blanks

1. A continuous load is a load for which the maximum current is expected to continue for 3 *or* three hours or more.

2. When the nameplate or terminal box of a heating appliance does not specify a temperature rating for the branch-circuit conductors, then conductors rated 60°F *or* 140°F *or* 60°C (140°F) are permitted to be used.

3. A room air conditioner is required to have factory-installed LCDI or AFCI protection built into the power supply cord or attachment plug.

4. When it is unlikely that two loads will operate at the same time (such as heating and air-conditioning), they are called non-coincident loads.

5. Providing a device in a power supply cord or cordset that senses leakage current flowing between or from the cord conductors and interrupts the current at a predetermined level of leakage is known as a leakage current detection and interruption *or* LCDI.

6. In residences, the service disconnect is permitted to serve as the required second disconnecting means for fixed electric space-heating equipment.

7. Low-voltage thermostats are typically used to control electric baseboard heating units when the load rating of the baseboard heaters to be controlled exceeds that of a line-voltage thermostat *(or equivalent wording)*.

8. When a piece of utilization equipment isn't marked to indicate a temperature rating for the supply conductors, then conductors rated 60°C *or* 140°F *or* 60°C (140°F) can normally be used.

9. The attachment plug and receptacle *(or equivalent wording)* normally serves as the required disconnect for room air conditioners.

10. Fixed electric space-heating equipment in dwellings is required to be supplied by an individual branch circuit rated 50 amperes or less.

True or False

1. Thermostats can always serve as the required disconnecting means for electric space-heating equipment.
   False

2. Fixed gas space-heating equipment is required to have integral electrical backup protection, such as supplemental electric heating coils.

   False

3. A single receptacle installed on an individual branch circuit must have a rating not exceeding 80 percent of the branch circuit on which it is installed.

   False

4. A unit switch built into fixed electric space-heating equipment that has a marked OFF position is permitted to serve as the required disconnecting means.

   True

5. Line-voltage thermostats are required to have integral overcurrent protection unless they are inherently self-limiting.

   False

6. If a manufacturer's label or nameplate on air-conditioning equipment specifies maximum fuse size, then either fuses or HACR circuit breakers may be permitted to provide the required overcurrent protection.

   False

7. A cord-connected electric water heater must be supplied through a single receptacle installed on an individual branch circuit.

   False

8. Some baseboard heaters are suitable for use at more than one voltage.

   True

9. A 120- or 240-volt, single-phase room air conditioner equipped with an attachment cord and plug is required to have either LCDI or AFCI protection, but not both.

   True

10. Class 2 wiring can be installed in the same raceway with power wiring only when the two wiring systems are connected to the same piece of equipment.

    False

Challenge Question Answers

1. When they start up, hermetic refrigerant motor-compressors have higher inrush currents than other types of motorized appliances. For this reason, conventional circuit breakers might trip when the motor starts. HACR circuit breakers can withstand higher momentary starting currents without tripping.

2. *NEC* 110.3(B) states that "listed or labeled equipment shall be installed and used in accordance with any instructions included in the listing or labeling." This means that equipment listing or labeling instructions are, in effect, part of the *Code* rules for that equipment.

3. Flexible cords are permitted to be used for appliances only where the fastening means and connections are specifically designed to permit ready removal for maintenance and repair. Kitchen waste disposers, which are small appliances, are designed to be removed for repair or maintenance, while larger water heaters aren't. Another reason is that waste disposers come supplied with attachment cords and plugs, while water heaters don't.

4. Heating appliances may reach temperatures that could damage the insulation on conductors rated 60°C (140°F), thus exposing live conductors and creating a shock hazard.

5. The safety reasons for this requirement is to ensure that a person working on the appliance can easily verify that the power is disconnected and to prevent another person from accidentally re-energizing the appliance while exposed parts are being worked on.

CHAPTER 15

Multiple Choice

1. In dwelling units, smoke detectors must be located
   D. All of the above

2. Residential telephone outlets are required to be
   B. Listed

3. Coaxial cables are not permitted to be installed in the same raceways as
   B. Electric light and power wiring

4. The communications (telephone) primary protector grounding conductor must be
   B. 14 AWG

5. A two-story house with three bedrooms requires how many smoke detectors?
   D. Cannot be determined from the available information

6. All communications (telephone) cables installed in new dwellings are required to be at least
   C. Category 3

7. The *NEC* requires cable television outlets in the following rooms:
   D. None of the above

8. A single receptacle is defined as
   C. A single contact device

9. Smoke detectors are permitted to be wired using
   D. All of the above

10. The telephone company's wiring typically terminates at a component called the
    C. NIU/NID

Fill in the Blanks

1. Smoke alarms and smoke detectors located within <u>20</u> ft of a cooking appliance should be of the photoelectric type, which are less likely to be sensitive to alarms caused by cooking vapors, or they should have a temporary silencing means ("hush button").

2. Occupants in two-family dwellings are required to have <u>ready access</u> to the overcurrent devices protecting conductors in their occupancy.

3. All communications (telephone) and coaxial cables installed in dwelling units are required to be <u>listed</u>.

4. *NEC* 820.24 requires that coaxial cable be installed in a "neat and <u>workmanlike</u> manner."

5. *NFPA 72* requires that smoke detectors be located not closer than <u>3</u> ft from the door to a bathroom or kitchen.

6. The device used in place of a junction box to join coaxial cables is called a <u>splitter</u>.

7. When the metal sheath of incoming coaxial cable is connected to a separate grounding electrode, a bonding jumper of minimum size <u>6 AWG</u> or equivalent is required to be connected to the power grounding electrode system.

8. Smoke detectors must be placed to avoid <u>dead-air spaces</u> *or* <u>dead air</u>, which smoke and heat have difficulty penetrating.

9. Battery-powered smoke detectors are permitted only in <u>existing</u> dwelling units.

10. Low-voltage wiring and power wiring cannot be installed in the same outlet box or other enclosure unless there is a <u>barrier</u> separating the two types of wiring.

True or False

1. When the length of the grounding conductor for a telephone primary protector would otherwise exceed 15 ft, a separate ground rod must be driven.

    False

2. Smoke detectors in new construction are permitted to be either direct-connected or stand-alone units with battery power.

    False

3. If a bedroom is located in a basement or in a habitable attic, the smoke detector located "in the immediate vicinity" can also serve as the required unit for the "additional story."

    True

4. Coaxial cable and power cables, such as Type AC or Type NM, are not permitted to occupy the same bored hole through a wood framing member.

    False

5. Raceways cannot be used to support communications (telephone) wiring.

    True

6. In one- and two-family dwellings, smoke detectors or smoke alarms are required in all habitable rooms.

    False

7. Smoke detectors in unfinished basements should be mounted between joists rather than on them, because smoke collects in these "dead-air spaces."

    False

8. Cooking vapors or bathroom moisture can cause unwanted operation of smoke detectors, often called "nuisance alarms."

    True

9. The *Code* classifies door chime wiring as Class 1, power-limited.

    False

10. Smoke detectors aren't required in attics, unless the attic is habitable and considered a "story."

    True

Challenge Question Answers

1. Both the service point and the demarcation point are found where the wiring owned by the service or utility provider ends, and the premises wiring covered by the *National Electrical Code* begins.

2. The *NEC* is a safety code, not a design guide. Unlike locations of lighting and receptacle outlets, there aren't important safety considerations that affect locations of telephone and cable TV outlets. The *Code* doesn't even require that dwellings have a telephone or cable TV outlet.

3. The reason for prohibiting telephone and cable TV conductors in the same raceways or enclosures as power wiring is to prevent the low-voltage conductors from inadvertently being energized at 120 volts.

4. *NFPA 72®*, *National Fire Alarm Code®*, provides the basic rules for smoke detectors. The *NEC* covers branch-circuit wiring practices and power supply rules for residential smoke detectors.

5. There isn't a firm *yes* or *no* answer to this question. *NEC* 90.3 states that "Chapters 1, 2, 3, and 4 apply generally; Chapters 5, 6, and 7 apply to special occupancies . . . Chapters 1–4 apply except as amended by Chapters 5, 6, and 7 for the particular conditions."

6. No, they cannot. Section 210.25 of the *Code* only permits residential branch circuits to supply loads within, or associated with, each individual dwelling unit.

CHAPTER 16

Multiple Choice

1. Generator sets installed at residences are normally rated:

   B. 120/240-volt, single-phase, 3-wire

2. When sizing a generator to supply essential loads, the following must be included:

   D. Both A and C

3. Residential generator installers are typically responsible for providing the following non-electrical item(s):

   A. Fuel

4. Sizing a standby generator to supply only "essential circuits" has the following advantages:

   D. All of these

5. Residential generators may need optional enclosures, mufflers, or fences in order to meet local codes governing

   B. Generator sound level

Fill in the Blanks

1. Loads should be balanced to the extent possible on generator phases. This is because the generator must be sized to supply the <u>largest phase current</u> of the connected load.

2. The <u>generator disconnecting means</u> is installed between the generator set and the transfer switch or transfer panel, and is used to disconnect the generator set for servicing.

3. Transfer switches mounted indoors normally have NEMA/UL Type <u>1</u> enclosures.

4. *NEC* <u>110.26</u> specifies minimum working space requirements around residential generator sets.

5. The ampacity of field-installed conductors from the generator output terminals to the first overcurrent device must be at least <u>115</u> percent of the nameplate current rating of the generator.

True or False

1. "Essential loads" are required by 760.13 to include fire alarm and burglar alarm systems.

   False

2. Generators with dual 120-volt windings are used only to supply extremely small dwellings with two 120-volt branch circuits.

   False

3. A transfer panel combines a transfer switch with a circuit breaker panelboard for supplying essential loads.

   True

4. A generator is one example of a separately derived system.

   True

5. Vibration isolation is used to prevent generator damage due to unbalanced loads on generator phases.

   False

Challenge Question Answers

1. The whole-house design is most convenient for homeowners because all electrical outlets and equipment are usable during a utility power outage. But this approach is more expensive because it requires a larger generator, generator output conductors, transfer switch, and utility disconnect switch. The essential loads design offers more limited electric service during an outage. But this approach is less expensive because it requires a smaller generator with a lower kVA rating. It also allows use of smaller generator output conductors, transfer switch, and utility disconnect switch.

2. A *transfer switch* transfers a load from one power source (such as a utility service) to an alternate source (emergency generator). Typically, a transfer switch feeds an emergency loads panelboard that then distributes the alternate power to individual branch circuits. A *transfer panel* combines a transfer switch with circuit breakers for feeding the emergency loads.

3. Yes. Since a two-family dwelling has a single utility service, a generator can be installed to supply the service equipment for both dwellings (and the house loads panel, if one exists) through a transfer switch.

4. According to the definition in Article 100, a separately derived system has the following two major characteristics:

   a. Its power is derived from a source other than a service.

   b. It has no direct connection, including a solidly connected grounded (neutral) conductor, to supply conductors of another system.

5. The electrical contractor is responsible for installing all components of the emergency generator system.

CHAPTER 17

Multiple Choice

1. A dwelling built to *NEC* rules is considered to be *Code*-compliant for how long?

   C. Indefinitely

2. Fished cables are required to be secured and supported

   D. None of these

3. Boxes that can be installed in finished surfaces without being attached to structural members are known as

   C. Old work boxes

4. In old work, an existing nongrounding-type receptacle is permitted to be replaced by

   D. Any of these

5. A surface extension from an existing outlet box can be installed using

   C. Both A and B

6. The *National Electrical Code* specifies intervals (distances) at which the following conductors must be secured and supported

   D. None of these

7. In old work, when a grounding means exists in the box, a general-use snap switch must be replaced by

   C. A nongrounding-type snap switch

8. Surface metal raceways and surface nonmetallic raceways are permitted to be installed

   C. As surface extensions from existing outlet boxes

9. The commonest wiring method(s) used in residential old work is (are)

   D. Both A and B

10. The following technique(s) is (are) permitted in old work but not in new construction

    D. Any of these

Fill in the Blank

1. Old work boxes are designed to be installed in flat surfaces without being attached to framing members. (*studs* or *joists* are also considered correct answers)

2. Pushing and pulling cables through concealed spaces in existing walls and ceilings is called fishing.

3. Type NM and AC cables must normally be supported and secured at intervals not exceeding 4½ ft and within 12 in. of every termination.

4. When new cables are fished in the course of doing old work, they are not required to be secured or supported. (Similar language such as *secured to studs* is also considered a correct answer.)

5. Surface raceways are manufactured in both metal and nonmetallic versions.

6. Surface metal and nonmetallic raceways are permitted to extend through partitions so long as the conductors are accessible. (Similar language such as *walls and ceilings* is also considered a correct answer.)

7. Typically, surface raceways are used to extend power from an existing outlet.

8. Wiring can be installed in surface metal and nonmetallic raceway only after the raceway installation is complete. (*finished* is also considered a correct answer)

9. The *Code* requires that general-use snap switches and dimmers be effectively grounded.

10. If a grounding means doesn't exist in a switch box or the wiring method doesn't provide an equipment ground, then an existing snap switch can be replaced by a new snap switch without a grounding terminal. This type of switch is sold for replacement purposes only.

TRUE OR FALSE

1. When existing outlet boxes are supplied by an old-style Type NM cable that doesn't have a separate equipment-grounding conductor, no wire nut exists in the receptacle enclosure.

   False

2. When an existing nongrounding-type receptacle is replaced with a new GFCI receptacle, the new GFCI receptacle is required to be marked "Caution — No GFCI."

   False

3. The steel or aluminum armor of Type AC cable is permitted to serve as an equipment-grounding conductor in accordance with 250.118(6).

   False

4. Nongrounding-type receptacles have two slots that accept receptacle blades.

   True

5. When a nongrounding snap switch is installed for replacement purposes and is located within reach of earth or conducting surfaces, the faceplate isn't permitted to be made of metal.

   True

6. The *Code* requires that low-voltage cables be secured at intervals not exceeding 4½ ft and within 12 in. of every bracket or other termination.

   False

7. A raceway with conductors to the cover of an existing flush box is called a *fixture whip*.

   False

8. When a GFCI receptacle is used to protect new grounding-type receptacles installed without a ground conductor in old work, the *Code* doesn't permit an equipment-grounding conductor to be connected between the GFCI and non-GFCI receptacles.

   True

9. Type AC cable can't be used to make surface extensions because the wiring method doesn't contain an anti-short bushing ("redhead").

    False

10. When existing outlet boxes are supplied by old-style Type NM cable that doesn't have a separate equipment-grounding conductor, no grounding means exists in the receptacle enclosure.

    True

Challenge Question Answers

1. The reason is to avoid potential safety problems caused by confusing homeowners and other electricians who might later assume that the equipment-grounding conductor was actually grounded at the service entrance.

2. Requiring that all wiring methods follow normal *Code* rules for securing and supporting in old work would re-quire a great deal of damage to existing structures. Permitting fishing in old work is a practical adaptation to real-life construction conditions. Also, fishing usually accounts for only a limited amount of wiring.

3. So that grounding or bonding continuity will not be interrupted by a loose box cover or by removing the cover to access the conductors inside.

4. *NEC* 300.15 lists the types of wiring methods for which boxes are required. Article 725 (Class 2 and 3 cables), Article 800 (communications conductors), and Article 820 (coaxial cables) don't contain any requirements for boxes at outlets, splices, and terminations.

5. Electrical work in exposed areas of existing structures, such as unfinished basements and attics, is essentially the same as new work. Thus, all normal *Code* rules for securing and supporting wiring methods, installing grounding-type receptacles and snap switches, providing GFCI protection for receptacles in certain locations, etc., apply just as they would in new construction.

Index